Analysis and Applications of Artificial Neural Networks

To Rutger, Wendy, Irene and Gerrie

Analysis and Applications of Artificial Neural Networks

L. P. J. Veelenturf

Prentice Hall
London New York Toronto Sydney Tokyo Singapore Madrid Mexico City Munich

First published 1995 by
Prentice Hall International (UK) Limited
Campus 400, Maylands Avenue
Hemel Hempstead
Hertfordshire, HP2 7EZ
A division of
Simon & Schuster International Group

Typeset in 10 on 12 pt Times by P&R Typesetters Ltd, Salisbury, Wiltshire, UK

Printed and bound in Great Britain by Bookcraft, Midsomer Norton

---

Library of Congress Cataloging-in-Publication Data

Veelenturf, L. P. J. (Leo P. J.)
Analysis and applications of artificial neural networks / by L. P. J. Veelenturf
p. cm.
Includes index.
ISBN 0-13-489832-X (cased)
1. Neural networks (Computer science) I. Title.
QA76.87.V42 1995
006.3--dc20 94-40215
CIP

British Library Cataloguing in Publication Data

A catalogue record for this book is available from the British Library

ISBN 0-13-489832-X (hbk)

---

1 2 3 4 5 99 98 97 96 95

Contents

Preface

'Any process that works can be understood; what cannot be understood is suspect.' This is effectively what Marvin Minsky and Seymour Papert stated twenty years ago during the first wave of interest in artificial neural networks. This critical remark could be made again in the recent second wave of interest in the topic, because for many people neural networks still seem something of a black art. Perhaps this attitude has been engendered by the tremendous amount of recent publications on neural networks, where it seems that every author invents his or her own type of neural network, which is frequently only justified by some particular application. For the reader it must be a chaotic collection of seemingly unrelated questions without a consistent framework and a unifying perspective. We certainly do not claim that we can give such a unified framework at this stage of development of this new scientific field, but we can at least open the black boxes of the main types of neural networks that can be thoroughly understood and that have turned out to be very useful in a broad range of applications.

A reason for the excitement about neural networks might be that in the literature on artificial neural networks one frequently encounters very promising and attractive statements about the generalization capability of neural networks, like: 'Neural networks are capable of adapting themselves with the aid of a learning rule and a set of examples to model relationships among the data without any *a priori* assumptions about the nature of the relationships.' A similar statement is: 'After learning neural networks they may be used to predict characteristics of new samples or to derive empirical models from examples in situations in which no theoretically based model is known.' Although to a certain extent these types of statements are true, one must be careful with the substatements that no model or *a priori* information about the nature of the relationship between examples is assumed.

Generalization is the process of inductive inference of general relationships from a finite number of samples. An example is the inference of a new number in a finite sequence of numbers: one is inclined to say that the next example in sequence 1, 4, 9, 16,... will be the number 25, because we observe a simple regularity in the sequence: the kth element in the sequence is k^2. However, if no prejudice in favour of some type of 'model' exists, any number may follow the given sequence. For example, one might as well say that the next number is 27 because one is in favor of the regularity where the nth number $y(n)$ in the sequence is defined by $y(n)$ = 'sum of first *n* uneven primes'. If one asks many people (who know elementary calculus!) to guess the next number in the sequence given above, almost all will say that the next number is 25. This phenomenon reveals the human attitude to select always the most 'simple' model to explain a sequence of experimental observations.

A neural network is capable of modelling relationships among data by learning from the examples but is always using some *a priori* set of models. 'Modelling' can be defined as the process of formulating a finite set of interrelated rules (or the construction of a finite set of interconnected mechanisms) by which one can generate or explain the (potentially infinite) set of observed data. The simplicity of the model is very subjective because it depends on the domain of knowledge of the person (or of the mechanical process) that is doing the modelling.

If one is using complex neural networks to 'model' the relationships behind a given set of data, it is hard to demonstrate that one is using (or assuming) *a priori* information about the kind of relationship. We will, however, demonstrate in several sections of this book that one needs, or assumes, *a priori* information about the relationship between the given data in order to be justified in accepting the outcome of the learning process of a neural network. The *a priori* information one is using is determined by the type of neural network, the configuration of the neural network and the type of neural transfer functions. In the application of neural networks to the solution of real-life problems it is important to be aware of this phenomenon. For example, if we want to infer with a two-layer continuous Perceptron from a given data set an unknown functional relationship and select the number of first-layer neurons in a Perceptron too low, then the unknown function will be underfitted (i.e. will not go through all data points); if we select the number of neurons too high, then the unknown function will be overfitted (the realized function will go through the data points but will fluctuate wildly in between).

This example, more extensively discussed in Section 3.13, shows that generalization by learning from examples and counterexamples is in general impossible without utilizing *a priori* knowledge about the properties of the function to be identified.

If one uses a neural network to find the relation behind the data in a data set, one is in addition assuming that the data set is representative of the (unknown!) relation. Nevertheless, it must be said that by using neural networks one can solve in an optimal way certain problems that are hard to tackle by conventional methods. For instance, we will show that among the set of all classifiers that divide the n-dimensional input space by a $n-1$ dimensional hyperplane, the single-neuron Perceptron is an optimal classifier.

We will concentrate in this book on these kinds of limitations and capabilities of the main type of neural networks, rather than giving a review of the multitude of different neural networks. Our aim is that the reader should profoundly understand and be able to apply artificial neural networks to the solution of practical problems.

Almost 70 per cent of all publications deal with the type of networks we will analyze in this book: the binary Perceptron, the continuous Perceptron and the self-organizing neural network, and when we consider the applications that turn out to be useful, this percentage is even higher.

The book is self-contained, which means that we will deliberately avoid the fashionable vicious circle of proving propositions by quoting the propositions of other authors. The methods of analyzing neural networks are largely of a mathematical nature but the level of mathematical rigour in our exposition will nevertheless be low because we are far more concerned with providing insight and understanding than establishing a rigorous mathematical foundation. The required mathematical maturity is that of a typical final-year undergraduate student in electrical engineering or computer science.

Understanding is the ability to transform new phenomena to a coherent simple structure of already well understood phenomena. For this reason we will give many illustrative examples and plausible arguments in terms of what is supposed to be well known by the reader. A great number of real-life applications will also contribute to this understanding and will show at the same time the powerful usefulness of neural networks.

For several years we have taught courses at the University of Twente based on the material in this book. The book can be covered in a one-semester course.

Each chapter concludes with some exercises. The lists of literature are far from complete because we only want to give the reader a map for the main routes in the bewildering and chaotic landscape of published material. The exercises are meant as a means to check one's own understanding of the presented theory.

While writing this book we benefited from the comments of many colleagues and from the experiments performed by our students. We would like to thank especially Philip de Bruin, Mark Bentum, Andre Beltman and Cuun Krugers-Dagneaux-Rikkers for their suggestions for improving the manuscript.

Leo P. J. Veelenturf

Acknowledgements

While writing the first chapter of this book we benefited from the contribution of several of our students. We would like to thank Arend Lammertink for his investigations on hyphenation with a binary Perceptron and especially for his idea to decompose the neural network into many subnetworks in order to reduce the computation time; Erik Coelingh for the realization of a software simulation program for the adaptive recruitment learning rule as well as for the recruitment and reinforcement learning rule; and Herco Reinders for his experiments on hyphenation with the recruitment and reinforcement learning rule.

Thanks are due to all the students who performed many experiments with the continuous multi-layer Perceptron. I want to mention Ronald van der Zee, Sander Lokerse, Berend Jan van der Zwaag, Peter Degen, Robert de Jong, Jan Wijnholt, Rolf van der Wal, Hans van't Spijker and Henk Martijn for their experiments on classification and function identification. Ronald Edens and Remco Couwenberg performed experiments on speech recognition with time delay and recurrent Perceptrons. A. Rensink, Manjo Nijrolder and Stephan ten Hage did the experiments with the harmonic signal detector. Peter Spekreijse and Geert Pronk used the multi-layer Perceptron for machine condition monitoring. Guido Frederiks made a useful contribution to determining the configuration of a neural net. Marcel Budding demonstrated that a neural net can be used for principal component analysis. John Kroeze made a contribution by analyzing neural networks with a piecewise linear approximation. Theo Roelands investigated the relation between the size of the data set and the complexity of the neural network. Special thanks are due to Erik Tromp for his proof that a two-layer neural network can approximate any continuous function that is arbitrarily close.

I want to express my gratitude to Jan Zandhuizen who made me aware of the intriguing properties of Kohonen's self-organizing neural network. In order to obtain a better understanding of the self-organizing neural network and especially to evaluate the learning process parameters that are theoretically difficult to establish, a lot of experiments were performed by our students.

Investigations on anthropomorphic pattern recognitions were done by Gerrit van Brakel, Ruud Coppoolse, Barthold Lichtenbelt, Walter Jansen, Frank Staijen, and last but not least by Ronald Muller.

Experiments on classification with the self-organizing neural network were performed by Ronald van der Zee, Wieant Nielander, Geert Pronk and Sander Lookerse. Paul Buisman made an analysis of the behaviour of the original neural network introduced by Kohonen.

The vector quantization properties were investigated by Hans Neggers. Marco Bloemendaal, Ruurd de Vries, Ruud Busschers, Erik Groot, Patrick Wilmerink, Stefan van Hal and Klaas Scheppink applied the self-organizing algorithm to spike-wave detection in EEG signals.

Arjan Draaijer investigated the application to adaptive system control. Experiments on speech recognition were performed by Ronald Edens, Paul de Haan, Andre Beltman, Guido de Jong, Herman Woudstra and Stefan ten Hage.

Sake Buwalda developed an VHDL implementation of the self-organizing algorithm.

Finally I owe a debt of gratitude to Andre Beltman who was involved in almost all experiments mentioned above and who developed very practical measures for the evaluation of ordering and topology preservation with a self-organizing neural algorithm.

1

INTRODUCTION

1.1 Machines and brains

For several years the author has been involved in research on pattern recognition. In that period he became aware of the tremendous range of sophisticated methods used to analyze and to recognize pictures by machines. The pattern recognition machines were equipped with large numbers of vast and complicated algorithms. The most advanced machines could, for instance, recognize a certain class of handwritten digits, but in spite of the sophisticated nature of these machines they were limited to recognizing those pictures that had been foreseen by the system builders as potential elements to be recognized in the future. For example, one can build machins to recognize handwritten capital 'A's but the system will fail to recognize a capital 'A' as given in Figure 1.1. It is surprising that human beings can recognize the letter in Figure 1.1 as an 'A' as it is very unlikely that one has ever seen the figure before.

It is very unlikely that human beings compare the handwritten 'A' to some reference picture stored in their brain. Probably they know the characteristic features of an 'A' and their perception is not disturbed by artefacts in the picture. The interference of artefacts in a picture will, however, destroy the correct classification by a programmed machine, which is probably not able to judge the importance of deviations from the preprogrammed standard features.

The way people have acquired the ability to recognize pictures can only be by experience. By trial and error they have learned to perform certain tasks. Machines do not learn, they are preprogrammed, and if they can learn they are restricted to certain classes of preprogrammed methods of learning.

The lesson seems to be that the capability of learning is essential for more advanced and intelligent artificial machines.

Another striking difference between machines and human beings is the 'computation' time required for complicated tasks such as pattern recognition. Computers are extremely fast but it is hard to design machines that can recognize three-dimensional objects in real time, whereas humans, whose brains are composed of neurons switching about a million times slower than electronic components, can recognize old friends almost instantaneously. We know that computers perform their computations sequentially, step by step, whereas the human brain is processing the information in parallel.

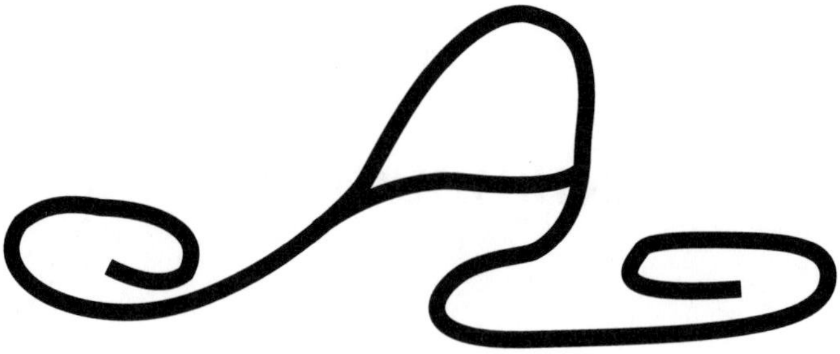

Figure 1.1 A handwritten capital 'A'

The second lesson for designing more intelligent systems seems to be the use of the parallel processing of information.

Man-made machines are built with a large number of different complicated functional building blocks; if one unit fails, the whole system collapses. The brain, however, is built out of a large number of, at least from a functional point of view, almost identical building bricks: the neurons. Many units may be destroyed without significantly changing the behaviour of the total system.

This comparison between the behaviour and construction of artificial machines and the behaviour and physiological configuration of the human brain might give new ideas for developing more intelligent machines. Artificial neural networks are the results of the first steps in this new direction for intelligent system design.

1.2 The artificial neural network

The building unit of a neural network is a simplified model of what is assumed to be the functional behaviour of an organic neuron. The human brain contains about 10^{11} neurons. For almost all organic neurons one can distinguish anatomically roughly three different parts: a set of incoming fibers (the dendrites), a cell body (the soma) and one outgoing fiber (the axon). For a simplified configuration see Figure 1.2. The axons divide up into different endings, each of which makes contact with other neurons. A neuron can receive up to 10 000 inputs from other neurons. The bulb-like structures where fibers contact are called synapses. Electrical pulses can be generated by neurons (so-called neuron firing) and are transmitted along the axon to the synapses. When the electrical activity is transferred by the synapse to another neuron, it may contribute to the excitation or inhibition of that neuron. The synapses play an important role because their transmission efficiency for electrical pulses from an axon to the dendrites (or somas) of other neurons can be changed depending on the 'profitability' of that alteration.

The learning ability of human beings is probably incorporated in the facility of changing the transmission efficiency of those synapses. Donald O. Hebb was among the first who postulated this mechanism in his book *Organization of Behavior* (1949):

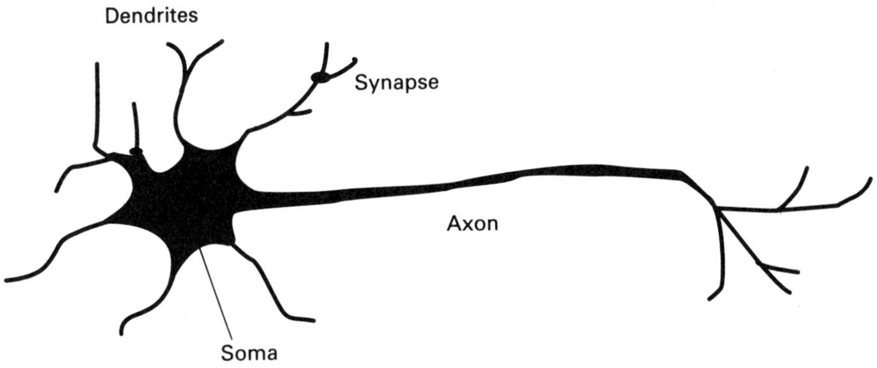

Figure 1.2 Simplified configuration of an organic neuron

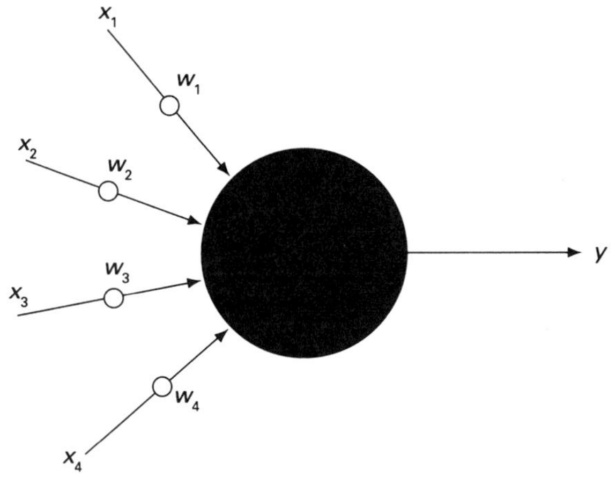

Figure 1.3 Artificial model of a neuron.

'When an axon of cell A is near enough to excite a cell B and repeatedly or persistently takes part in firing it, some growth process or metabolic change takes place in one or both cells such that A's efficiency, as one of the cells firing B, is increased.' The change of the synaptic transmission efficiency acts as a memory for past experiences.

In a simplified artificial model of a neuron (see Figure 1.3), the synaptic transmission efficiency is translated into a real number w_i by which an input x_i is multiplied before entering the neuron cell. The number w_i is called the weight of input x_i. The absence or presence of a train of electrical pulses in a real neural fiber is modelled by a variable x_i which respectively may have the value zero or one. In that case we will say that we have a binary artificial neuron representing the 'one-or-zero' behaviour of a real neuron.

In reality, neurons may fire up to about 100 pulses in a second. We can model this gradual change in pulse train frequency by a variable x_i which may have any value between zero and one. In that case we have a continuous artificial neuron.

In almost all artificial models of neurons, all inputs x_i are weighted by the synaptic transmission efficiency and are summed to one number $s = \sum x_i w_i$. This weighted input s determines in some way or another the output value y of the artificial neuron. In a binary neuron the output y will be one (the neuron is firing) if the weighted input exceeds some threshold T, and will be zero (the neuron is silent) if the weighted input is below the threshold. In a continuous artificial neuron, the output y may be some monotone increasing function (the frequency of the pulse train on the outgoing axon is gradually changing depending on s) of the weighted input s.

The model of an artificial neuron outlined above was first introduced by the neurophysiologist Warren McCulloch and the logician Walter Pitts in 1943. In a famous paper by the mathematician S. C. Kleene in 1951, it was shown that with the artificial neurons introduced by McCulloch and Pitts, one can build a system that behaves in the same way as a computer. Although it is important to know that artificial neural nets are not inferior in their computation capabilities to computers, it is of no practical use to mimic computers. The benefits of an artificial neural network are mainly the results of the modifiability of behaviour by changing the weights w_i in a learning process.

The learning behaviour of artificial neural nets was first treated extensively in a book by Frank Rosenblatt in 1962, *Principles of Neurodynamics.* He introduced a learning algorithm by which the weights can be changed such that a desired computation was performed. The wave of activity on artificial neural networks in the mid-1960s was, however, challenged in 1969 by Marvin Minsky and Seymour Papert, who showed in their book *Perceptrons* that some simple computations cannot be done with a one-layer neural net, and doubted that a learning algorithm could be found for multi-layer neural networks. At that time many scientists left the field of artificial neural networks.

A second upheaval took place in the mid-1980s when several people found a learning algorithm, called the *back-propagation algorithm*, that could adjust the weights in multi-layer neural nets. About that time also new types of neural net with dynamic behaviour were introduced. We mention the neural net of Hopfield (1982) (not treated in this book) and the self-organizing neural net of Kohonen (1982).

Dynamic neural nets are characterized by feedback: the output of a neuron depends, after some delay, on its own output because of the fully interconnected structure of the neural nets. The binary Perceptron we will discuss in Chapter 2 and the continuous Perceptron of Chapter 3 are not dynamic systems: after learning, the output $\mathbf{y}(t)$ of the neural net will only depend on the actual input $\mathbf{x}(t)$ and not on previous inputs. However, the input–output behaviour F: $\mathbf{y}(t) = F\{\mathbf{x}(t)\}$, learned by the neural net in the training phase, will depend on the sample input–output behaviour of the training set.

For a dynamic system the actual output $\mathbf{y}(t)$ depends not only on the actual input $\mathbf{x}(t)$ but also on the actual state $\mathbf{q}(t)$: $\mathbf{y}(t) = F\{\mathbf{q}(t), \mathbf{x}(t)\}$, whereas the state depends on inputs in the past. There exist neural networks that can learn dynamic behaviour from sequences of inputs and corresponding sequences of outputs (Veelenturf, 1981). We will not discuss these networks because they are a typical example of solving problems by using a neural network, whereas, except for special cases, there are more efficient methods of finding the solutions with conventional methods (Veelenturf, 1978). This observation leads to the warning that one must be aware that using neural networks is not a panacea; it is frequently better to have recourse to more conventional methods.

2

THE BINARY PERCEPTRON

2.1 Introduction

The behaviour of an artificial neuron is inspired by the assumed behaviour of a real neuron in organic neural networks. A simplified model of a real neuron is composed of a cell body or soma, a set of fibers entering the cell body, called the dendrites, and one special fiber leaving the soma, called the axon. The dendrites transmit trains of electrical pulses towards the soma and the axon conducts trains of pulses away from the soma. The axon terminates by branching into many filaments. The filaments end in bulb-like structures called synapses that make contact with dendrites or somas of other neurons. The transfer of electrical pulses from the final filaments of some axon to the dendrites or soma of another neuron depends on the *synaptic transmission efficiency*, represented by the variable w. If the synaptic transmission efficiency is positive, the synapse is said to be *excitatory*, if negative, the synapse is called *inhibitory*. The positive or negative transmission efficiency may vary between small and large values. Only when the sum of 'synaptic weighted' incoming pulses is greater than some threshold is a train of pulses generated by the soma and transmitted by the axon (see Figure 2.1). In organic neural tissue the pulse frequency may vary between a few pulses per second up to twenty pulses per second. An additional simplification is to disregard the frequency of the pulse trains and to consider only the presence, represented by the number 1, or absence represented by the number 0, of a pulse train. This simplified model of a real neuron is called the *one-or-zero behaviour* of a neuron.

This simplified model of a neuron can easily be simulated by an *artificial neuron* (see Figure 2.2). Dendrites are represented by input lines and a variable x_i represents the presence $[x_i(t)=1]$ or absence $[x_i(t)=0]$ of a pulse train on fiber i at time t. Every artificial neuron has one output line representing the axon of the neuron and the presence or absence of a pulse train at an axon is presented by the value 1 or 0 of the variable $y(t)$. There will be one special input line with a constant input $x_0=1$, and a weight w_0. This constant input, $x_0=1$, and the weight w_0 realize a threshold equal to $-w_0$. When the 'weighted' sum of incoming signals is greater than the *threshold* $T=-w_0$ the output y will become 1 after a delay τ. The input–output behaviour of

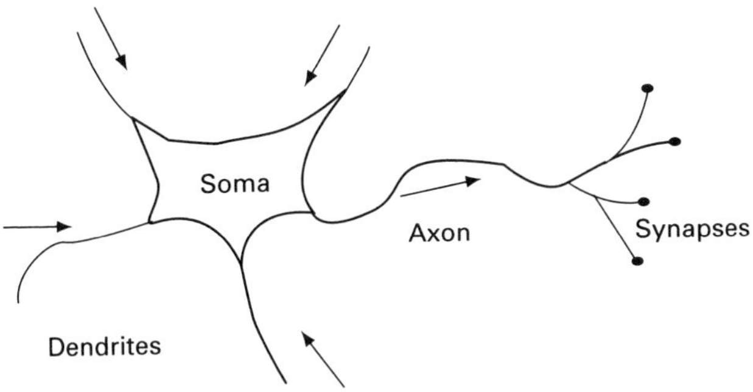

Figure 2.1 Simplified configuration of an organic neuron

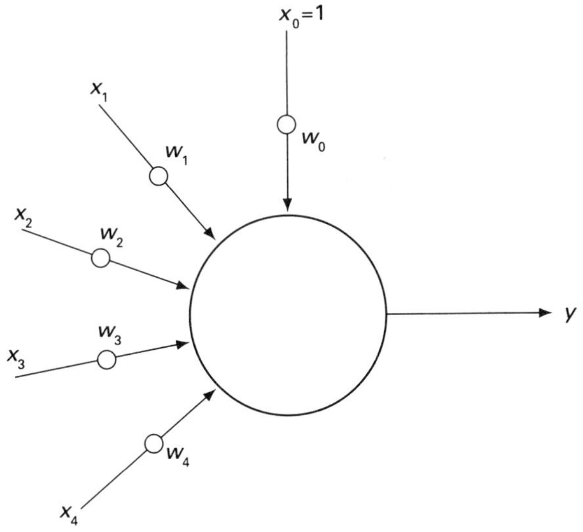

Figure 2.2 Artificial model of a neuron

an artificial neuron is now specified by:

$$\ddot{y}(t+\tau)=\begin{cases}=1 & \text{if} \sum_{i}^{n} w_i x_i(t) > -w_0 \\ =0 & \text{otherwise}\end{cases}$$

The variable w_i is called the *weight* of input line i and represents the synaptic transmission efficiency of the synapse between the final filament of a neuron and the dendrite i (or the soma) of a particular neuron. The threshold $T=-w_0$, the weights w_i and the delay τ are real valued. If there is no feedback in the neural network we may take $\tau=0$, and the time dependency of x_i and y can be ignored. So the previous

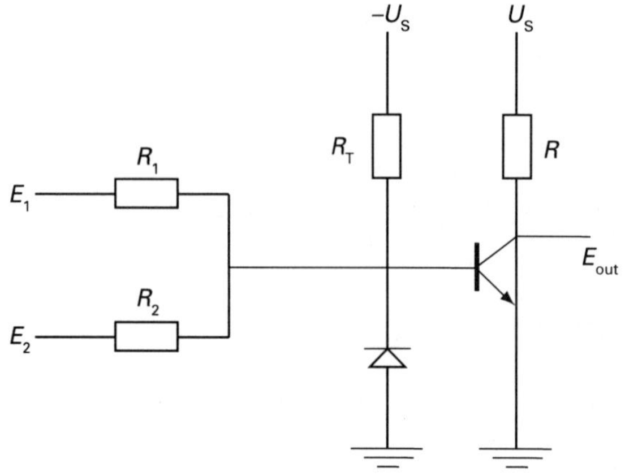

Figure 2.3 Electronic implementation of an artificial neuron

formulation of the input–output behaviour can be replaced by:

$$y = \begin{cases} =1 & \text{if} \sum_i^n w_i x_i > -w_0 \\ =0 & \text{otherwise} \end{cases}$$

An artificial neuron can easily be implemented in a simple electronic circuit (see Figure 2.3). Those acquainted with electronics will understand that the transistor will be open if:

$$\frac{E_1}{R_1} + \frac{E_2}{R_2} > \frac{-U_S}{R_T}$$

If the voltage E_1 represents x_1, E_2 represents x_2, and $-U_s/R_T$ represents the threshold $-w_0$, then we obtain with $1/R_1 = w_1$ and $1/R_2 = w_2$ that the transistor is open if:

$$w_1 x_1 + w_2 x_2 > -w_0$$

Networks composed of layers of interconnected artificial neurons have been studied extensively by many authors. The analysis of neural networks is attractive because all the building units, the neurons, are the same and the transfer function of such a unit is quite simple. More important, however, is that we can alter the behaviour of a neuron by changing in a learning process the weights w_i in the input lines. Changing weights is the artificial counterpart of the adaptation of the synaptic efficiency in real organic neural networks. Before examining this learning behaviour of a neural network, we consider the 'zero-or-one behaviour' of just one single artificial 'binary' neuron.

With a single neuron, for example, we can realize some restricted class of *predicate logic*. Consider the statement: 'John is going out for a walk if and only if the sun is shining or if it is cold and the wind is blowing west.' The predicate 'John is going

Table 2.1

x_1	x_2	x_3	y
0	0	0	0
0	0	1	0
0	1	0	0
0	1	1	1
1	0	0	1
1	0	1	1
1	1	0	1
1	1	1	1

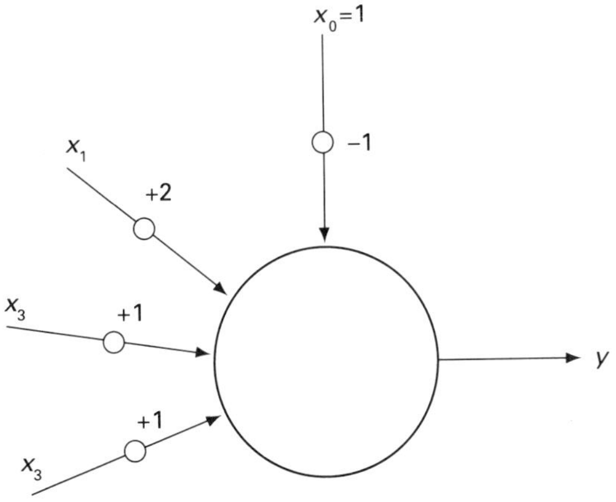

Figure 2.4 Neuron illustrating Table 2.1

out for a walk' is only TRUE if the conditions mentioned are TRUE. Now we can represent the 'truth value' TRUE by the number 1 and FALSE by the number 0. We represent the truth value of the predicate 'The sun is shining' by x_1, the truth value of 'It is cold' by x_2, the truth value of 'The wind is blowing west' by x_3, and the truth value of 'John is going out for a walk' by y. With this notation we can enumerate all possible situations in a simple truth table, as shown in Table 2.1.

If we now consider the values of x_1, x_2 and x_3 as the inputs of a single neuron and y as the output, we can select the weights w_0, w_1, w_2 and w_3 in such a way that the output behaviour of that neuron yields the truth value of the predicate 'John is going out for a walk' (Figure 2.4). Methods for finding the appropriate weights analytically, or by a learning process, will be discussed later.

Pioneers in this field of research, like Rosenblatt (1962) and Minsky and Papert (1969), investigated neural networks with the aim of using such networks mainly for

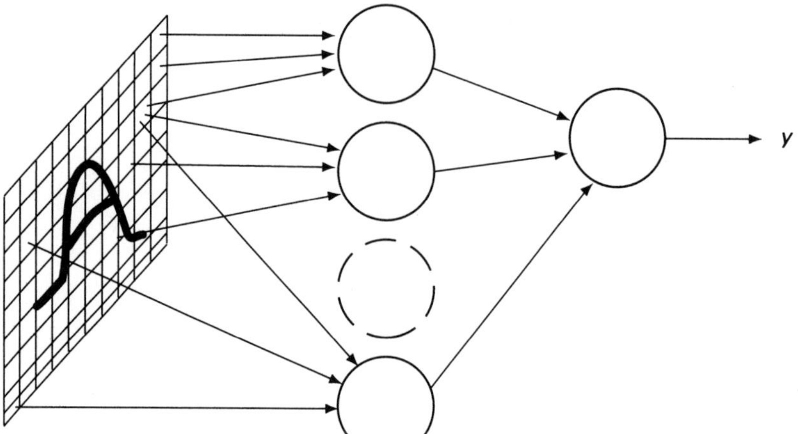

Figure 2.5 Neural network for pattern recognition

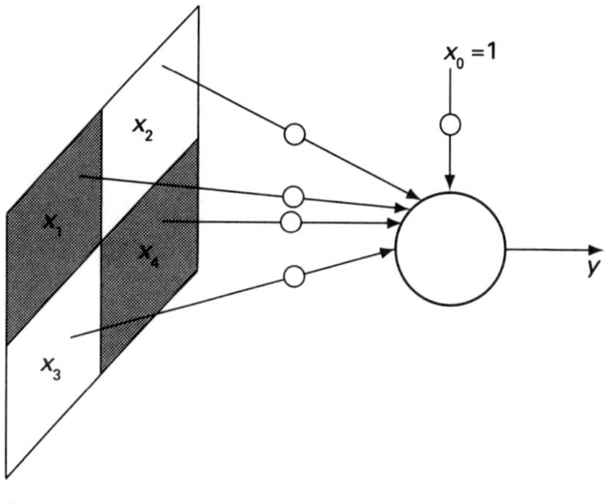

Figure 2.6 Classification of sixteen patterns by a neural network

pattern recognition problems. For this reason they called those networks *Perceptrons* (from perception). In honour of Rosenblatt, who used the term first, we will call the networks discussed in Chapters 2 and 3 Perceptrons.

As an example we consider a pattern recognition problem. Some pattern is projected onto a grid of small squares. A variable x_i is assigned to each square. The variable x_i will have the value 1 if the pattern is covering that particular square, and 0 if the pattern is not covering that square (see Figure 2.5). The values of the variables x_i constitute the inputs of a binary neural network. The output of the neural network will classify patterns as belonging to some predefined class ($y=1$) or not ($y=0$).

For example, if we have a very small grid of four squares and one single neuron (see Figure 2.6), one may wish to classify the sixteen different artificial patterns (see

Table 2.2

x_1	x_2	x_3	x_4	y	x_1	x_2	x_3	x_4	y
0	0	0	0	0	1	0	0	0	0
0	0	0	1	0	1	0	0	1	0
0	0	1	0	0	1	0	1	0	1
0	0	1	1	1	1	0	1	1	1
0	1	0	0	0	1	1	0	0	1
0	1	0	1	1	1	1	0	1	1
0	1	1	0	0	1	1	1	0	1
0	1	1	1	1	1	1	1	1	1

Table 2.2) as whether patterns of at least two black squares are connected (i.e. the black squares are adjacent), $y=1$, or not, $y=0$.

Although there are many pattern classification problems that can be solved with a single neuron, we will demonstrate in the next chapter that there exists no set of weights w_0, w_1, w_2, w_3 and w_4 such that Table 2.2 is realized by a single neuron classifier. It turns out that we need a two-layer neural network with at least two neurons in the first layer. We can see directly that the problem can also be solved with four neurons in the first layer (one neuron for detecting that x_1 and x_2 are both 1, one neuron for detecting that x_1 and x_3 are both 1, one neuron for detecting that x_2 and x_4 are both 1, and one neuron to detect that x_3 and x_4 are both 1), and one neuron in a second layer to detect that least one neuron in the first has an output of 1. One might suspect that a single neuron is not able to solve a classification problem if there are a great number of input variables. There are, however, problems with only two input variables that are also not solvable with a single neuron.

Consider, for example, the Boolean 'exclusive-or' function: $y=x_1 \otimes x_2$, i.e. the output of the single neuron must be 1 if and only if $x_1=1$ or $x_2=1$ but not both. We will see in the next chapter that we cannot solve this problem with a single neuron but we will demonstrate, on the other hand, that any Boolean function can be realized with a two-layer Perceptron. This example indicates at the same time a third application area for the use of binary Perceptrons: the realization of *Boolean functions.*

Because we can realize any Boolean function with a binary Perceptron and because every neuron can be implemented with an electronic circuit, we have the fourth application area: *switching circuits.*

In a subsequent section of this chapter we will study two-layer binary neural networks composed of interconnected artificial neurons without feedback connections between neurons.

Different Boolean functions can be realized in parallel, e.g. with the two-layer neural network given in Figure 2.7. The neural net of Figure 2.7 classifies simple patterns consisting of three pixel points in three classes as specified by Table 2.3.

If a pattern $\mathbf{p}=\langle p_1 p_2 p_3 \rangle$ is a member of the class $K_1=\{\langle 000\rangle, \langle 001\rangle, \langle 100\rangle, \langle 111\rangle\}$,

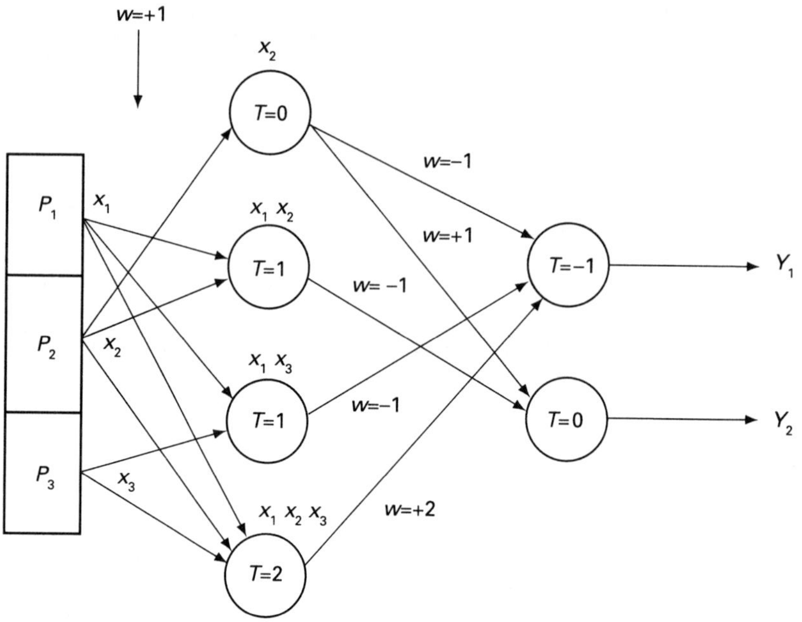

Figure 2.7 Neural net for classification with two-dimensional output

then $y_1=1$ and $y_2=0$. If the pattern is a member of the class $K_2=\{\langle 010\rangle, \langle 011\rangle\}$, then $y_1=0$ and $y_2=1$. If the pattern is not a member of K_1 or K_2, then $y_1=0$ and $y_2=0$.

2.2 The performance of a single-neuron binary Perceptron

In the previous section we saw that a single neuron performs a kind of 'weighted voting' on variables x_i: the output y of the neuron will be 1 if and only if $w_1x_1+w_2x_2+\cdots+w_nx_n$ is greater than some threshold T.

Example 2.1

Consider the balance of Figure 2.8. At equally spaced points there might be objects with some weight g_k at the balance pole. At the left-hand side there is one fixed weight g_0 attached to the balance at a unit distance from the suspension point. We use the variable x_k to indicate whether ($x_k=1$) or not ($x_k=0$) if there is an object placed at distance k from the point of suspension.

Now 'The balance will tip to the right' if and only if:

$$\sum kg_kx_k>g_0$$

or with kg_k replaced by w_k, $g_0=T$, and when the predicate 'The balance will tip to

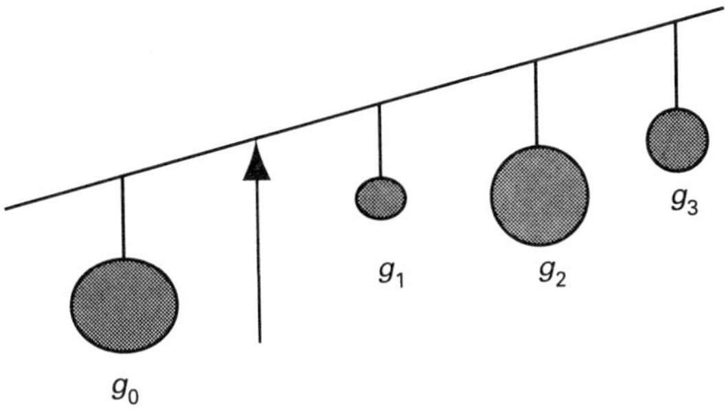

Figure 2.8 The mechanical equivalent of a threshold function

the right' is replaced by the binary variable y, we obtain:

$$y=1 \text{ if and only if } \sum w_k x_k > T \quad \blacksquare$$

We will now investigate the properties of threshold functions like the one used above. We define first: a function $y=f\{x_1, x_2,\ldots, x_n)$ is called a *binary linear threshold function with respect to the binary-valued variables* $x_1, x_2,\ldots, x_n$ if there exist a number T and a set of numbers $\{w_1, w_2,\ldots, w_n\}$ such that $y=1$ if and only if $\Sigma w_i x_i > T$. We usually will drop the adjective 'binary' and, although important, we will frequently drop the phrase 'with respect to the binary valued variables $x_1, x_2,\ldots, x_n$'.

With the use of the *step function* $S(z)$ defined by: $S(z)=1$ if $z>0$ and $S(z)=0$ if $z\leqslant 0$ we can equivalently say that $y=f(x_1, x_2,\ldots, x_n)$ is a linear threshold function if $y=S(\Sigma w_i x_i - T)$.

If the weights w_i constitute the components of a so-called *weight vector* $\mathbf{w}$ and the variables x_i are the components of the *input vector* $\mathbf{x}$, we can also write a linear threshold function as $y=S(\mathbf{w}^t\mathbf{x}-T)$, with $\mathbf{w}^t$ the transpose of the vector $\mathbf{w}$.

Note that a linear threshold function is not a linear function in the ordinary sense, because for a linear threshold function we have $f(\alpha\mathbf{x})\neq\alpha f(\mathbf{x})$.

In the introduction to section 2.1 we have demonstrated that a linear threshold function can be realized by a single-neuron binary Perceptron with the threshold $T=-w_0$ realized by a constant input $x_0=1$ and a corresponding weight w_0. If we introduce a so-called *extended weight vector* $\hat{\mathbf{w}}$ with $\hat{\mathbf{w}}^t=[w_0, w_1, w_2,\ldots, w_n]$, and a so-called *extended input vector* $\hat{\mathbf{x}}$ with $\hat{\mathbf{x}}^t=[1, x_1, x_2,\ldots, x_n]$, then we can compactly write a linear threshold function realized by a single neuron Perceptron as $y=S(\hat{\mathbf{w}}^t\hat{\mathbf{x}})$.

Now we concentrate on the class of logical functions that can be realized by a single-neuron binary Perceptron. If $y=f(\mathbf{x})$ is a logical function of two variables x_1 and x_2 we can have sixteen different functions. (NB: there are 2^2 different arguments and for each argument the function value can be either 1 or 0.) Of those sixteen functions, fourteen can be realized by a single-neuron binary Perceptron. The two functions that cannot be realized are specified by Table 2.3. The first function is called the *exclusive-or function* and the second the *identity function*.

We will prove now that the first function is not a linear threshold function and

Table 2.3

x_1	x_2	y	x_1	x_2	y
0	0	0	0	0	1
0	1	1	0	1	0
1	0	1	1	0	0
1	1	0	1	1	1

thus it cannot be realized by a single-neuron binary Perceptron. Assuming, however, that we can realize the exclusive-or function, then there must be weights w_0, w_1 and w_2 such that:

For the first argument [0, 0] we must have:

$$w_0+w_1x_1+w_2x_2\leqslant 0 \text{ thus: } w_0\leqslant 0$$

For the second argument [0, 1] we must have:

$$w_0+w_1x_1+w_2x_2>0 \text{ thus: } w_0+w_2>0$$

For the third argument [1, 0] we must have:

$$w_0+w_1x_1+w_2x_2>0 \text{ thus: } w_0+w_1>0$$

For the fourth argument [1, 1] we must have:

$$w_0+w_1x_1+w_2x_2\leqslant 0 \text{ thus: } w_0+w_1+w_2\leqslant 0$$

One can easily verify that the four inequalities for the weights cannot be satisfied simultaneously. This completes our proof.

Our conclusion that the exclusive-or function cannot be realized by a single-neuron Perceptron also becomes clear when we consider the realization of the exclusive-or function as a classification problem. We have two classes of points in a two-dimensional input space (see Figure 2.9). For one class of points $\{(0, 1), (1, 0)\}$ the output of the neuron must be 1, and for the other class $\{(0, 0), (1, 1)\}$ the output must be 0. Thus for the first class we must have $w_0+w_1x_1+w_2x_2>0$, and for the second class we must have $w_0+w_1x_1+w_2x_2\leqslant 0$. The set of points for which $w_0+w_1x_1+w_2x_2=0$ represents a separating line in the two-dimensional input space. On one side of this line we will have for every point (x_1, x_2) that $w_0+w_1x_1+w_2x_2>0$ and thus the output of the neuron will be equal to 1. On the other side of the line we will have $w_0+w_1x_1+w_2x_2\leqslant 0$ and the output of the neuron will be 0. Now one can easily check that we cannot locate a line between the two sets of points that must be separated; thus there exists no solution for our classification problem.

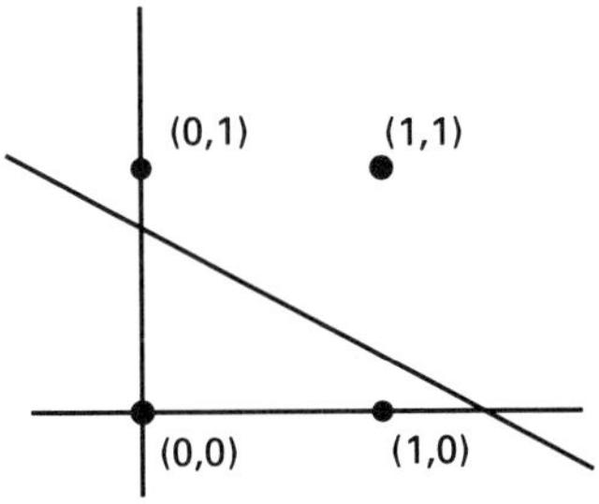

Figure 2.9 Exclusive-or as a classification problem

Table 2.4

	x_1	x_2	x_3	y
1	0	0	0	0
2	0	0	1	0
3	0	1	0	0
4	0	1	1	1
5	1	0	0	1
6	1	0	1	1
7	1	1	0	1
8	1	1	1	1

Fortunately we can show in a later section that the problem can be solved with a two-layer network.

We will now discuss the more general situation of the realization of logical (or binary) functions with n arguments with a single-neuron Perceptron. There are 2^{2^n} different functions with n arguments. A large number of these functions are not linear threshold functions with respect to the variables $x_1, x_2, \ldots, x_n$ and thus cannot be written as $y = S(\Sigma\, w_i x_i - T)$ and thus cannot be realized with a single-neuron binary Perceptron. We will give an example of how to find the linear threshold function (if it exists) given the truth table of a logical function with three arguments.

Example 2.2

Let a function be specified by the truth table given in Table 2.4 (see also Figure 2.4). We want to obtain a function of the form $y = S(\Sigma\, w_i x_i - T)$, or with the threshold T replaced by the weight $-w_0$ we want to find the expression $y = S(w_0 + w_1 x_1 + w_2 x_2 + w_3 x_3)$.

For the successive arguments we must have the following:

1. $w_0 \leqslant 0$
2. $w_0 + w_3 \leqslant 0$
3. $w_0 + w_2 \leqslant 0$
4. $w_0 + w_2 + w_3 > 0$
5. $w_0 + w_1 > 0$
6. $w_0 + w_1 + w_3 > 0$
7. $w_0 + w_1 + w_2 > 0$
8. $w_0 + w_1 + w_2 + w_3 > 0$

From (1) and (4) we conclude $w_2 + w_3 > 0$. Take $w_2 = 1$ and $w_3 = 1$, then from (2) and (3) we conclude $w_0 \leqslant -1$. Take $w_0 = -1$. From (5) we infer $w_1 = 2$. With this selection of weights all inequalities are satisfied and we obtain the following linear threshold function: $y = S(-1 + 2x_1 + x_2 + x_3)$. ■

From this example it becomes clear that there exists, within the bounds given by the set of inequalities, a certain amount of freedom to select the weights. For instance we could select as well: $w_0 = -2$, $w_1 = 4$, $w_2 = 2$ and $w_3 = 2$; another selection could be: $w_0 = -1$, $w_1 = 4$, $w_2 = 1$ and $w_3 = 1$. We will return to this topic in Section 2.3.

As stated before, not all logical functions are linear threshold functions with respect to the variables $x_1, x_2, \ldots, x_n$. It is not easy to determine whether a given function is a linear threshold function or not. At present there is only one more or less practical method by which this can be done, and that is by determining whether or not the set of inequalities associated with the logical function contains a contradiction.

Example 2.3

Consider the logical function $y = f(x_1, x_2, x_3)$ defined by $y = 1$ if and only if the number of 1 in the argument is odd. This function is known as the *parity problem* and is specified in Table 2.5.

Table 2.5

	x_1	x_2	x_3	y
1	0	0	0	0
2	0	0	1	1
3	0	1	0	1
4	0	1	1	0
5	1	0	0	1
6	1	0	1	0
7	1	1	0	0
8	1	1	1	1

We want to find an expression of the form: $y=S(w_0+w_1x_1+w_2x_2+w_3x_3)$. For the successive arguments we must have the following:

1. $w_0 \leqslant 0$
2. $w_0+w_3>0$
3. $w_0+w_2>0$
4. $w_0+w_2+w_3 \leqslant 0$
5. $w_0+w_1>0$
6. $w_0+w_1+w_3 \leqslant 0$
7. $w_0+w_1+w_2 \leqslant 0$
8. $w_0+w_1+w_2+w_3>0$

In this case we can see immediately that the parity problem cannot be solved with a linear threshold function, because from (1) and (2) we conclude that $w_3>0$ and from (3) and (4) we conclude that $w_3<0$; hence we have a contradiction.

In the general case of n variables we have to investigate 2^n inequalities. We can save ourselves a great deal of effort if we eliminate redundant equations and simplify expressions by using the following set of properties.

Properties of inequalities

1. $a>0$ and $b>0 \Rightarrow a+b>0$
2. $a \leqslant 0$ and $b \leqslant 0 \Rightarrow a+b \leqslant 0$
3. $a>0$ and $a+b \leqslant 0 \Rightarrow b<0$
4. $a \leqslant 0$ and $a+b>0 \Rightarrow b>0$
5. $a+b>0$ and $a+c \leqslant 0 \Rightarrow b>c$
6. $\Sigma a_i>0$ and $\lambda>0 \Rightarrow \Sigma \lambda a_i>0$
7. $\Sigma a_i<0$ and $\lambda>0 \Rightarrow \Sigma \lambda a_i<0$

From rules (1) and (2) we can derive a property that we can sometimes use to check whether a logical function can be a linear threshold function or not without writing down the set of all inequalities. Let a in rule (1) represent the sum of some set of weights associated with some input vector $\mathbf{x}_i$ with $f(\mathbf{x}_1)=1$. Thus the first inequality in rule (1) with $a=\hat{\mathbf{w}} \cdot \hat{\mathbf{x}}_1>0$ must hold. Let b in rule (1) be the sum of weights associated with some vector $\mathbf{x}_2$ with $f(\mathbf{x}_2)=1$. Thus the second inequality in rule (1) with $b=\hat{\mathbf{w}} \cdot \hat{\mathbf{x}}_2>0$ must hold. According to rule (1) we must have $\hat{\mathbf{w}} \cdot \hat{\mathbf{x}}_1+\hat{\mathbf{w}} \cdot \hat{\mathbf{x}}_2>0$. Assume we have for the zero vector $\mathbf{0}$: $f(\mathbf{0})=0$. This implies $w_0 \leqslant 0$ and thus $(-w_0)+\hat{\mathbf{w}} \cdot \hat{\mathbf{x}}_1+\hat{\mathbf{w}} \cdot \hat{\mathbf{x}}_2>0$. Let $\mathbf{x}_1$ and $\mathbf{x}_2$ be vectors with no 1s in the same position, we write $\mathbf{x}_1 \cap \mathbf{x}_2=\mathbf{0}$. Let $\mathbf{z}$ be a vector obtained from the two input vectors $\mathbf{x}_1$ and $\mathbf{x}_2$ such that $z_i=1$ if $x_{1i}=1$ or $x_{2i}=1$ otherwise $z_i=0$, we write $\mathbf{z}=\mathbf{x}_1 \cup \mathbf{x}_2$. Vector $\mathbf{z}$ has a corresponding inequality $\hat{\mathbf{w}} \cdot \hat{\mathbf{z}}=(-w_0)+\hat{\mathbf{w}} \cdot \hat{\mathbf{x}}_1+\hat{\mathbf{w}} \cdot \hat{\mathbf{x}}_2$. Because $(-w_0)+\hat{\mathbf{w}} \cdot \hat{\mathbf{x}}_1+\hat{\mathbf{w}} \cdot \hat{\mathbf{x}}_2>0$ we must have $f(\mathbf{z})=1$. The same kind of reasoning holds if $f(\mathbf{x}_1)=0$, $f(\mathbf{x}_2)=0$ and $f(\mathbf{0})=1$. In that case $f(\mathbf{z})=0$ must hold. Thus we can write the *consistency property of the binary linear threshold function*: If the logical function $y=f(\mathbf{x})$ is a binary linear threshold function

Table 2.6

x_1	x_2	y
0	0	1
0	1	0
1	0	0
1	1	1

with respect to $x_1, x_2, \ldots, x_n$ and $f(\mathbf{x}_i)=u$, $f(\mathbf{x}_j)=u$, $f(\mathbf{0})=\bar{u}$, then $f(\mathbf{z})=u$ must hold, with $\mathbf{z}=\mathbf{x}_i \cup \mathbf{x}_j$ and $\mathbf{x}_i \cap \mathbf{x}_j=\mathbf{0}$, and $u=1$ ($\bar{u}=0$) or $u=0$ ($\bar{u}=1$).

Example 2.4

For the identity function we have Table 2.6. From Table 2.6 we find:

$$f(0, 0)=1 \text{ and thus } w_0>0$$

$$f(0, 1)=0 \text{ and thus } w_0+w_2 \leqslant 0$$

$$f(1, 0)=0 \text{ and thus } w_0+w_1 \leqslant 0$$

For f to be a linear threshold function we must have $f(1, 1)=0$, which implies $w_0+w_1+w_2 \leqslant 0$. However, $f(1, 1)=1$, and thus f cannot be a linear threshold function with respect to x_1 and x_2. ■

Many logical functions of n arguments cannot be realized by a single-neuron Perceptron. This also becomes clear when we consider the determination of a logical function as a *classification problem.* We have two classes of points in a n-dimensional input space (see Figure 2.10) for the parity problem as presented in Example 2.3. For one class of points $A=\{(0, 0, 1), (0, 1, 0), (1, 0, 0), (1, 1, 1)\}$ the number of 1s is odd and the output of the neuron must be 1, and for the other class $B=\{(0, 0, 0), (0, 1, 1), (1, 1, 0), (1, 0, 1)\}$ the number of 1s is even and the output must be 0. Thus for the first class we must have $w_0+w_1x_1+w_2x_2+w_3x_3>0$ and for the second class we must have $w_0+w_1x_1+w_2x_2+w_3x_3 \leqslant 0$. The set of points (x_1, x_2, x_3) for which $w_0+w_1x_1+w_2x_2+w_3x_3=0$ represents a two-dimensional separating plane H in the three-dimensional input space. On one side of this plane H we must have for every point (x_1, x_2, x_3) that $w_0+w_1x_1+w_2x_2+w_3x_3>0$ and thus the output of the neuron will become equal to 1. On the other side of the plane H we want to have $w_0+w_1x_1+w_2x_2+w_3x_3 \leqslant 0$ and the output of the neuron must be 0. Now one can easily check that we cannot locate a plane between the two sets of points that must be separated, thus there exists no solution for our classification problem and thus the parity function cannot be represented by a linear threshold function. Or equivalently one can say: both sets of points *are not linearly separable.*

In the n-dimensional case we must have a $(n-1)$-dimensional *separating hyperplane*

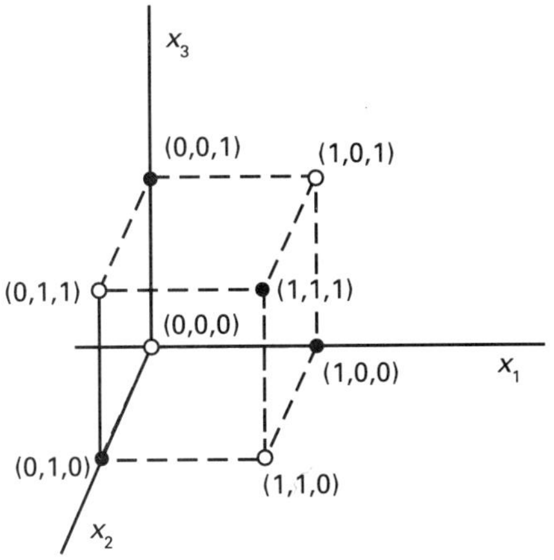

Figure 2.10 The parity problem as a classification problem

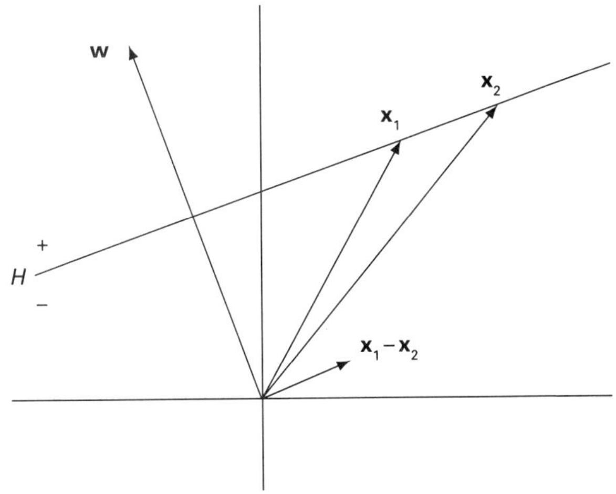

Figure 2.11 One-dimensional separating hyperplane H in a two-dimensional space

H in the n-dimensional input space X if the logical function is a linear threshold function with respect to $x_1, x_2, \ldots, x_n$. The separating hyperplane is defined by $w_1x_1 + w_2x_2 + \cdots + w_nx_n = -w_0$. The weight vector $\mathbf{w} = [w_1, w_2, \ldots, w_n]$ is orthogonal to the separating hyperplane. This becomes clear when we take two n-dimensional input vectors $\mathbf{x}_1$ and $\mathbf{x}_2$ located on the hyperplane. Figure 2.11 shows the two-dimensional case. For these vectors $\mathbf{x}_1$ and $\mathbf{x}_2$ we have that $\mathbf{w}^t(\mathbf{x}_1 - \mathbf{x}_2) = 0$ and thus $\mathbf{w}$ and the hyperplane are orthogonal.

The separating hyperplane H divides the n-dimensional input space X into two half-spaces, the region X^+ where $\Sigma\, w_ix_i > -w_0$ (i.e. $y = 1$) and the region X^- where

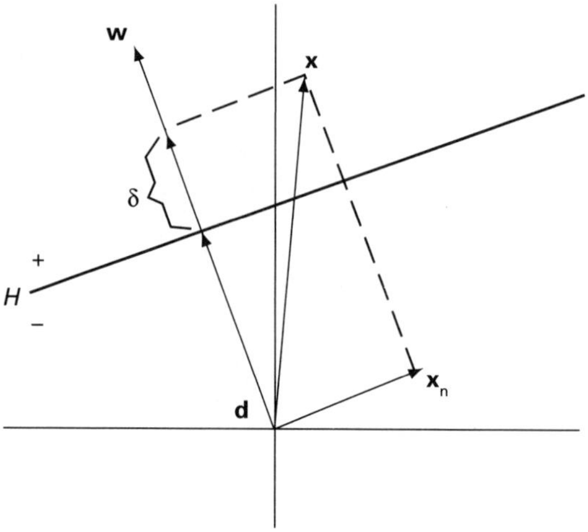

Figure 2.12 Decomposition of the input vector **x**

$\Sigma w_i x_i \leqslant -w_0$ (i.e. $y=0$). Since a vector **x** in the region X^+ will give $\mathbf{w}^t\mathbf{x} > -w_0$, the weight vector **w** points into the region X^+. It is said that **x** is on the *positive side of the hyperplane H* if **x** is in X^+, and **x** is on the negative side of H if **x** is in X^-.

The distance d from the origin to the hyperplane H is equal to the projection of a vector **x** in H (i.e. $\mathbf{w}^t\mathbf{x} = -w_0$) on the unit vector $\mathbf{w}/|\mathbf{w}|$ normal to H. Thus $d = \mathbf{w}^t\mathbf{x}/|\mathbf{w}|$, or with $\mathbf{w}^t\mathbf{x} = -w_0$ we find for the distance along **w** from the origin to the hyperplane:

$$d = \frac{-w_0}{|\mathbf{w}|}$$

The distance δ from the hyperplane H to an input vector **x** is proportional to $\mathbf{w}^t\mathbf{x} + w_0$. The easiest way to see this is by decomposing the vector **x** into a component in the direction of **w** and another component $\mathbf{x}_n$ orthogonal to **w** (see Figure 2.12):

$$\mathbf{x} = \mathbf{x}_n + \lambda \frac{\mathbf{w}}{|\mathbf{w}|}$$

By forming the vector product $\mathbf{w}^t\mathbf{x}$ and noting that $\mathbf{w}^t\mathbf{x}_n = 0$ we obtain for the length of the component **x** along **w**:

$$\lambda = \frac{\mathbf{w}^t\mathbf{x}}{|\mathbf{w}|}$$

By subtraction of the distance d from the origin to the hyperplane we obtain for the distance δ from the hyperplane to **x**:

$$\delta = \frac{\mathbf{w}^t\mathbf{x} + w_0}{|\mathbf{w}|}$$

We finally mention that one can prove (see Cover, 1965) that in the n-dimensional case the number of linear threshold functions is equal to:

$$C(n)=2\sum_{i=0}^{n}\binom{2^n-1}{i}$$

For partial functions, 2^n must be replaced by the number of samples.

2.3 Equivalent linear threshold functions

This section on equivalent linear threshold functions can be omitted on a first reading of this book. For a more profound understanding of the learning capabilities of a binary Perceptron, however, one must know that in general a binary Perceptron yields an infinite number of solutions to the same problem.

In the next chapter we will give a learning procedure of how to find from a set of samples of a logical function, a single-neuron binary Perceptron that will realize that logical function, if the logical function is a linear threshold function with respect to the variables $x_1, x_2, \ldots, x_n$. It will turn out that the weights of the linear threshold function, obtained after learning by the neuron, will depend on the particular sequence of applied samples. Different threshold functions can, however, realize the same logical function, known as equivalent linear threshold functions.

In addition, it is worth determining which features of linear threshold functions are essential and which are arbitrary. It frequently occurs that the descriptions of linear threshold functions are different, i.e. the weights w_i are different, whereas the realized logical function is the same. We will now give four different simple theorems on equivalent linear threshold functions. The first theorem is important, though its proof is trivial.

Theorem 2.1

If $y=S(\Sigma\, w_i x_i+w_0)$ is a linear threshold function, then $y'=S(\Sigma\, \lambda w_i x_i+\lambda w_0)$ with $\lambda>0$ is an equivalent linear threshold function.

Example 2.5

Let $y=S(x_1+x_2-x_3)$ and thus $w_1=1$, $w_2=1$ and $w_3=-1$, then with $\lambda=2$ the linear threshold function $y'=S(2x_1+2x_2-2x_3)$ is equivalent to y. ■

Theorem 2.2a

If $y=S(\Sigma\, w_i x_i+w_0)$ is a linear threshold function and $y'=\Sigma\, w_i' x_i+w_0'$ is a linear function with respect to the binary variables $x_1, x_2, \ldots, x_n$ such that if $y'>0$

Table 2.7

x_1	x_2	x_3	y	y'	y''
0	0	0	0	-1	0
0	0	1	0	-1	0
0	1	0	1	0	1
0	1	1	0	0	1
1	0	0	1	0	1
1	0	1	0	0	0
1	1	0	1	1	1
1	1	1	1	1	1

then $y=1$, and if $y'=0$ then $y=1$ or $y=0$, and if $y'<0$ then $y=0$, then $y''=S(\Sigma(w_i+\lambda w_i')x_i+w_0+\lambda w_0')$ with $\lambda \geqslant 0$ is equivalent to $y=S(\Sigma\, w_ix_i+w_0)$.

Proof

Because we add to the value of the argument of the step function in y a positive or zero value when $y=1$ and we add to the value of the argument of the step function in y a negative or zero value when $y=0$, it will be clear that $y=y''$ for all the values of the argument. QED

Example 2.6

Let $y=S(x_1+x_2-x_3)$ as specified in Table 2.7. Let $y'=x_1+x_2-1$ as specified in column 5 of Table 2.7. The conditions on y' are satisfied so the linear threshold function $y''=S(3x_1+3x_2-x_3-2)$ (with $\lambda=2$) is equivalent to y.

The trivial complement of the previous theorem is as follows:

Theorem 2.2b

If $y=S(\Sigma\, w_ix_i+w_0)$ is a linear threshold function and $y'=\Sigma\, w_i'x_i+w_0'$ is a linear function with respect to the binary variables $x_1, x_2, \ldots, x_n$ such that if $y'<0$ then $y=1$, and if $y'=0$ then $y=1$ or $y=0$, and if $y'>0$ then $y=0$, then $y''=S(\Sigma(w_i+\lambda w_i')x_i+w_0+\lambda w_0')$ with $\lambda \leqslant 0$ is equivalent to $y=S(\Sigma\, w_ix_i+w_0)$.

For the next theorem we need some auxiliary concepts and lemmas. We first define X^+ as the set of argument values of a logical function for which $y=1$; in the same way we have X^- as the set of argument values for which $y=0$.

Lemma 2.1

1. If $y=S(\Sigma\, w_ix_i+w_0)$ is a linear threshold function, then $y'=S(\Sigma\, w_ix_i+w_0-\Delta^+)$ with $0 \leqslant \Delta^+ < \min(\Sigma\, w_ix_i+w_0)$ over X^+ is an equivalent linear threshold function. If

$y'' = S(\Sigma w_i x_i + w_0 - \Delta^-)$ with $0 \geqslant \Delta^- \geqslant \max(\Sigma w_i x_i + w_0)$ over X^-, then y'' is also an equivalent linear threshold function.

Proof

When we subtract from the argument of the step function in y a positive constant Δ less than $\min(\Sigma w_i x_i + w_0)$ over X^+, then the argument of the step function will remain positive for all elements of X^+ and the argument of the step function will be negative or zero for all elements of X^- thus y will remain the same for all elements. If we add to the argument a positive constant value Δ smaller than or equal to $-\max(\Sigma w_i x_i + w_0)$ over X^-, then y will also remain the same for all values of the argument. QED

Example 2.7

The linear threshold function $y = S(x_1 + x_2 - x_3)$ is equivalent to $y' = S(x_1 + x_2 - x_3 - 0.5)$ because $\Delta^+ = 0.5$ and $0 \leqslant 0.5 < \min(\Sigma w_i x_i + w_0) = 1$ over X^+ (see Table 2.7). Note: $\Delta^- = 0$. ■

Lemma 2.2

If $y = S(\Sigma w_i x_i + w_0)$ is a linear threshold function w.r.t. $x_1, x_2, \ldots, x_n$, then $y' = S(\Sigma(w_i - \delta_i)x_i + w_0)$ is an equivalent linear threshold function if $0 \leqslant \delta_i < (1/n)\min_i(\Sigma w_i x_i + w_0)$ over X^+ or if $0 \geqslant \delta_i \geqslant (1/n)\max_i(\Sigma w_i x_i + w_0)$ over X^-.

Proof

If, in the worst case, for all x_i we have $x_i = 1$, we subtract in the first case from the argument of the step function in y the constant $n\delta_i$ with $0 \leqslant n\delta_i < \min_i(\Sigma w_i x_i + w_0)$ over X^+. According to Lemma 2.1, we obtain in that case an equivalent linear threshold function. The same holds if $0 \geqslant \delta_i \geqslant (1/n)\max_i(\Sigma w_i x_i + w_0)$ over X^-. QED

Example 2.8

The linear threshold function $y = S(x_1 + x_2 - x_3)$ is equivalent to $y = S[(1 - 1/6)x_1 + (1 - 1/6)x_2 - (1 - 1/6)x_3]$ because $\delta_i = 1/6$ and $0 \leqslant 1/6 < 1/3\min_i(\Sigma w_i x_i + w_0)$ over x^+. ■

Theorem 2.3

If $y = S(\Sigma w_i x_i + w_0)$ is a linear threshold function, then there exists a linear threshold function $y' = \Sigma w'_i x_i + w'_0$ such that all weights are integers.

Proof

If all weights in y are rational numbers, we can form the product D of all denominators of the weights and multiply all weights by D. All weights will become an integer and the obtained threshold function y' is, according to Theorem 2.1 equivalent to y. If a weight w_i is a real number, we can replace weight w_i, according to Lemma 2.2, by a rational weight w'_i such that $w'_i = w_i - \delta_i$ and rational weights can be replaced by integer weights. This completes our proof. QED

2.4 Learning a single-neuron binary Perceptron with the reinforcement rule

Although a single-neuron binary Perceptron is not of great practical use, because only a few of the logical functions are linear threshold functions with respect to the variables $x_1, x_2, \ldots, x_n$, we can nevertheless gain much understanding of more complicated networks by investigating the learning behaviour of one building unit.

In Section 2.2 we demonstrated how to determine the weights of a single-neuron binary Perceptron from a set of inequalities. Now we investigate how we can adapt step by step in a learning process the weights of a neuron in order to identify some logical function. In fact we can use the learning process as an algorithm to solve a set of inequalities. We assume that the function to be realized by the Perceptron is a linear threshold function.

At a given step of the learning process we have some extended weight vector $\hat{\mathbf{w}} = [w_0, w_1, \ldots, w_n]$; the output will be correct for a subset of all arguments of the function to be identified, and for the remaining arguments the output will be wrong. The set of arguments for which the target output is equal to 1, whereas the actual output is equal to 0, will be denoted by $T^+_{\mathbf{w}}$, the set of arguments for which the target output is equal to 0 and the actual output is 1 will be denoted by $T^-_{\mathbf{w}}$. The arguments $x_1, x_2, \ldots, x_n$ of the function y will be extended with the constant internal input $x_0 = 1$ of the neuron. It will turn out that we have to change the weights proportional to the elements of $T^+_{\mathbf{w}}$ and negative proportional to the elements of $T^-_{\mathbf{w}}$.

Example 2.9

The function y to be realized is specified in Table 2.8. Note that in Table 2.8 the extended inputs $[x_0, x_1, x_2]$ with $x_0 = 1$ are given, whereas y is a logical function of x_1 and x_2. The neuron in the initial learning state (i.e. at step $k=0$) has the weight vector $\hat{\mathbf{w}}(0) = [w_0(0), w_1(0), w_2(0)]^t = [0.5, 1, -1]^t$ (see Figure 2.13). For the output $y(0)$ of the neuron at step $k=0$ we have (see Table 2.8) the following:

$$y(0) = S[w_0(0) + w_1(0)x_1 + w_2(0)x_2] = S(0.5 + x_1 - x_2)$$

We see that for the extended input vectors $[1, 0, 0]^t$ and $[1, 1, 0]^t$ the output is wrong. The only way to improve the output for the vector $[1, 0, 0]^t$ is by decreasing

Table 2.8

x_0	x_1	x_2	y	y^0
1	0	0	0	1
1	0	1	0	0
1	1	0	0	1
1	1	1	1	1

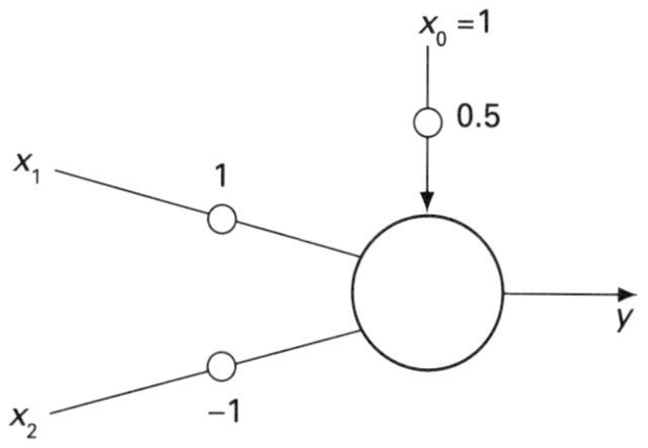

Figure 2.13 Single neuron with initial weights

the value of w_0. So we can change the weight vector $\hat{\mathbf{w}}=[w_0, w_1, w_2]^t$ by an amount $\Delta\hat{\mathbf{w}}$ proportional to $-1[1, 0, 0]^t$, i.e. minus the first extended input vector. To improve the output for the vector $[1, 1, 0]^t$ we have to decrease the values of w_0 and w_1. So we can change the weight vector by an amount $\Delta\hat{\mathbf{w}}$ proportional to $-1[1, 1, 0]^t$, i.e. minus the third extended input vector.

We can take the proportionality equal to 1 and add both increments $\Delta\hat{\mathbf{w}}$. Thus we obtain $\Delta\hat{\mathbf{w}}=[-2, -1, 0]^t$. The new weight vector becomes $\hat{\mathbf{w}}(1)=[w_0(1), w_1(1), w_2(1)]^t=[-1.5, 0, -1]^t$. Because:

$$y(1)=S[w_0(1)+w_1(1)x_1+w_2(1)x_2]=S(-1.5-x_2)$$

we observe that now we obtain the wrong output only for the input vector $[1, 1, 1]^t$. To improve for that argument the output of the neuron we have to increase w_0, w_1 and w_2. Thus we can take $\Delta\hat{\mathbf{w}}$ proportional to the corresponding misclassified vector $[1, 1, 1]^t$. We can proceed in the same way as above and after a finite number of steps we will obtain a correct response of the single-neuron Perceptron, if the original function is a linear threshold function with respect to $x_1, x_2, \ldots, x_n$. In Table 2.9 the results for six learning steps are given. The final linear threshold function is equal to:

$$y=S(-1.5+x_1+x_2)$$ ∎

Let in general X^+ be the set of argument values with target output $y=1$ and X^- the set of argument values with target output $y=0$. For a given value of the weight vector $\hat{\mathbf{w}}$, a subset $T^+_{\mathbf{w}}$ of X^+ and a subset $T^-_{\mathbf{w}}$ of X^- are misclassified.

Table 2.9

Step		w_0	w_1	w_2	T^+	T^-
0	$\hat{\mathbf{w}}$	0.5	1	−1	—	[1, 0, 0], [1, 1, 0]
	$\Delta\hat{\mathbf{w}}$	−2	−1	0		
1	$\hat{\mathbf{w}}$	−1.5	0	−1	[1, 1, 1]	—
	$\Delta\hat{\mathbf{w}}$	1	1	1		
2	$\hat{\mathbf{w}}$	−0.5	1	0	—	[1, 1, 0]
	$\Delta\hat{\mathbf{w}}$	−1	−1	0		
3	$\hat{\mathbf{w}}$	−1.5	0	0	[1, 1, 1]	—
	$\Delta\hat{\mathbf{w}}$	1	1	1		
4	$\hat{\mathbf{w}}$	−0.5	1	1	—	[1, 1, 0], [1, 0, 1]
	$\Delta\hat{\mathbf{w}}$	−2	−1	−1		
5	$\hat{\mathbf{w}}$	−2.5	0	0	[1, 1, 1]	
	$\Delta\hat{\mathbf{w}}$	1	1	1		
6	$\hat{\mathbf{w}}$	−1.5	1	1	—	—

We will show that for a convergent learning process we can modify the weight vector $\hat{\mathbf{w}}$ by adding $\Sigma\, \varepsilon\hat{\mathbf{x}}_i$ summed over $\hat{\mathbf{x}}_i \in \hat{T}_{\mathbf{w}}^{+}$, where $\varepsilon > 0$ is a proportionality constant, and subtracting from $\hat{\mathbf{w}}$ the value of $\Sigma\, \varepsilon\hat{\mathbf{x}}_j$ for $\hat{\mathbf{x}}_j \in \hat{T}_{\mathbf{w}}^{-}$. Because at each step we use the total set $T_{\mathbf{w}}^{+} \cup T_{\mathbf{w}}^{-}$ for modifying the weight vector we will call this way of learning *global learning.* However, we can also modify at each step the weight vector with only one element of $\hat{T}_{\mathbf{w}}^{+} \cup \hat{T}_{\mathbf{w}}^{-}$. This way of learning will be called *local learning.* The learning rule is called *reinforcement learning* for both cases. The proportionality constant ε is called the *learning rate.*

We see that the output of a single-neuron binary Perceptron at a certain step k depends on the actual input and on the weight vector $\hat{\mathbf{w}}(k)$ at step k. We can consider the value of the weight vector at step k as the *state* of the learning system. This enables us to describe the single-neuron binary Perceptron within the framework of the theory of *finite sequential machines.* A finite sequential machine is described by a *state transition function* δ and an *output function* λ. Given the state $\hat{\mathbf{w}}(k)$ and the input $\mathbf{x}$, the next state is defined by $\hat{\mathbf{w}}(k+1) = \delta(\hat{\mathbf{w}}(k), \hat{\mathbf{x}})$ and the output at step k is defined by $y(k) = \lambda(\hat{\mathbf{w}}(k), \hat{\mathbf{x}})$. The behaviour of a sequential machine can be graphically represented by a so-called *state diagram.* In a state diagram, states are represented by circles and transitions between states by arrows pointing from the actual state to the next state. Each arrow has a label indicating the supplied input and a label corresponding to the output.

We can define the *output function* λ of the sequential machine corresponding to a single-neuron Perceptron as:

$$\lambda(\hat{\mathbf{w}}, \hat{\mathbf{x}}) = S(w_0 + w_1x_1 + w_2x_2 + \cdots + w_nx_n)$$

In case of local learning and with the learning rate ε equal to 1, we can define the

Table 2.10

$\mathbf{p}$	x_0	x_1	x_2	y	y^0
$\mathbf{p}_0$	1	0	0	0	1
$\mathbf{p}_1$	1	0	1	0	1
$\mathbf{p}_2$	1	1	0	0	1
$\mathbf{p}_3$	1	1	1	1	1

state transition function δ as:

$$\delta(\mathbf{w}, \mathbf{x}) = \begin{cases} \mathbf{w} & \text{if } \lambda(\mathbf{w}, \mathbf{x}) \text{ is correct} \\ \mathbf{w}+\mathbf{x} & \text{if } \mathbf{x} \in T_{\mathbf{w}}^{+} \\ \mathbf{w}-\mathbf{x} & \text{if } \mathbf{x} \in T_{\mathbf{w}}^{-} \end{cases}$$

Example 2.10

The function y to be realized is specified in Table 2.10. We use the extended input, i.e. with the constant $x_0 = 1$. The function y to be identified is a logical function of x_1 and x_2. The different input values will be denoted by $\mathbf{p}_0$, $\mathbf{p}_1$, $\mathbf{p}_2$ and $\mathbf{p}_3$. The neuron in the initial learning state (i.e. at step $k=0$) has the weight vector $\hat{\mathbf{w}}(0) = [w_0(0), w_1(0), w_2(0)]^t = [1, 1, 1]^t$. The output is defined by:

$$y(0) = S(w_0(0) + w_1(0)x_1 + w_2(0)x_2) = S(1 + x_1 + x_2)$$

Only for input $\mathbf{p}_3$ is the output correct, and so we remain in the same state only for that input. For input $\mathbf{p}_0 \in T_{\mathbf{w}}^{-}$ the output is 1 and the next state is $\mathbf{w}(0) - \mathbf{p}_0 = [0, 1, 1]^t$. For input $\mathbf{p}_1 \in T_{\mathbf{w}}^{-}$ the output is 1 and the next state is $\mathbf{w}(0) - \mathbf{p}_1 = [0, 1, 0]^t$. For input $\mathbf{p}_2 \in T_{\mathbf{w}}^{-}$ the output is 1 and the next state is $\mathbf{w}(0) - \mathbf{p}_2 = [0, 0, 1]^t$ (see Figure 2.14). We can continue this process in the same way, with the final result as shown in Figure 2.15.

■

In order to find a weight vector for a correct realization of a logical function, we found intuitively that in the learning phase the weight vector must be increased or decreased with vectors proportional to the misclassified extended input vectors.

We can see the same in a more formal way. In the case of local learning the adaptation at step k becomes:

$$\hat{\mathbf{w}}(k+1) = \hat{\mathbf{w}}(k) + \varepsilon \hat{\mathbf{x}}_i \quad \text{if } \hat{\mathbf{x}}_i \in \hat{T}_{\mathbf{w}}^{+}$$

and

$$\hat{\mathbf{w}}(k+1) = \hat{\mathbf{w}}(k) - \varepsilon \hat{\mathbf{x}}_i \quad \text{if } \hat{\mathbf{x}}_i \in \hat{T}_{\mathbf{w}}^{-}$$

In the case $\hat{\mathbf{x}}_i \in T_{\mathbf{w}}^{+}$ we have prior to adaptation that the inner product $\hat{\mathbf{w}}(k) \cdot \hat{\mathbf{x}}_i \leqslant 0$. After adaptation we have $\hat{\mathbf{w}}(k+1) \cdot \hat{\mathbf{x}}_i = \hat{\mathbf{w}}(k) \cdot \hat{\mathbf{x}}_i + \varepsilon |\hat{\mathbf{x}}_i|^2$. So we add a positive number

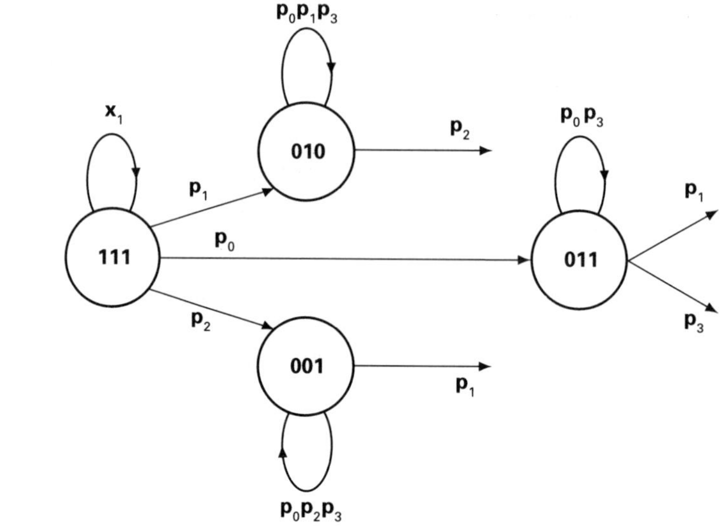

Figure 2.14 The initial state diagram of Example 2.10

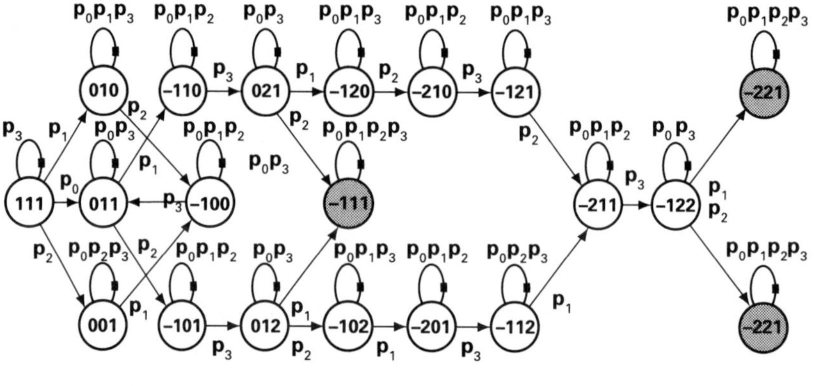

Figure 2.15 The complete state diagram of Example 2.10

$\varepsilon|\hat{\mathbf{x}}_i|^2$ to the old inner product, and the inner product is changed into the desired direction.

In the case $\hat{\mathbf{x}}_i \in \hat{T}_{\mathbf{w}}^-$ we have prior to adaptation that the inner product $\hat{\mathbf{w}}(k)\cdot\hat{\mathbf{x}}_i > 0$. After adaptation we have $\hat{\mathbf{w}}(k+1)\cdot\hat{\mathbf{x}}_i = \hat{\mathbf{w}}(k)\cdot\hat{\mathbf{x}}_i - \varepsilon|\hat{\mathbf{x}}_i|^2$. So the inner product is again changing into the desired direction.

In cases of global learning the adaptation at step k becomes:

$$\hat{\mathbf{w}}(k+1) = \hat{\mathbf{w}}(k) + \varepsilon(\Sigma\,\hat{\mathbf{x}}_i - \Sigma\,\hat{\mathbf{x}}_j) \text{ with } \hat{\mathbf{x}}_i \in \hat{T}_{\mathbf{w}}^+ \text{ and } \hat{\mathbf{x}}_j \in \hat{T}_{\mathbf{w}}^-$$

We can conceive $\Sigma\,\hat{\mathbf{x}}_i - \Sigma\,\hat{\mathbf{x}}_j$ as one misclassified correction vector $\hat{\mathbf{c}}$ of $\hat{T}_{\mathbf{w}}^+$. We have prior to adaptation that the inner product $\hat{\mathbf{w}}(k)\cdot\hat{\mathbf{c}} \leqslant 0$. After adaptation we have

$\hat{\mathbf{w}}(k+1)\cdot\hat{\mathbf{c}} = \hat{\mathbf{w}}(k)\cdot\hat{\mathbf{c}} + \varepsilon|\hat{\mathbf{c}}|^2$. Thus in the case of global learning the inner product is also changing into the desired direction. We see that at each step the behaviour of the single-neuron binary Perceptron is improved with respect to the misclassified input vectors at that step. However, in general after an adaptation step the set of misclassified vectors is changed, so we have to adapt the weight vector for that new set of misclassified vectors.

One may wonder whether we can always be sure that we will finally arrive at a state where the outputs are correct. It might be supposed, for instance, that we may never finish updating the weight vector because the process can enter a loop of states. However, we will show that when at a certain step a state (i.e. a weight vector) $\hat{\mathbf{w}}(k)$ is obtained, we will never return to the same state if there exists a solution.

Theorem 2.4

During the learning process a single-neuron binary Perceptron will never enter the same state more than once if there exists a solution space.

Proof

Let $\hat{\mathbf{w}}(k+n)$ be a state reached from $\hat{\mathbf{w}}(k)$ after n additional adaptations; we then have a sum or vectors added to $\hat{\mathbf{w}}(k)$:

$$\varepsilon(\Sigma\, \hat{\mathbf{x}}_i - \Sigma\, \hat{\mathbf{x}}_j) \text{ with } \hat{\mathbf{x}}_i \in \hat{T}^{+}_{\mathbf{w}(k+p)} \text{ and } \hat{\mathbf{x}}_j \in \hat{T}^{-}_{\mathbf{w}(k+p)} \text{ with } p=(0, 1, \dots, n-1)$$

Let $\hat{\mathbf{s}}$ (a *solution vector*) be a weight vector for which a correct solution is obtained. We consider the inner product of $\hat{\mathbf{s}}$ and $\hat{\mathbf{w}}(k+n)$:

$$\hat{\mathbf{s}}\cdot\hat{\mathbf{w}}(k+n) = \hat{\mathbf{s}}\cdot\hat{\mathbf{w}}(k) + \varepsilon(\Sigma\, \hat{\mathbf{s}}\cdot\hat{\mathbf{x}}_i - \Sigma\, \hat{\mathbf{s}}\cdot\hat{\mathbf{x}}_j)$$

summed over $\hat{\mathbf{x}}_i \in \hat{T}^{+}_{\mathbf{w}(k+p)}$ and $\hat{\mathbf{x}}_j \in \hat{T}^{-}_{\mathbf{w}(k+p)}$ with $p=(0, 1, \dots, n-1)$.

For every $\hat{\mathbf{x}}_i \in \hat{T}^{+}_{\mathbf{w}(k+p)}$ we have $\hat{\mathbf{s}}\cdot\hat{\mathbf{x}}_i > 0$ and for every $\hat{\mathbf{x}}_j \in \hat{T}^{-}_{\mathbf{w}(k+p)}$ we have $\hat{\mathbf{s}}\cdot\hat{\mathbf{x}}_j \leqslant 0$. Thus $\hat{\mathbf{s}}\cdot\hat{\mathbf{w}}(k+n) \neq \hat{\mathbf{s}}\cdot\hat{\mathbf{w}}(k)$ and hence $\hat{\mathbf{w}}(k+n) \neq \hat{\mathbf{w}}(k)$.

(It may happen that $\hat{T}^{+}_{\mathbf{w}(k+p)}$ is empty for all p and that $\hat{\mathbf{s}}\cdot\hat{\mathbf{x}}_j = 0$, thus in that case $\hat{\mathbf{s}}\cdot\hat{\mathbf{w}}(k+n) = \hat{\mathbf{s}}\cdot\hat{\mathbf{w}}(k)$. However, in that case we can take another solution vector $\mathbf{s}'$ from the solution space such that $\hat{\mathbf{s}}'\cdot\hat{\mathbf{x}}_j \neq 0$.) QED

Although we will never return in the same state during learning, this does not guarantee that the learning process will stop because the number of states in the space containing incorrect states is infinite. However, the quotient of the number of correct states and the number of incorrect states is finite. The subspace $\hat{S}$ of the extended weights space $\hat{W}$ containing all correct weight vectors is called the *solution space*. The solution space also contains an infinite number of weight vectors, as becomes clear from the following arguments.

If $\hat{\mathbf{w}}$ is a solution vector, then the inner product $\hat{\mathbf{w}}\cdot\hat{\mathbf{x}} > 0$ for all $\hat{\mathbf{x}} \in \hat{X}^{+}$ and $\hat{\mathbf{w}}\cdot\hat{\mathbf{x}} \leqslant 0$ for all $\hat{\mathbf{x}} \in \hat{X}^{-}$. The same holds for $\beta\hat{\mathbf{w}}$ with $\beta > 0$, thus $\beta\hat{\mathbf{w}}$ is also a solution vector. If $\hat{\mathbf{w}}_i$ and $\hat{\mathbf{w}}_j$ are solution vectors, then one easily verifies that $\beta_i\hat{\mathbf{w}}_i + \beta_j\hat{\mathbf{w}}_j$ with β_i and

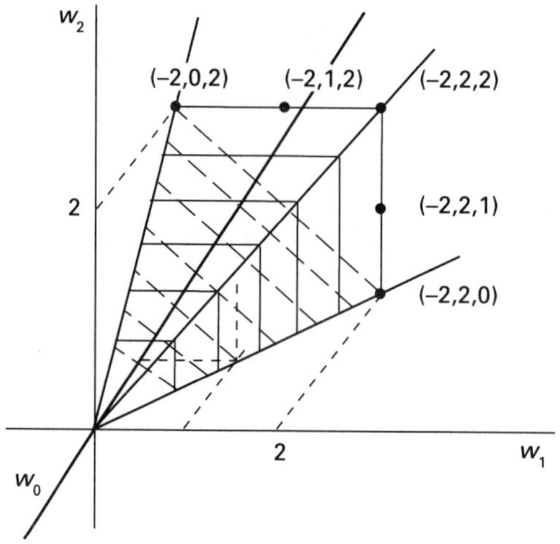

Figure 2.16 The solution space of Example 2.9

$\beta_j > 0$ is also a solution vector. A space having these properties is called a *convex cone.* In Figure 2.16 we have given the solution space of Example 2.9.

The quotient of the 'volume' of the solution space $\hat{S}$ and the volume of its complement $\hat{W} - \hat{S}$ is finite; thus if we could select weight vectors randomly we would have a finite probability of selecting a correct solution vector. In the learning process we do better because we never select a weight vector that has been chosen previously (Theorem 2.4), and moreover after each adaptation we move with the new weight vector in the direction of the solution space, as should become clear from the next example. In our previous discussion it should have been noticeable that we did not build our theory explicitly on binary valued input vectors. It turns out that linear threshold functions can also be functions of real-valued input vectors. In the next example we will use real valued inputs.

Example 2.11

Assume we want to learn a simple threshold function with one-dimensional real-valued input vectors such that the output equals 1 for $\mathbf{x}_1 = (1)$ and for $\mathbf{x}_2 = (1.5)$, whereas the output must be 0 for the inputs $\mathbf{x}_3 = (0.25)$ and $\mathbf{x}_4 = (-0.5)$.

In Figure 2.17 we have given for the extended input space the solution space for $\hat{X}^+ = \{\hat{\mathbf{x}}_1, \hat{\mathbf{x}}_2\} = \{[1, 1]^t, [1, 1.5]^t\}$ and $\hat{X}^- = \{\hat{\mathbf{x}}_3, \hat{\mathbf{x}}_4\} = \{[1, 0.25]^t, [1, -0.5]^t\}$. Because the output for $\hat{\mathbf{x}}_1$ must be 1, we must have for the solution weight vector **s**: $\mathbf{s} \cdot \hat{\mathbf{x}}_1 > 0$, and thus **s** must be located to the right of line l_1. For input $\hat{\mathbf{x}}_3$ the output must be 0 and thus $\mathbf{s} \cdot \hat{\mathbf{x}}_3 \leqslant 0$ and thus **s** must be located to the left of the line l_3 or on the line l_3. We can do the same for the other inputs. The intersection of the separate solution spaces gives the solution space, indicated by the shaded area in Figure 2.17.

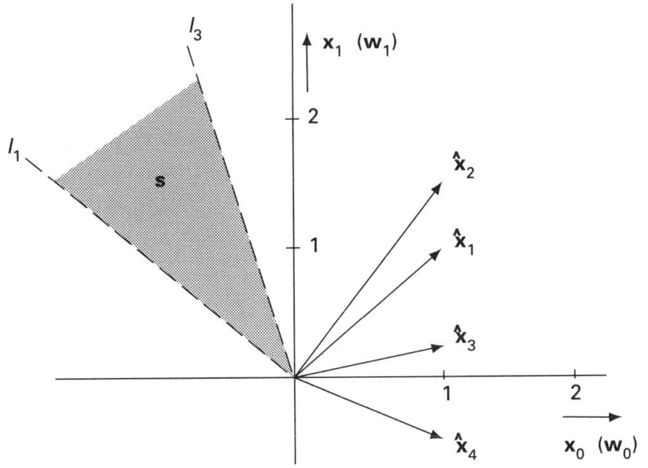

Figure 2.17 Construction of the solution space of Example 2.11

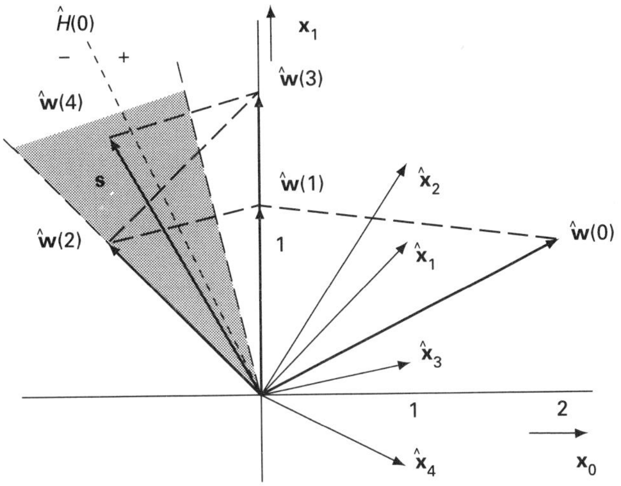

Figure 2.18 The construction of the sequence of weight vectors of Example 2.11

Let the initial weights be $[w_0(0), w_1(0)] = (2, 1)$, then the initial separating hyperplane $\hat{H}(0)$ is defined by $2x_0 + x_1 = 0$ (see Figure 2.18). With this initial hyperplane the extended inputs $\hat{\mathbf{x}}_3$ and $\hat{\mathbf{x}}_4$ give the wrong output. If we use the procedure for global learning we have to subtract from $\hat{\mathbf{w}}(0)$ the vector $\varepsilon(\hat{\mathbf{x}}_3 + \hat{\mathbf{x}}_4)$. With $\varepsilon = 1$ the new weight vector becomes $\hat{\mathbf{w}}(1) = (2, 1) - (2, -0.25) = (0, 1.25)$. We see that the weight vector is changed in the direction of the solution space. The corresponding separating hyperplane is now identical with the x_0-axis, and only for the input vector $\hat{\mathbf{x}}_3=(1, 0.25)$ do we obtain a wrong output because $w_0(1)x_0+w_1(1)x_1=0+(1.25)(0.25)>0$ and hence $y=1$. Now we subtract from $\hat{\mathbf{w}}(1)$ the input vector $\hat{\mathbf{x}}_3=(1, 0.25)$. The next weight vector becomes $\hat{\mathbf{w}}(2)=(-1, 1)$ at the border of the solution space. The output for the input vector $\hat{\mathbf{x}}_1$ is now just wrong: $y=0$. Thus in the next step we add $\hat{\mathbf{x}}_1=(1, 1)$ to the weight vector and we obtain $\hat{\mathbf{w}}(3)=(0, 2)$. Input vector $\hat{\mathbf{x}}_3$ is now misclassified, so we have to subtract from $\hat{\mathbf{w}}_3$ the vector $\hat{\mathbf{x}}_3=(1, 0.25)$; the next weight vector $\hat{\mathbf{w}}(4)=(-1, 1.75)$ is in the solution space. ■

The question arises of whether we will always enter, in a finite number of adaptation steps, the solution space. One can prove that for any constant learning rate ε (called *fixed increment learning*) this will be the case – even for a time-varying learning rate like $\varepsilon(k)=1/k$ or $\varepsilon(k)=k$. These statements are consequences of the *Perceptron convergence theorem*, which will be discussed in the next section.

2.5 The Perceptron convergence theorem

This section mainly deals with the formal statement of the Perceptron convergence theorem and its proof. Because we have already outlined the theorem in the previous section in an informal way, this section can be omitted on a first reading of the book without loss of continuity.

The Perceptron convergence theorem concerns the convergence of the learning procedure to find, from samples of correct behaviour, the linear threshold function $y=S(w_0+w_1x_1+w_2x_2+\cdots+w_nx_n)$ realized by a single binary Perceptron, if the function $y=f(x_1, x_2,\ldots, x_n)$ to be identified is a linear threshold function (the symbol S represents the step function).

In the previous section the variables y and x_i were binary valued; the Perceptron convergence theorem is, however, also applicable if the variables x_i are real valued.

The *reinforcement learning rule* is given by:

Let $\hat{\mathbf{w}}(0)=(w_0(0), w_1(0),\ldots, w_n(0))$ be any initial weight vector.

Let $\hat{\mathbf{w}}(k)=(w_0(k), w_1(k),\ldots, w_n(k))$ be the weight vector at step k.

Let $\varepsilon(k)$ be a variable *learning rate*.

Local learning

If at step k: $w_0(k)+w_1(k)x_{i1}+w_2(k)x_{i2}+\cdots+w_n(k)x_{in}\leqslant 0$ and it is given that $y=f(\mathbf{x}_i)=1$ for some input vector $\mathbf{x}_i=(x_{i1}, x_{i2},\ldots, x_{in})$, then change $\hat{\mathbf{w}}(k)$ into:

$$\hat{\mathbf{w}}(k+1)=\hat{\mathbf{w}}(k)+\varepsilon(k)\hat{\mathbf{x}}_i$$

If at step k: $w_0(k)+w_1(k)x_{i1}+w_2(k)x_{i2}+\cdots+w_n(k)x_{in}>0$ and it is given that

$y = f(\mathbf{x}_i) = 0$ for some input vector $\mathbf{x}_i = (x_{i1}, x_{i2}, \ldots, x_{in})$, then change $\hat{\mathbf{w}}(k)$ to:

$$\hat{\mathbf{w}}(k+1) = \hat{\mathbf{w}}(k) - \varepsilon(k)\hat{\mathbf{x}}_i$$

Global learning

If $\hat{\mathbf{c}}(k) = \Sigma\,\hat{\mathbf{x}}_i - \Sigma\,\hat{\mathbf{x}}_j$ with $\hat{\mathbf{x}}_i \in \hat{T}^+(k)$ and $\hat{\mathbf{x}}_j \in \hat{T}^-(k)$ with $\hat{T}^+(k)$ the set of extended input vectors with target value 1 and actual output of the neuron at step k equal to 0, and $\hat{T}^-(k)$ the set of extended input vectors with target value 0 and actual output of the neuron at step k equal to 1, then change the weight vector into:

$$\hat{\mathbf{w}}(k+1) = \hat{\mathbf{w}}(k) + \varepsilon(k)\hat{\mathbf{c}}(k)$$

The Perceptron convergence theorem states: in the case of local or global learning the weight vector $\hat{\mathbf{w}}(k)$ converges to a solution vector $\hat{\mathbf{s}}$ if the samples are linearly separable and if the following conditions on the learning rate $\varepsilon(k)$ are satisfied:

1. $\varepsilon(k) \geqslant 0$
2. $\lim_{m\to\infty} \sum_{k=1}^{m} \varepsilon(k) = \infty$
3. $\lim_{m\to\infty} \dfrac{\sum_{k=1}^{m} (\varepsilon(k))^2}{\left(\sum_{k=1}^{m} \varepsilon(k)\right)^2} = 0$

The conditions imply that convergence occurs for any positive constant learning rate ε or if $\varepsilon(k) = 1/k$, or even if it increases like $\varepsilon(k) = k$. If the set of samples is not linear separable, then the 'separating' hyperplane defined by $\Sigma_{i=0}^{n} w_i x_i = 0$ will 'oscillate' between several positions if the learning rate is constant or increasing. The same occurs when the data set contains contradicting samples, so we can formulate the following practical statement:

Practical statement 2.1

A decreasing value of the learning rate $\varepsilon(k)$ is particularly advisable if the set of samples may not be linear separable or contains contradictions, because in that case the effect of 'disruptive' samples will be reduced.

(Note: For a formal proof of the theorem it is required that during learning for correct classification the inner product $\hat{\mathbf{w}} \cdot \hat{\mathbf{x}} > \delta$ for $\mathbf{x} \in \mathbf{T}_\mathbf{w}^+$ and $\mathbf{w} \cdot \mathbf{x} \leqslant \delta$ for $\mathbf{x} \in T_\mathbf{w}^-$ with δ a certain small positive constant.)

We will not give a general proof of the theorem and restrict it to the case where $\varepsilon(k) = 1/|\mathbf{c}(k)|$ for global learning.

Proof

If the learning process converges, then there exists a solution vector $\hat{\mathbf{s}}$ such that after some finite time m: $\hat{\mathbf{w}}(m)=\hat{\mathbf{s}}$. If $\hat{\mathbf{s}}$ is a solution vector, then the unit vector $\hat{\mathbf{u}}=\hat{\mathbf{s}}/|\hat{\mathbf{s}}|$ is also a solution vector. Thus $\hat{\mathbf{w}}(m)=\lambda\hat{\mathbf{u}}$ for some $\lambda>0$. If $\hat{\mathbf{w}}(m)=\lambda\hat{\mathbf{u}}$, then $\hat{\mathbf{u}}\cdot\hat{\mathbf{w}}(m)/|\hat{\mathbf{w}}(m)|=1$. At every step k we have for the value of the cosine of the angle between $\hat{\mathbf{u}}$ and $\mathbf{w}(k)$: $\hat{\mathbf{u}}\cdot\hat{\mathbf{w}}(k)/|\hat{\mathbf{w}}(k)|\leqslant 1$. We will show that at each step $\hat{\mathbf{u}}\cdot\hat{\mathbf{w}}(k)/|\hat{\mathbf{w}}(k)|$ will increase with a positive amount and thus after a finite number of steps must become equal to 1.

Consider $\hat{\mathbf{u}}\cdot\hat{\mathbf{w}}(k+1)=\hat{\mathbf{u}}\cdot\{\hat{\mathbf{w}}(k)+\varepsilon(k)\hat{\mathbf{c}}(k)\}$. Because we take $\varepsilon(k)=1/|\mathbf{c}(k)|$ we obtain $\hat{\mathbf{u}}\cdot\hat{\mathbf{w}}(k+1)=\hat{\mathbf{u}}\cdot\hat{\mathbf{w}}(k)+\hat{\mathbf{u}}\cdot\hat{\mathbf{c}}(k)/|\hat{\mathbf{c}}(k)|$. Because $\hat{\mathbf{c}}(k)=\Sigma\,\hat{\mathbf{x}}_i-\Sigma\,\hat{\mathbf{x}}_j$ with $\hat{\mathbf{x}}_i\in\hat{T}_{\mathrm{w}}{}^{+}(k)$ and $\hat{\mathbf{x}}_j\in\hat{T}_{\mathrm{w}}{}^{-}(k)$ we have $\hat{\mathbf{u}}\cdot\hat{\mathbf{c}}(k)=\delta(k)>0$. Let $\delta'=\min_k \delta(k)/|\mathbf{c}(k)|$. So at each step the inner product $\hat{\mathbf{u}}\cdot\mathbf{w}(k)$ is increased with at least the value δ', thus $\hat{\mathbf{u}}\cdot\mathbf{w}(k)>k\delta'$. (Note that if we take $\varepsilon(k)>1/|\mathbf{c}(k)|$, then the increments at each step will be larger.)

Now consider $|\hat{\mathbf{w}}(k+1)|^2=|\hat{\mathbf{w}}(k)+\varepsilon(k)\hat{\mathbf{c}}(k)|^2=|\hat{\mathbf{w}}(k)|^2+2\hat{\mathbf{w}}(k)\cdot\hat{\mathbf{c}}(k)/|\hat{\mathbf{c}}(k)|+1$. Because $\hat{\mathbf{w}}(k)\cdot\hat{\mathbf{c}}(k)\leqslant 0$ we have $|\hat{\mathbf{w}}(k+1)|^2<|\hat{\mathbf{w}}(k)|^2+1$. This implies $|\hat{\mathbf{w}}(k)|^2<|\hat{\mathbf{w}}(0)|^2+k$ and thus $|\hat{\mathbf{w}}(k)|<(|\hat{\mathbf{w}}(0)|^2+k)^{1/2}$. The expression $\hat{\mathbf{u}}\cdot\hat{\mathbf{w}}(k)/|\hat{\mathbf{w}}(k)|$ can thus be approximated by: $\hat{\mathbf{u}}\cdot\hat{\mathbf{w}}(k)|/\hat{\mathbf{w}}(k)|>\delta' k/(|\mathbf{w}(0)|^2+k)^{1/2}$ and thus after a finite number of steps k the value of $\hat{\mathbf{u}}\cdot\hat{\mathbf{w}}(k)/|\hat{\mathbf{w}}(k)|$ must become equal to 1 and thus a solution vector must be reached.

QED

We have found that after a finite number of steps the reinforcement learning rule will provide a correct solution if certain conditions are satisfied. There are, however, other learning rules that will give correct solutions under certain conditions. These rules are also based on the *gradient descent procedures* for minimizing certain *criterion functions*, like *cost functions* (for misclassification), sometimes called *error functions* or *energy functions.*

The criterion function $E(\hat{\mathbf{w}})$ is a positive scalar function which is zero if $\hat{\mathbf{w}}$ is equal to a solution vector. With the gradient descent procedure we move through a sequence of weight vectors $\hat{\mathbf{w}}_0$, $\hat{\mathbf{w}}_1$, $\hat{\mathbf{w}}_2,\ldots$ such that $E(\hat{\mathbf{w}}_0)>E(\hat{\mathbf{w}}_1)>E(\hat{\mathbf{w}}_2)>\cdots$ and finally end up with $E(\hat{\mathbf{w}}_f)=0$. The procedure starts with some arbitrary weight vector $\hat{\mathbf{w}}_0$, then computes the gradient vector $\nabla E(\hat{\mathbf{w}}_0)$. The next weight vector is obtained by moving in the direction of the steepest descent, i.e. along the negative of the gradient vector $\nabla E(\hat{\mathbf{w}}_0)$.

An obvious choice for $E(\hat{\mathbf{w}})$ would be the number of misclassified input vectors for the weight vector $\hat{\mathbf{w}}$. But in that case $E(\hat{\mathbf{w}})$ would be piecewise constant and the gradient of $E(\hat{\mathbf{w}})$ is zero or undefined, so this would be a poor candidate for the criterion function.

The reinforcement learning rule used before is obtained if we minimize the *Perceptron criterion function,* i.e.

$$E(\hat{\mathbf{w}})=\sum_{\hat{\mathbf{x}}\in\hat{T}_{\mathrm{w}}}-\hat{\mathbf{w}}\cdot\hat{\mathbf{x}}\ \text{with}\ T_{\mathrm{w}}=T_{\mathrm{w}}^{+}\cup\{-T_{\mathrm{w}}^{-}\}$$

Because $\hat{\mathbf{w}}\cdot\hat{\mathbf{x}}$ is negative for each $\mathbf{x}\in\hat{T}_{\mathrm{w}}$ the criterion function $E(\hat{\mathbf{w}})$ is always positive if there are misclassified input vectors, and it will only be zero for a solution vector $\mathbf{s}$. Because $\Delta E(\hat{\mathbf{w}}) = \nabla\{E(\hat{\mathbf{w}})\}\Delta\hat{\mathbf{w}}$ we have the situation that $E(\hat{\mathbf{w}})$ will reduce if $\Delta\hat{\mathbf{w}}$ is proportional to the negative value of the gradient vector $\nabla\{E(\hat{\mathbf{w}})\}$. The ith component of the gradient vector $\partial E(\hat{\mathbf{w}})/\partial w_i$ is equal to x_i, thus we obtain for $\Delta\hat{\mathbf{w}}$:

$$\Delta\hat{\mathbf{w}} = \varepsilon \cdot \sum_{\hat{\mathbf{x}} \in \hat{T}_{\mathbf{w}}} \hat{\mathbf{x}}$$

as we used in the reinforcement learning rule.

Another criterion function could be, for example:

$$E'(\hat{\mathbf{w}}) = \sum_{\hat{\mathbf{x}} \in \hat{T}_{\mathbf{w}}} (\hat{\mathbf{w}}^t\hat{\mathbf{x}})^2 \text{ with } T_{\mathbf{w}} = T_{\mathbf{w}}^+ \cup \{-T_{\mathbf{w}}^-\}$$

Although this criterion function can be used, it turns out that the corresponding learning rule is inferior to the reinforcement learning rule.

2.6 Performance of a two-layer binary Perceptron

In the previous sections we found that a single-neuron binary Perceptron can realize linear threshold functions with respect to the binary variables $x_1, x_2, \ldots, x_n$. However, most logical functions $f: \{0, 1\}^n \rightarrow \{0, 1\}$ are not linear threshold functions with respect to $x_1, x_2, \ldots, x_n$. In this section, we will show that any logical function can be realized with a two-layer binary Perceptron with one neuron in the second layer.

First we define some concepts, using a terminology related to pattern recognition.

The argument values of logical functions will be called patterns, so we have a *pattern set*: $P = \{0, 1\}^n$; a *pattern* $\mathbf{p}_i$ is an element of P, i.e. $\mathbf{p}_i = [p_{i_1}, p_{i_2}, \ldots, p_{i_n}]$ with $p_{i_k} \in \{0, 1\}$.

The *intersection* $\mathbf{r}_k$ of a pattern $\mathbf{p}_i \in \{0, 1\}^n$ and a pattern $\mathbf{q}_j \in \{0, 1\}^n$ is defined by $\mathbf{p}_i \cap \mathbf{q}_j = \mathbf{r}_k$, with $r_{k_m} = 1$ if $p_{i_m} = 1$ and $q_{j_m} = 1$ and with $r_{k_m} = 0$ otherwise.

The *union* $\mathbf{r}_k$ of a pattern $\mathbf{p}_i \in \{0, 1\}^n$ and a pattern $\mathbf{q}_j \in \{0, 1\}^n$ is defined by $\mathbf{p}_i \cup \mathbf{q}_j = \mathbf{r}_k$. With $r_{k_m} = 1$ if $p_{i_m} = 1$ or $q_{j_m} = 1$ or both and with $r_{k_m} = 0$ otherwise.

The *order* of a pattern $\mathbf{p}_i$ is defined as the number of 1s occurring in $\mathbf{p}_i$, and is denoted by $|\mathbf{p}_i|$.

In the previous section we treated x_j as a binary variable; however, it is helpful to consider x_j as a function: $x_j: P \rightarrow \{0, 1\}$ defined by:

$$x_j(\mathbf{p}_i) = p_{i_j}$$

Because $x_j(\mathbf{p}_i)$ depends on one component, or one pixel, from the pattern $\mathbf{p}_i$ we will call $\mathbf{x}_j$ a *pixel function.*

We introduce a special logical function which acts as a mask by which we observe a pattern. Let $\mathbf{q} \in P$, then $\mathbf{x}_\mathbf{q}: P \rightarrow \{0, 1\}$ is a so-called *mask function* defined by:

$$\mathbf{x}_\mathbf{q}(\mathbf{p}_i) = 1 \text{ if } \mathbf{q} \cap \mathbf{p}_i = \mathbf{q} \text{ and } \mathbf{x}_\mathbf{q}(\mathbf{p}_i) = 0 \text{ otherwise}$$

Example 2.12

$$\mathbf{x}_{1100}(1110)=1 \text{ and } \mathbf{x}_{1100}(1001)=0$$ ■

A mask function can be written as a product of pixel functions. Let $\mathbf{q}\in P$, then $\mathbf{x_q}(\mathbf{p_h})=\tilde{x}_{q_1}(\mathbf{p_h})\tilde{x}_{q_2}(\mathbf{p_h})\cdots\tilde{x}_{q_n}(\mathbf{p_h})$

with

$$\tilde{x}_{q_i}(\mathbf{p_h})=\begin{cases}=x_i(\mathbf{p_h}) & \text{if } q_i=1\\ =1 & \text{if } q_i=0\end{cases}$$

We will call $\mathbf{q}$ in $\mathbf{x_q}(\mathbf{p_h})$ the *mask* of the mask function $\mathbf{x_q}$.

Example 2.13

$$\text{For } n=3 \quad \mathbf{x}_{101}=x_1x_3$$ ■

We define the *substratum* $S_{\mathbf{p}_i}$ of a pattern $\mathbf{p}_i$ as the set of patterns: $S_{\mathbf{p}_i}=\{\mathbf{q}_k|\mathbf{x}_{\mathbf{q}_k}(\mathbf{p}_i)=1\}$.

Example 2.14

$$S_{0011}=\{0000, 0010, 0001, 0011\}$$ ■

We define the *cover* $C_{\mathbf{p}_i}$ of a pattern $\mathbf{p}_i$ as the set of patterns: $C_{\mathbf{p}_i}=\{\mathbf{q}_k|\mathbf{x}_{\mathbf{p}_i}(\mathbf{q}_k)=1\}$.

Example 2.15

$$C_{0011}=\{0011, 0111, 1011, 1111\}$$ ■

If a logical function y: $\{0, 1\}^n\rightarrow\{0, 1\}$ is written as an arithmetical linear combination of the mask function:

$$y(\mathbf{p}_j)=\Sigma\, w_i\mathbf{x}_{\mathbf{q}_i}(\mathbf{p}_j) \text{ with } w_i\in\mathbb{R}$$

then we will call such a form an *arithmetical conjunctive normal form.*

In the case where the logical function is written as:

$$y(\mathbf{p}_j)=S(\Sigma\, w_i\mathbf{x}_{\mathbf{q}_i}(\mathbf{p}_j)) \text{ with } w_i\in\mathbb{R} \text{ and } S \text{ the step function}$$

then we call this form an *indirect arithmetical conjunctive normal form.*

Now we can state the following theorem:

Theorem 2.5

Any logical function y: $\{0, 1\}^n\rightarrow\{0, 1\}$ can be written in an arithmetical conjunctive

Table 2.11

x_1	x_2	y
0	0	0
1	0	1
0	1	1
1	1	0

normal form:

$$y=\sum_{\mathbf{q}_j} w_{\mathbf{q}_j}\cdot \mathbf{x}_{\mathbf{q}_j} \quad \mathbf{q}_j\in\{0,1\}^n$$

with $w_{\mathbf{q}_j}$ an integer such that for each pattern $\mathbf{p}_i\in P$:

$$\sum_{\mathbf{q}_j\in S_{\mathbf{p}_i}} w_{\mathbf{q}_j}=y(\mathbf{p}_i) \quad \text{with } S_{\mathbf{p}_i} \text{ the substratum of } \mathbf{p}_i$$

Before proving the theorem we will give an example of the theorem.

Example 2.16

The exclusive-or function defined by Table 2.11 can be written as:

$$y=w_{00}\mathbf{x}_{00}+w_{10}\mathbf{x}_{10}+w_{01}\mathbf{x}_{01}+w_{11}\mathbf{x}_{11}$$

It turns out that $w_{00}=0$, $w_{10}=1$, $w_{01}=1$ and $w_{11}=-2$. Thus:

$$y=x_1+x_2-2x_1x_2$$

■

Proof of Theorem 2.5

We have to prove that for any logical function y: $\{0,1\}^n\rightarrow\{0,1\}$ we can find a unique set of coefficients $w_{\mathbf{q}_j}$ such that:

$$\sum_{\mathbf{q}_j\in P} w_{\mathbf{q}_j}\cdot \mathbf{x}_{\mathbf{q}_j}(\mathbf{p}_i)=y(\mathbf{p}_i) \text{ for each } \mathbf{p}_i\in P=\{0,1\}^n$$

Because $\mathbf{x}_{\mathbf{q}_j}(\mathbf{p}_i)=1$ if $\mathbf{q}_j\in S_{\mathbf{p}_i}$ and 0 otherwise, we have for every $\mathbf{p}_i$:

$$\sum_{\mathbf{q}_j\in P} w_{\mathbf{q}_j}\cdot \mathbf{x}_{\mathbf{q}_j}(\mathbf{p}_i)=\sum_{\mathbf{q}_j\in S_{\mathbf{p}_i}} w_{\mathbf{q}_j}\cdot \mathbf{x}_{\mathbf{q}_j}(\mathbf{p}_i)$$

For $\mathbf{q}_j\in S_{\mathbf{p}_i}$ we have $\mathbf{x}_{\mathbf{q}_j}(\mathbf{p}_i)=1$, thus:

$$\sum_{\mathbf{q}_j\in S_{\mathbf{p}_i}} w_{\mathbf{q}_j}\cdot \mathbf{x}_{\mathbf{q}_j}(\mathbf{p}_i)=\sum_{\mathbf{q}_j\in S_{\mathbf{p}_i}} w_{\mathbf{q}_j}$$

Thus if:

$$\sum_{\mathbf{q}_j \in S_{\mathbf{p}_i}} w_{\mathbf{q}_j} = y(\mathbf{p}_i) \text{ for each } \mathbf{p}_i \in P$$

we obtain:

$$\sum_{\mathbf{q}_j \in P} w_{\mathbf{q}_j} \cdot \mathbf{x}_{\mathbf{q}_j}(\mathbf{p}_i) = y(\mathbf{p}_i)$$

Moreover we have for a total function 2^n patterns and thus 2^n independent linear equations of the form:

$$\sum_{\mathbf{q}_j \in S_{\mathbf{p}_i}} w_{\mathbf{q}_j} = y(\mathbf{p}_i)$$

Because the number of coefficients $w_{\mathbf{q}_j}$ is 2^n, the solution is unique. QED

Example 2.17

Given the function y: $\{0, 1\}^n \rightarrow \{0, 1\}$ specified by Table 2.12, with the conditions for $\Sigma\, w_{\mathbf{q}}$ given in the last column. The solution for the set equations is:

$$\begin{aligned} w_{000} &= 1 & w_{100} &= 0 \\ w_{001} &= 0 & w_{101} &= -1 \\ w_{010} &= -1 & w_{110} &= 0 \\ w_{011} &= 0 & w_{111} &= 2 \end{aligned}$$

Thus the function y can be written as a linear combination of mask functions:

$$y = 1 - x_2 - x_1 x_3 + 2 x_1 x_2 x_3$$ ■

In Section 2.2 we defined a linear threshold function with respect to the binary

Table 2.12

	x_1	x_2	x_3	y	
$\mathbf{p}_0$	0	0	0	1	$= w_{000}$
$\mathbf{p}_1$	0	0	1	1	$= w_{000} + w_{001}$
$\mathbf{p}_2$	0	1	0	0	$= w_{000} + w_{010}$
$\mathbf{p}_3$	0	1	1	0	$= w_{000} + w_{001} + w_{010} + w_{011}$
$\mathbf{p}_4$	1	0	0	1	$= w_{000} + w_{100}$
$\mathbf{p}_5$	1	0	1	0	$= w_{000} + w_{001} + w_{100} + w_{101}$
$\mathbf{p}_6$	1	1	0	0	$= w_{000} + w_{100} + w_{010} + w_{110}$
$\mathbf{p}_7$	1	1	1	1	$= w_{000} + w_{001} + w_{010} + w_{011} + w_{100} + w_{101} + w_{110} + w_{111}$

variables $x_1, x_2, \ldots, x_n$ as a function that can be written as $y = S(\Sigma\, w_i x_i - T)$, with S the step function.

From Theorem 2.5 the next theorem follows immediately.

Theorem 2.6

Any logical function y: $\{0, 1\}^n \to \{0, 1\}$ is a linear threshold function with respect to the set of binary mask functions, i.e.

$$y = \sum_{\mathbf{q}_j} w_{\mathbf{q}_j} \mathbf{x}_{\mathbf{q}_j}$$

or

$$y = S\left(\sum_{\mathbf{q}_j \neq \mathbf{q}_0} w_{\mathbf{q}_j} \cdot \mathbf{x}_{\mathbf{q}_j} - T \right) \text{ with threshold } T = -w_{00\ldots0}$$

Example 2.18

The exclusive-or function of Example 2.16 is a linear threshold function with respect to the binary mask functions: $\mathbf{x}_{00} = 1$, $\mathbf{x}_{10} = x_1$, $\mathbf{x}_{01} = x_2$ and $\mathbf{x}_{11} = x_1 x_2$:

$$y = S(x_1 + x_2 - 2x_1 x_2 - T) \text{ with } T = w_{00} = 0$$ ■

A central theorem is as follows:

Theorem 2.7

Any logical function f: $\{0, 1\}^n \to \{0, 1\}$ can be realized by a simple two-layer binary Perceptron.

Proof

Any logical function y: $\{0, 1\}^n \to \{0, 1\}$ is a linear threshold function with respect to the binary valued mask functions:

$$y = S\left(\sum_{\mathbf{q}_j \neq \mathbf{q}_0} w_{\mathbf{q}_j} \cdot \mathbf{x}_{\mathbf{q}_j} - T \right) \text{ with threshold } T = -w_{00\ldots0}$$

Given the binary valued mask functions $\mathbf{x}_\mathbf{q}$, one single binary output neuron can realize the linear threshold function with respect to the mask functions $\mathbf{x}_\mathbf{q}$. The threshold of the second-layer neuron is equal to $-w_{00\ldots0}$ of the linear threshold function y.

Any mask function $\mathbf{x}_\mathbf{q} = x_i x_j \cdots x_k$ is a linear threshold function with respect to $x_i, x_j, \ldots, x_k$, because $\mathbf{x}_\mathbf{q} = S(x_i + x_j + \cdots + x_k - T)$ with T equal to the number of variables in the product $\mathbf{x}_q = x_i x_j \cdots x_k$ minus 1, and thus it can be realized by a single binary neuron in the first layer. Such a first-layer neuron has an input x_i for every x_i occurring in the product $\mathbf{x_q}$ with a corresponding weight $w_i=1$. The output of such a first-layer neuron equals 1 if $\mathbf{x_q}(\mathbf{p}_i)=1$, and 0 otherwise.

All neurons realizing mask functions constitute the first layer of the Perceptron. The output of a first-layer neuron realizing a mask function $\mathbf{x_q}$ is multiplied by the *synaptic weight* $w_\mathbf{q}$ of the connection to the output neuron.

Thus the output of the second-layer neuron equals 1 if and only if:

$$\sum_\mathbf{q} w_\mathbf{q} \cdot \mathbf{x_q} > -w_{00\ldots0} \qquad \text{QED}$$

Figure 2.19 gives the two-layer binary Perceptron which realizes the logical function described in Example 2.17. Figure 2.20 gives an alternative realization of the same function.

In Theorem 2.5 we found that any logical function can be written in an arithmetical conjunctive normal form. The proof of the theorem also revealed a method of how to find that arithmetical conjunctive normal form. There is, however, another method to find the arithmetical conjunctive normal form. If we have written a logical function as a Boolean function then we can convert it in a systematic way into an equivalent arithmetical function using the following rules:

1. If x_i is Boolean function (a single variable), then x_i is an equivalent arithmetical function.

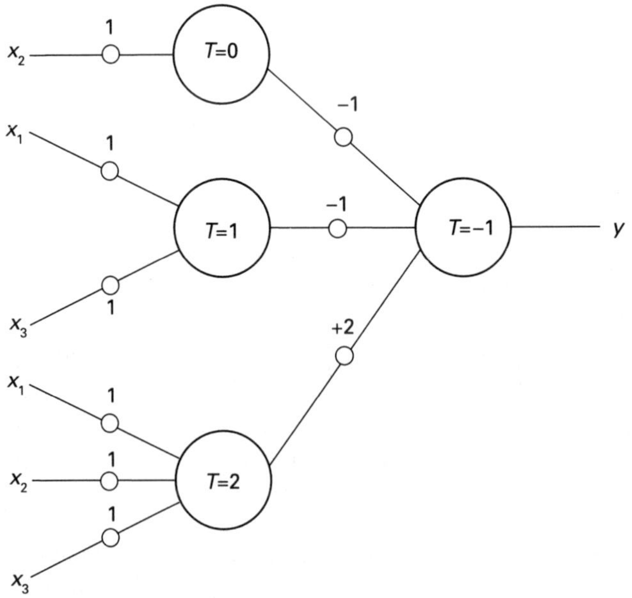

Figure 2.19 Two-layer binary Perceptron illustrating Table 2.11

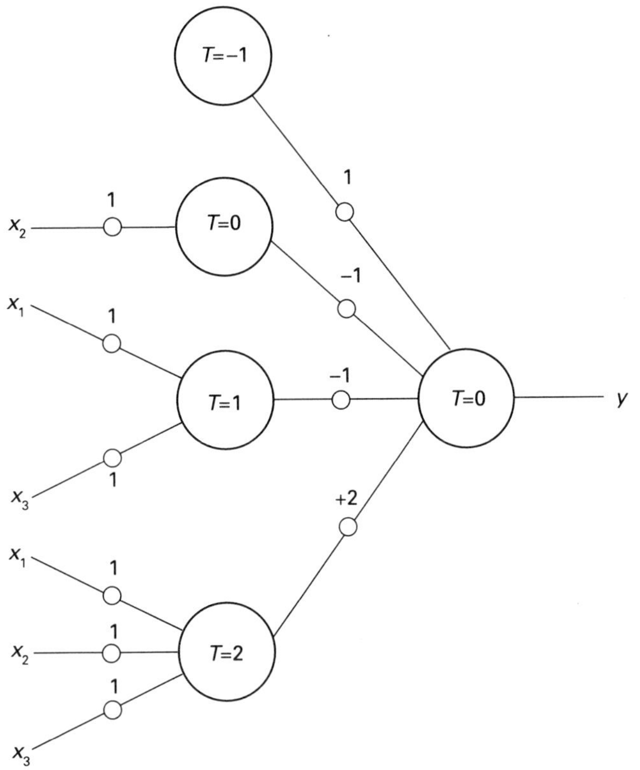

Figure 2.20 Alternative configuration of the binary Perceptron of Figure 2.19

2. If $\bar{x}_i$ is Boolean function (the inverse of x_i), then $1-x_i$ is an equivalent arithmetical function.
3. If the Boolean function f_1 is equivalent with the arithmetical function f'_1 and the Boolean function f_2 is equivalent with the arithmetical function f'_2, then the Boolean function $f_1 \wedge f_2$ is equivalent with the arithmetical function $f'_1 f'_2$.
4. If the Boolean function f_1 is equivalent to the arithmetical function f'_1 and the Boolean function f_2 is equivalent to the arithmetical function f'_2, then the Boolean function $f_1 \vee f_2$ (also written as $f_1 + f_2$) is equivalent to the arithmetical function $S(f'_1 + f'_2)$, with S the step function.
5. If the Boolean function f_1 is equivalent to the arithmetical function f'_1 and the Boolean function f_2 is equivalent to the arithmetical function f'_2 and the Boolean functions f_1 and f_2 are not true for the same arguments, then the Boolean function $f_1 \vee f_2$ is equivalent to the arithmetical function $f'_1 + f'_2$.

Table 2.13

x_1	x_2	x_3	y
0	0	0	1
0	0	1	1
0	1	0	0
0	1	1	0
1	0	0	1
1	0	1	0
1	1	0	0
1	1	1	1

Example 2.19

Consider the logical function specified by Table 2.13. This function can be specified by the Boolean function: $y=\bar{x}_1\bar{x}_2\bar{x}_3+\bar{x}_1\bar{x}_2x_3+x_1\bar{x}_2\bar{x}_3+x_1x_2x_3$. Because one and only one argument of y can be true we can use rule (5) above and obtain the equivalent arithmetical conjunctive normal form:

$$y=(1-x_1)(1-x_2)(1-x_3)+(1-x_1)(1-x_2)x_3+x_1(1-x_2)(1-x_3)+x_1x_2x_3$$

or

$$y=1-x_2-x_1x_3+2x_1x_2x_3 \text{ as found before in Example 2.17}$$ ■

A logical function can also be considered as a *characteristic function of some set $Q\subseteq P$*. A characteristic function of a set $Q\subseteq P$ is a logical function f_Q such that $f_Q(\mathbf{p}_i)=1$ if $\mathbf{p}_i\in Q$ and $f_Q(\mathbf{p}_i)=0$ otherwise. As a consequence of the conversion rules stated above, we have the next theorem. In Theorem 2.8 we will use the *complement* of $\mathbf{p}_i$ denoted by $\bar{\mathbf{p}}_i$ (e.g. if $\mathbf{p}_i=\langle 010\rangle$ then $\bar{\mathbf{p}}_i=\langle 101\rangle$).

Theorem 2.8

A characteristic function $f_{\mathbf{p}_i}$: $P\rightarrow\{0, 1\}$ for the singleton $K=\{\mathbf{p}_i\}$ can always be written in the arithmetical conjunctive normal form:

$$f_{\mathbf{p}_i}=\sum_{\mathbf{q}_m\in C_{\mathbf{p}_i}}(-1)^{|\mathbf{q}_m\cap\bar{\mathbf{p}}_i|}\mathbf{x}_{\mathbf{q}_m}$$

with $C_{\mathbf{p}_i}$ the cover of pattern $\mathbf{p}_i$ and $|\mathbf{q}_m\cap\bar{\mathbf{p}}_i|$ the order of $\mathbf{q}_m\cap\bar{\mathbf{p}}_i$ (i.e. the number of ones occurring in $\mathbf{q}_m\cap\bar{\mathbf{p}}_i$).

Proof

Consider x_i as a Boolean function, then any characteristic function $f_{\mathbf{p}_i}$ of the singleton $\{\mathbf{p}_i\}$ can be written as a Boolean product $f^B_{\mathbf{p}_i}=\tilde{x}_1\tilde{x}_2\cdots\tilde{x}_n$ with $\tilde{x}_j=x_j$ if $\mathbf{p}_{i_j}=1$ and

$\tilde{x}_j = \bar{x}_j$ (the negation of x_j) if $\mathbf{p}_{i_j} = 0$. (For example for $n=3$ and pattern $\mathbf{p}_i = \langle 101 \rangle$: $f^B_{\mathbf{p}_i} = x_1 \bar{x}_2 x_3$.)

The Boolean function x_i is equivalent to the arithmetical pixel function x_i. The Boolean function $\bar{x}_i$ is equivalent to the arithmetical function $(1 - x_i)$, thus if we replace the Boolean variables in the Boolean product by the corresponding arithmetical functions we obtain an equivalent arithmetical product. The elimination of the brackets in this product (and some thinking) gives the desired result. QED

Example 2.20

Let $f^B_{\mathbf{p}_i} = \bar{x}_1 \bar{x}_2 x_3 x_4$ be the Boolean form of the characteristic function of the singleton $\{\mathbf{p}_i\} = \{0011\}$. The complement of $\mathbf{p}_i$ is: $\bar{\mathbf{p}}_i = 1100$. The cover of $\mathbf{p}_i$ is: $C_{\mathbf{p}_i} = \{0011, 1011, 0111, 111\}$. Thus we can replace the Boolean function by the equivalent arithmetical function:

$$f_{0011} = (-1)^{|0000|} \mathbf{x}_{0011} + (-1)^{|1000|} \mathbf{x}_{1011} + (-1)^{|0100|} \mathbf{x}_{0111} + (-1)^{|1100|} \mathbf{x}_{1111}$$

or simply:

$$f_{\mathbf{p}_i} = x_3 x_4 - x_1 x_3 x_4 - x_2 x_3 x_4 + x_1 x_2 x_3 x_4 \text{ (note: } \bar{x}_1 \bar{x}_2 x_3 x_4 = (1 - x_1)(1 - x_2) x_3 x_4)$$

■

Theorem 2.8 can be extended to sets of patterns as follows:

Theorem 2.9

A characteristic function f_K: $P \to \{0, 1\}$ for a class of patterns $K \subseteq P$ can always be written in the arithmetical conjunctive normal form:

$$f_K = \sum_{\mathbf{p}_i \in K} f_{\mathbf{p}_i}$$

with:

$$f_{\mathbf{p}_i} = \sum_{\mathbf{q}_m \in C_{\mathbf{p}_i}} (-1)^{|\mathbf{q}_m \cap \bar{\mathbf{p}}_i|} \mathbf{x}_{\mathbf{q}_m}$$

Proof

If f_S is the characteristic function of a set of patterns S (e.g. a singleton $\{\mathbf{p}_i\}$) and $f_{\mathbf{p}_j}$ the characteristic function of the singleton $\{\mathbf{p}_j\}$, then one easily verifies that $f_S + f_{\mathbf{p}_j}$ is the characteristic function of the set $S \cup \{\mathbf{p}_j\}$. QED

Example 2.21

For $n=2$ let $K = \{00, 10\}$. The characteristic function of $\{00\}$ is $1 - x_1 - x_2 + x_1 x_2$

and the characteristic function of $\{10\}$ is $x_1 - x_1 x_2$. The characteristic function of $K = \{00, 10\}$ becomes $1 - x_2$. ■

A mask function $\mathbf{x_q}$ is the characteristic function of a set of patterns $K_\mathbf{q}$ defined by:

$$K_\mathbf{q} = \{\mathbf{p}_i | \mathbf{x_q}(\mathbf{p}_i) = 1\}$$

Because this definition of $K_\mathbf{q}$ is identical with the definition of the cover $C_\mathbf{q}$ of pattern $\mathbf{q}$, we have the following lemma:

Lemma 2.3

The mask function $\mathbf{x_q}$ is the characteristic function of the *cover* $C_\mathbf{q}$ of pattern $\mathbf{q}$.

This lemma gives us a convenient set-theoretical interpretation of the arithmetical conjunctive normal form of a logical function y, and a way to determine the arguments for which $y = 1$. In a subsequent section we will see how it can also give us the means to show that a two-layer binary Perceptron is able to generalize from samples of desired behaviour.

Example 2.22

The arithmetical conjunctive normal form of the exclusive-or function is as follows:

$$y = x_1 + x_2 - 2x_1 x_2$$

Thus $y = 1$ for the elements of the set specified by:

$$C_{10} + C_{01} - 2C_{11} = \{10, 11\} + \{01, 11\} - 2 \cdot \{11\} = \{10, 01\}$$

In this example we used the operations of addition, subtraction and multiplication of sets. These operators must be defined in a formal way. We will come back to it in Section 2.8. ■

Because of Theorem 2.8 a characteristic function $f_{\mathbf{p}_i}$ of a pattern $\mathbf{p}_i$ can be written in an arithmetical conjunctive normal form:

$$f_{\mathbf{p}_i} = \sum_{\mathbf{q}_m \in C_{\mathbf{p}_i}} (-1)^{|\mathbf{q}_m \cap \bar{\mathbf{p}}_i|} \mathbf{x}_{\mathbf{q}_m}$$

and thus the class containing only pattern $\mathbf{p}_i$ can also be written as an arithmetical sum of the covers $C_{\mathbf{q}_m}$ (the set of patterns covering pattern $\mathbf{q}_m$) represented by the mask functions in the arithmetical conjunctive normal form:

$$\{\mathbf{p}_i\} = \sum_{\mathbf{q}_m \in C_{\mathbf{p}_i}} (-1)^{|\mathbf{q}_m \cap \bar{\mathbf{p}}_i|} C_{\mathbf{q}_m}$$

Example 2.23

For $n=4$ and pattern $\mathbf{p}_i=0011$ the characteristic function is as follows:

$$f_{\mathbf{p}_i}=x_3x_4-x_1x_3x_4-x_2x_3x_4+x_1x_2x_3x_4$$

Thus

$$\{\mathbf{p}_i\}=C_{0011}-C_{1011}-C_{0111}+C_{1111}$$

Hence

$$\{\mathbf{p}_i\}=\{0011\}=\{1111, 0111, 1011, 0011\}-\{1111, 1011\}-\{1111, 0111\}+\{1111\}$$

2.7 The adaptive recruitment learning rule

In the previous section we found that any logical function $y\colon \{0, 1\}^n \rightarrow \{0, 1\}$ can always be written in the arithmetical conjunctive normal form, and that every arithmetical conjunctive normal form can be realized by a two-layer binary Perceptron.

A two-layer neural network that realizes such a logical function by some linear threshold function with respect to the mask functions $\mathbf{x_q}$:

$$y(\mathbf{p}_i)=S\left(\sum_{\mathbf{q}} w_{\mathbf{q}}\cdot\mathbf{x_q}(\mathbf{p}_i)-T\right)$$

must have for each $\mathbf{q}\in\{0, 1\}^n$ for which $w_{\mathbf{q}}\neq 0$, an input neuron realizing $\mathbf{x_q}$, moreover the weight of the connections from the output of a first-layer neuron realizing $\mathbf{x_q}$ to the input of the output neuron must be equal to $w_{\mathbf{q}}$.

Assuming we do not have an explicit description of the threshold function but only a finite set D, the *training set*, of examples consisting of pairs $\langle \mathbf{p}_i, y(\mathbf{p}_i)\rangle$, then we investigate whether we can develop a learning rule leading to the recruitment of the required first-layer neurons and correct values of the weights $w_{\mathbf{q}}$ such that at the end of the learning process the two-layer binary Perceptron will at least give the correct response to all patterns of D.

A logical function $y\colon P\rightarrow\{0, 1\}$ with $P=\{0, 1\}^n$ can be considered as a *pattern classification function*, i.e. $y(\mathbf{p}_i)=1$ if $\mathbf{p}_i$ belongs to some subset K of P and $y(\mathbf{p}_i)=0$ if p_i belongs to the complement $\bar{K}=P-K$.

The subset of patterns in the given data set D that are elements of K will be called the *set of examples E*, and the subset of patterns of D that are elements of $\bar{K}$ will be called the *set of counterexamples.*

The learning rule we will present will give the correct response to the training set D after a finite learning time and will generalize in some sense on the basis of the training set to other inputs not present in the training set. In contradistinction with learning rules for other artificial neural nets, we have to present every element of the training set D just once. This implies that the learning time is proportional to the number of elements in D. As far as we know the learning rule presented here is new and has not been published before.

Assume we want to learn a logical function y: $\{0, 1\}^n \rightarrow \{0, 1\}$ and we have a finite set of examples (set E) and counterexamples (set F).

The adaptive recruitment learning rule

1. Given initially an arbitrary two-layer binary Perceptron with in the first layer an arbitrary number (it might be zero) of neurons realizing mask functions $\mathbf{x_q}$ with $\mathbf{q} \in \{0, 1\}^n$. The outputs $z_\mathbf{q}$ of input neurons are multiplied by arbitrary weights $w_\mathbf{q}$ and connected to one single output neuron with arbitrary valued threshold T.
2. Present all examples and counterexamples in the order of the number of 'ones' occurring in the set $D = E \cup F$.
3. If an example or counterexample is correctly classified, go to the next element of the ordered set D.
4. If a pattern $\mathbf{p}$ is presented and incorrectly classified and there exists no first-layer neuron that realizes the mask function $\mathbf{x_p}$, introduce such a neuron. Change the weight $w_\mathbf{p}$ to $w_\mathbf{q} + \Delta$ with Δ such that the output of the output neuron becomes correct. (Δ is positive if $\mathbf{p}$ belongs to E and the output was 0; Δ is negative if $\mathbf{p}$ belongs to F and the output was 1.)
5. Go to the next element of the training set D.

Before proving that after learning the set, D is correctly classified, we will give a simple example.

Example 2.24

Assume we want to identify the logical function such that $y = 1$ for the elements of $K = \{0100, 1001, 0101, 0110, 1101, 1011, 0111, 1111\}$ and thus $y = 0$ for the elements of $\bar{K} = \{0000, 0001, 0010, 0011, 1000, 1010, 1100, 1110\}$.

Assume we do not know K but only the set of examples: $E = \{0100, 1001\}$ and one counterexample $F = \{1100\}$.

Assume we start the learning process with a neural net without any first-layer neurons and only an output neuron with threshold $T = 0$.

We start the learning process with the example 0100. The output is incorrect so we have to introduce a first-layer neuron that realizes the mask function $\mathbf{x}_{0100}$. For the weight we will obtain $w_{0100} = 1$ (see Figure 2.21).

In the next learning step we take the counterexample: 1100. We observe that for the neural net obtained after the first step the output for input 1100 is wrong: $y = 1$. We have to introduce a second first-layer neuron that realizes the mask function $\mathbf{x}_{1100}$ with a weight $w_{1100} = -1$ (see Figure 2.22).

In the third step we take example 1001. For the neural net of Figure 2.22 we obtain for the input 1001 the output $y = 0$. Thus we have to add an additional first-layer

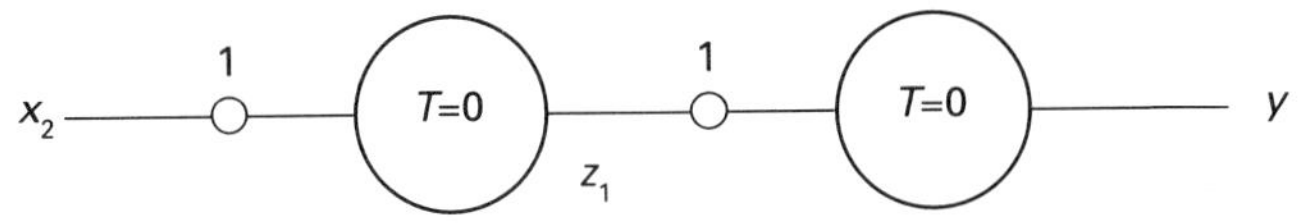

Figure 2.21 The neural net after the first learning step

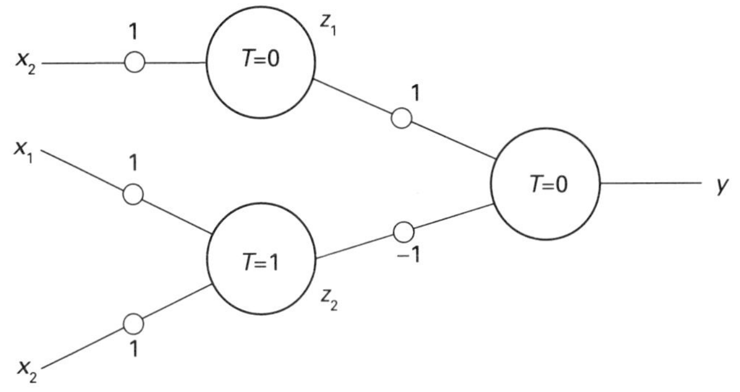

Figure 2.22 The neural net after the second learning step

neuron that realizes the mask function $\mathbf{x}_{1001}$ with a weight $w_{1001}=1$ (see Figure 2.23). Because we have presented all examples and counterexamples, we are at the end of the learning process. One easily verifies that now the output y of the final neural net is equal to 1 for all elements of K and $y=0$ for all elements of $\bar{K}$.

At a first glance one might be surprised that in the previous example we could identify the logical function y just with two examples and one counterexample. But the example was not fair because the unknown function could as well have been defined as $y=1$ for the set:

$$\begin{aligned} K' &= E \cup (\bar{K} - F) \\ &= \{0100, 1001\} \cup \{0000, 0001, 0010, 0011, 1000, 1010, 1100, 1110\} - \{1100\} \end{aligned}$$

and $y=0$ for the set:

$$\begin{aligned} \bar{K}' &= F \cup (K - E) \\ &= \{1100\} \cup \{0100, 1001, 0101, 0110, 1101, 1011, 0111, 1111\} - \{0100, 1001\} \end{aligned}$$

Learning with the same sets of examples E, and counterexamples F, would result in the same neural net but with a wrong response for all inputs except for the elements of E and F. ■

Although in an ideal learning situation one wishes to generalize from a restricted set of examples and counterexamples, the previous example gives ground to the

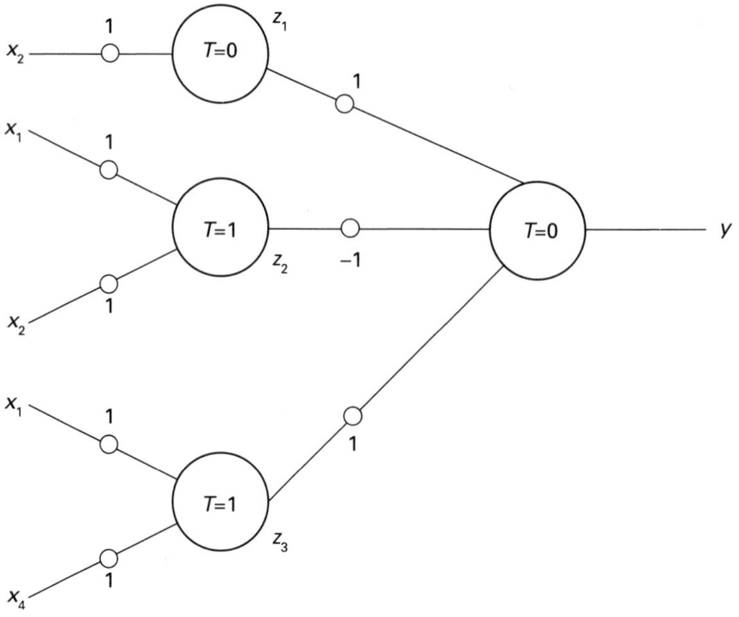

Figure 2.23 The neural net after the third learning step

following general hypothesis:

> Generalization by learning from examples and counterexamples is in general impossible without utilizing *a priori* knowledge about the properties of the function to be identified.

We will come back to this subject later. We will first present a proof of correctness of the adaptive recruitment learning rule.

Proof of correctness of the adaptive recruitment learning rule

We have to prove that after learning, the set $E \cup F$ of examples E and counterexamples F is correctly classified. This implies that if $E = K$ and $F = \bar{K}$ we can identify any logical function exactly.

Let $R(k)$ be a subset of D which is correctly classified after step k. Assume we present at step $k+1$ an element $\mathbf{p}_j \in E$. After step $k+1$ the linear threshold function realized by the neural net will have the form:

$$y(k+1) = S\left(\sum_{\substack{\mathbf{q} \in Q \\ \mathbf{q} \neq \mathbf{p}_j}} w_{\mathbf{q}} \mathbf{x}_{\mathbf{q}} + w_{\mathbf{p}_j} \mathbf{x}_{\mathbf{p}_j} - T \right) \text{ for some set } Q$$

Due to the ordered presentation of examples and counterexamples during the learning process, we have for every $\mathbf{p}_i \in R(k)$ that the number of 1s occurring in $\mathbf{p}_i$ is smaller than the number of 1s occurring $\mathbf{p}_j$, or if that number is the same, then $\mathbf{p}_i \neq \mathbf{p}_j$ and thus $\mathbf{x}_{\mathbf{p}_j}(\mathbf{p}_i)=0$ for all $\mathbf{p}_i \in R(k)$. Thus $y(k+1)=y(k)$ for all $\mathbf{p}_i \in R(k)$ and hence $R(k) \subseteq R(k-1)$. After step $k+1$ the weight $w_{\mathbf{p}_j}$ will be such that $\mathbf{p}_j$ is correctly classified and thus $R(k+1)=R(k) \cup \{\mathbf{p}_j\}$. The same reasoning holds if we present at step $k+1$ a counterexample. So we finally will end up with the situation that D is correctly classified. QED

If the initial net contains no neurons in the first layer, and all initial weights are zero, and the initial threshold of the output neuron is zero, and if we take for Δ in rule 4 the smallest integer satisfying the condition mentioned, then we will call the applied learning rule the *proper adaptive recruitment learning rule.*

It may be worthwhile noting that the linear threshold function realized by the adaptive recruitment rule may contain less terms than the arithmetical conjunctive normal form obtained by the procedure mentioned in the proof of Theorem 2.5.

Example 2.25

For the binary 'or' function we will find with Theorem 2.5 the following conjunctive normal form:

$$Y=x_1+x_2-x_1x_2$$

The linear threshold function realized by the proper adaptive recruitment learning rule will be:

$$y=S(x_1+x_2)$$ ■

2.8 Generalizing with a two-layer binary Perceptron

An ideal learning performance in pattern classification would be when correct classification of the patterns of a class K occurs after a learning phase in which the patterns of a finite proper subset of K and patterns of a finite proper subset of the complement of K (the counterexamples) are presented to the learning system. This will frequently occur with the two-layer binary Perceptron, as in Example 2.24, but in general we cannot guarantee that the obtained classification for the set K is correct unless $E=K$ and $F=\bar{K}$.

When, however, some *a priori* knowledge about the relation between the set of examples E, the set of counterexamples F, and the class K of patterns to be identified can be taken into account, then correct classification can in general be learned from a proper subset of K and a proper subset of $\bar{K}$.

We investigate the properties of the class L for which the output y of the binary Perceptron will be 1 after the learning phase. In order to formulate a theorem relating to the properties of class L we have to introduce some new concepts.

In regular set theory a *set* A is defined as a collection of distinguishable objects

a_i, each object occurring once in the set. In the subsequent discussion we need sets in which an object a_i can occur $\alpha_0+\alpha_i$ times, with α_0 and $\alpha_i \in \mathbb{R}$. For this purpose we define an *extensive set* $\hat{A}$ of a regular set A: $\hat{A}=\{a_0, \alpha_1 a_1, \alpha_2 a_2, \ldots, \alpha_n a_n\}$ with the following properties:

If $\hat{A}=\{\alpha_0, \alpha_1 a_1, \alpha_2 a_2, \ldots, \alpha_n a_n\}$ and $\gamma \in \mathbb{R}$, then $\gamma\hat{A}=\{\gamma\alpha_0, \gamma\alpha_1 a_1, \gamma\alpha_2 a_2, \ldots, \gamma\alpha_n a_n\}$.

If $\hat{A}=\{\alpha_0, \alpha_1 a_1, \alpha_2 a_2, \ldots, \alpha_n a_n\}$ and $\hat{B}=\{\beta_0, \beta_1 b_1, \beta_2 b_2, \ldots, \beta_m b_m\}$, then $\hat{A}+\hat{B}=\{\alpha_0+\beta_0, \alpha_1 a_1, \alpha_2 a_2, \ldots, \alpha_n a_n, \beta_1 b_1, \beta_2 b_2, \ldots, \beta_m b_m\}$. Note that if $a_i=b_j=z$, then $(\alpha_i+\beta_j)z \in A+B$.

We can convert an extensive set into a regular set with the *set step function* $\mathbb{S}$ defined as:

$$\mathbb{S}(\hat{A})=\mathbb{S}\{\alpha_0, \alpha_1 a_1, \alpha_2 a_2, \ldots, \alpha_n a_n\}=\{a_i | \alpha_0+\alpha_i>0\}$$

We can now give a theorem concerning the set of patterns accepted by the binary Perceptron after learning.

Theorem 2.10

If E is the set of examples and F the set of counterexamples and we use the adaptive recruitment learning rule, then after learning, the output of the Perceptron will be equal to 1 for elements of some set L and will be zero for the set $P-L$, with L:

$$L=\mathbb{S}\left(\sum_{\mathbf{q}_i \in E'} \alpha_i C_{\mathbf{q}_i} - \sum_{\mathbf{q}_j \in F'} \alpha_j C_{\mathbf{q}_j}\right)$$

with $\mathbb{S}$ the set step function, $C_{\mathbf{q}_i}$ the cover of example $\mathbf{q}_i$ and $C_{\mathbf{q}_j}$ the cover of counterexample $\mathbf{q}_j$, with $\alpha_i>0$ and $\alpha_j>0$ and E' a subset of E and F' a subset of F.

Before presenting the proof of Theorem 2.10 we will illustrate this theorem with two examples.

Example 2.26

Let $E=\{010\}$ and $F=\{110\}$, then after learning the Perceptron will realize the following linear threshold function:

$$y=S(x_2-x_1x_2)$$

According to the theorem we obtain for the set L:

$$L=\mathbb{S}(C_{010}-C_{110})=\mathbb{S}(\{010, 011, 110, 111\}-\{110, 111\})=\{010, 011\} \quad \blacksquare$$

Example 2.27

Let for $n=3$ the set of patterns be: $K=\{\langle 010\rangle, \langle 011\rangle\}$. Assume we start learning with an initial neural network containing in the first layer a neuron that realizes the maskfunction $\mathbf{x}_{010}$ and that it is connected to the output neuron with a weight

$w_{010}=2$. Let the set of examples be $E=\{\langle 011\rangle\}$ and the set of counterexamples: $F=\{\langle 110\rangle, \langle 111\rangle\}$. After learning we obtain a neural net realizing the linear threshold function:

$$y=S\{2x_2-2x_1x_2+x_2x_3-x_1x_2x_3\}.$$

For the class L we obtain:

$$\begin{aligned}L&=\mathbb{S}(2C_{010}-2C_{110}+C_{011}-C_{111})\\&=\mathbb{S}(2\{010, 011, 110, 111\}-2\{110, 111\}+\{011, 111\}-\{111\})\\&=\mathbb{S}(2\{010\}, 3\{011\})\\&=\{010, 011\}\end{aligned}$$

■

Proof of Theorem 2.10

By inspection of the adaptive recruitment learning rule we see that for each example $\mathbf{q}_i\in E$, a first-layer neuron that realizes the mask function $\mathbf{x}_{\mathbf{q}_i}$ with a corresponding weight $w_{\mathbf{q}_i}>0$ will be introduced if the example does not already give the correct response. If the example has already been accepted, no first-layer neuron will be introduced and $w_{\mathbf{q}_i}=0$. The same holds for a counterexample $\mathbf{q}_j$, but now with $w_{\mathbf{q}_j}<0$ or $w_{\mathbf{q}_j}=0$. The output neuron realizes a step function with some threshold T. Thus after learning the Perceptron will realize a linear threshold function:

$$y(\mathbf{p})=S\left(\sum_{\mathbf{q}_i\in E'} w_{\mathbf{q}_i}\mathbf{x}_{\mathbf{q}_i}(\mathbf{p})+\sum_{\mathbf{q}_j\in F'} w_{\mathbf{q}_j}\mathbf{x}_{\mathbf{q}_j}(\mathbf{p})-T\right) \text{ with } E'\subseteq E \text{ and } F'\subseteq F$$

A mask function $\mathbf{x}_\mathbf{q}$ will have the value 1 for all elements of the cover $C_\mathbf{q}$. Thus for a pattern $\mathbf{p}$ accepted ($y=1$) by the Perceptron we have:

$$S\left(\sum_{\substack{\mathbf{q}_i\in E'\\ \mathbf{p}\in C_{\mathbf{q}_i}}} w_{\mathbf{q}_i}+\sum_{\substack{\mathbf{q}_j\in F'\\ \mathbf{p}\in C_{\mathbf{q}j}}} w_{\mathbf{w}_j}-T\right)=1$$

For a pattern $\mathbf{p}$ rejected ($y=0$) by the Perceptron we have:

$$S\left(\sum_{\substack{\mathbf{q}_i\in E'\\ \mathbf{p}\in C_{\mathbf{q}_i}}} w_{\mathbf{q}_i}+\sum_{\substack{\mathbf{q}_j\in F'\\ \mathbf{p}\in C_{\mathbf{q}_j}}} w_{\mathbf{q}_j}-T\right)=0$$

Thus if pattern $\mathbf{p}$ is accepted and is a member of some cover $C_{\mathbf{q}_i}$ (respectively $C_{\mathbf{q}_j}$), then we can equivalently count that pattern $w_{\mathbf{q}_i}$ (respectively $-w_{\mathbf{q}_j}$) times and add these numbers $w_{\mathbf{q}_i}$ (respectively $-w_{\mathbf{q}_j}$) to one total number β and subtract from β the threshold T. Now we can say that pattern $\mathbf{p}$ is $\beta-T$ times a member of the extended set $\hat{L}$. In the case that $\mathbf{p}$ is accepted we must have that $\beta-T>0$ because $S(\beta-T)>0$ and thus $\mathbf{p}\in\mathbb{S}(\hat{L})$. In case that pattern $\mathbf{p}$ is not accepted we obtain similarly $\beta - T \leqslant 0$. Thus with $\alpha_i = w_{\mathbf{q}_i}$ and $\alpha_j = -w_{\mathbf{q}_j}$ we obtain:

$$L = \mathbb{S}\left(\sum_{\mathbf{q}_i \in E'} \alpha_i C_{\mathbf{q}_i} - \sum_{\mathbf{q}_j \in F'} \alpha_j C_{\mathbf{q}_j} - T \right) \qquad \text{QED}$$

If the set L is identical with the set K that has to be identified we say that K is *coverable* by the set of examples and counterexamples.

We conclude that with the adaptive recruitment learning rule the Perceptron can be generalizing correctly from sets of examples and counterexamples. However, in general it might be hard to determine beforehand whether the (unknown) class K to be identified will be coverable by the given sets of examples E and counterexamples F.

Theorem 2.10 implies that any logical function can be written as:

$$y = S\left(\sum_{\mathbf{q}_i \in E} w_{\mathbf{q}_i} \mathbf{x}_{\mathbf{q}} + \sum_{\mathbf{q}_j \in F} w_{\mathbf{q}_j} \mathbf{x}_{\mathbf{q}} \right) \text{ with } y(\mathbf{p}) = 1 \text{ iff } \mathbf{p} \in K$$

$$\text{with } w_{\mathbf{q}_i} \geqslant 0 \text{ if pattern } \mathbf{q}_i \in E \subseteq K$$

$$\text{and } w_{\mathbf{q}_j} \leqslant 0 \text{ if pattern } \mathbf{q}_j \in F \subseteq \bar{K}$$

This statement can also be proven without relying on the adaptive recruitment learning rule, so we have the following theorem:

Theorem 2.11

For every class K of patterns there exists a linear threshold function:

$$y = S\left(\sum_{\mathbf{q}_i \in E} w_{\mathbf{q}_i} \mathbf{x}_{\mathbf{q}} + \sum_{\mathbf{q}_j \in F} w_{\mathbf{q}_j} \mathbf{x}_{\mathbf{q}} \right) \text{ with } y(\mathbf{p}) = 1 \text{ iff } \mathbf{p} \in K$$

$$\text{with } w_{\mathbf{q}_i} \geqslant 0 \text{ if pattern } \mathbf{q}_i \in F \subseteq K$$

$$\text{and } w_{\mathbf{q}_j} \leqslant 0 \text{ if pattern } \mathbf{q}_j \in F \subseteq \bar{K}$$

Proof

Let

$$y' = S\left(\sum_{\mathbf{q}_i} w'_{\mathbf{q}_i} \mathbf{x}_{\mathbf{q}_i} \right)$$

be some linear threshold function realizing the classification of class K. Assume $w'_{\mathbf{q}_i} \leqslant 0$, whereas pattern $\mathbf{q}_i \in K$. There exists, however, a characteristic function for pattern $\mathbf{q}_i$ of the form (Theorem 2.8):

$$f_{\mathbf{q}_i} = \sum_{\mathbf{q}_m \in C_{\mathbf{q}_i}} (-1)^{|\mathbf{q}_m \cap \bar{\mathbf{q}}_i|} \mathbf{x}_{\mathbf{q}_m}$$

Now $w_{\mathbf{q}_i} \mathbf{x}_{\mathbf{q}_i}$ is a term occurring in the expansion of $f_{\mathbf{q}_i}$ with $w_{\mathbf{q}_i} = 1$. The characteristic function $f_{\mathbf{q}_i}$ of $\mathbf{q}_i$ can be added to f' without changing the classification performed by f' (Theorem 2.3). The same holds for $\delta \cdot f_{\mathbf{q}_i}$ with $\delta \geqslant 0$.

We can always select δ such that the new synaptic weight of $\mathbf{x}_{\mathbf{q}_i}$ in f', $w_{\mathbf{q}_i} = w'_{\mathbf{q}_i} + \delta$, becomes zero or positive. If $w_{\mathbf{q}_i}$ becomes zero, then the mask $\mathbf{x}_{\mathbf{q}_i}$ can be eliminated from f' and hence $\mathbf{q}_i \in K$ does not occur in E.

Similarly, if $w'_{\mathbf{q}_i} \geqslant 0$ in f' whereas pattern $\mathbf{q}_i \in \bar{K}$, we can change the synaptic weight to a negative or zero value $w_{\mathbf{q}_i} = w'_{\mathbf{q}_i} + \delta$ with $\delta < 0$ without changing the classification function. If $w_{\mathbf{q}_i}$ becomes zero, then the mask $\mathbf{x}_{\mathbf{q}_i}$ can be eliminated from f' and hence $\mathbf{q}_i \in \bar{K}$ does not occur in F.

If we change the weights in such an order that $w'_{\mathbf{q}_i}$ is altered before $w'_{\mathbf{q}_j}$ if $|\mathbf{q}_i| < |\mathbf{q}_j|$, then we avoid the alteration of weights changed before. QED

Example 2.28

Let for $n = 3$ the class of patterns be $K = \{\langle 010 \rangle, \langle 011 \rangle\}$. A threshold function realizing the classification of class K is as follows:

$$y = S(2x_2 - 2x_1x_2 - x_2x_3 + x_1x_2x_3)$$

Note that the weight of $x_2x_3 = \mathbf{x}_{011}$ is negative, whereas $\langle 011 \rangle \in K$.

The characteristic function for the pattern 011 is $f = x_2x_3 - x_1x_2x_3$. We can add this function to the argument of the step function in y without changing the classification function y. We obtain:

$$y' = S(2x_2 - 2x_1x_2)$$

■

2.9 The recruitment and reinforcement learning rule

The adaptive recruitment learning rule discussed in Section 2.7 is very fast because we have to present all elements of the set of examples E and all elements of the set F of counterexamples just once, and we certainly obtain, after learning, the correct response for elements of E and F. A disadvantage of the adaptive recruitment learning rule is that we have to order the learning set $D = E \cup F$ according to the increasing number of 1s occurring in the binary vectors.

We can, however, also train a two-layer binary Perceptron with the reinforcement learning rule introduced in Section 2.4. In that case we do not have to present the samples in some fixed order, but on the other hand we must apply the whole set of samples many times during learning. The fact that we can use the reinforcement rule for training a two-layer binary Perceptron will become clear if we realize that any logical function y: $\{0, 1\}^n \to \{0, 1\}$ is a linear threshold function with respect to the set of all mask functions $\mathbf{x}_{\mathbf{q}}$ (see Theorem 2.6). However, if we want to train a two-layer binary Perceptron with the reinforcement learning rule we must have in the first layer a neuron for every mask function $\mathbf{x}_{\mathbf{q}}$ with $\mathbf{q} \in \{0, 1\}^n$. For any realistic application n will be large and will thus require a tremendous number of first-layer neurons (for $n=10$ the number of first-layer neurons will be 1024).

After learning with the reinforcement learning rule with a first-layer neuron for each mask function $\mathbf{x_q}$, it turns out that a great number of first-layer neurons can be removed because the corresponding weight $w_\mathbf{q}$ will be zero.

We can do better by not introducing first-layer neurons if that is not necessary. This principle is used in the next learning rule. The outputs of the first-layer neurons will be represented by the variables z_1, z_2, etc.

The recruitment and reinforcement learning rule

Step 0 There is no first-layer neuron and there is one output neuron with threshold zero. (The output neuron has a constant threshold input $z_0=1$ connected to some initial weight w_0.)

Step k Take randomly an element $\mathbf{q}$ of the training set $E \cup F$. If the output $y(k)$ of the output neuron is incorrect and there exists no input neuron that realizes the mask function $\mathbf{x_q}$ then introduce such a neuron first. If there exists an input neuron that realizes the mask function $\mathbf{x_q}$ and the output $y(k)$ is incorrect, then change the extended weight vector $\hat{\mathbf{w}}(k)$, composed of the ordered set of all weights $w_\mathbf{q}$ including the threshold weight $w_\mathbf{0}$, to:

$$\hat{\mathbf{w}}(k+1)=\hat{\mathbf{w}}(k)+\varepsilon(k)\hat{\mathbf{z}}(k) \quad \text{if } y(k)=0 \text{ whereas } \mathbf{q}\in E$$

$$\hat{\mathbf{w}}(k+1)=\hat{\mathbf{w}}(k)-\varepsilon(k)\hat{\mathbf{z}}(k) \quad \text{if } y(k)=1 \text{ whereas } \mathbf{q}\in F$$

with $\varepsilon(k)$ the learning rate (see Section 2.4) and $\hat{\mathbf{z}}(k)$ the extended vector composed of z_0 and the ordered set of all outputs of the first layer neurons introduced so far.

As stated before we do not have to order the learning set D when we use the *recruitment and reinforcement learning rule*, but we now have to present the whole set D many times in order to obtain the situation such that the set D is correctly classified.

We can use *local* and *global learning*, and the *learning rate* $\varepsilon(k)$ may be fixed or time varying, as discussed in Section 2.4. The correctness of the recruitment and reinforcement rule is based on the correctness of the reinforcement rule discussed in Sections 2.4 and 2.5. The difference is the recruitment during learning of first-layer neurons that realize mask functions. Every time we introduce a new first-layer neuron we can conceive that configuration as a new initial situation for learning a linear threshold function with the reinforcement learning rule. The Perceptron convergence theorem (see Section 2.5) does not, however, depend on the initial situation; we only have to guarantee that during learning at least all mask functions required for identification of the linear threshold function are realized by neurons in the first layer. If any of the required mask functions are not realized by the neurons in the first layer, then the output y cannot be correct, but in that case we introduce the missing

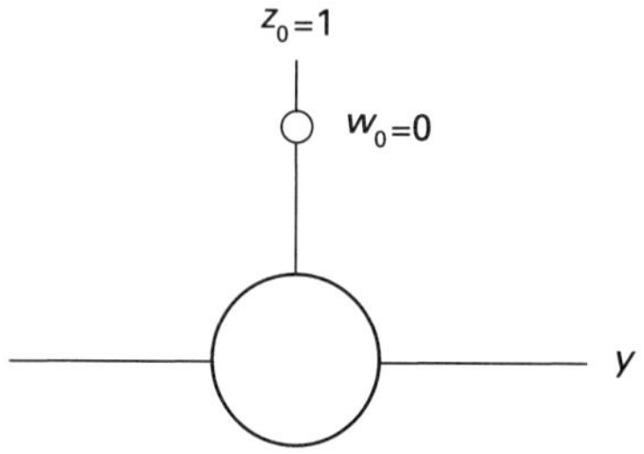

Figure 2.24 Initial configuration of Example 2.29

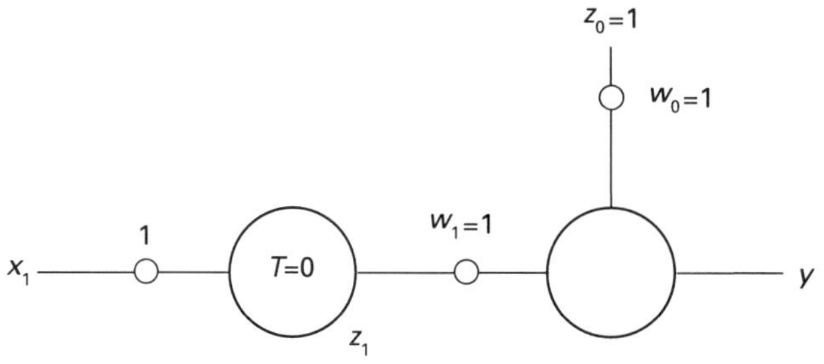

Figure 2.25 The neural net after the first learning step

neuron in the first layer with the recruitment and reinforcement rule. It might be that we introduce a redundant number of first-layer neurons but many of them will be eliminated during learning because the corresponding weights to the output neuron will become zero.

Example 2.29

Suppose we want to learn with the recruitment and reinforcement learning rule the exclusive-or function, with $E=\{[01], [10]\}$ and $F=\{[00], [11]\}$. We start with the neural net of Figure 2.24 and take the learning rate equal to 1. We apply example [10]. The output $y=0$ is not correct. We have to introduce a first-layer neuron that realizes the mask function $\mathbf{x}_{10}$. The output of the neural net for example [10] is now still incorrect because the introduced neuron is not connected to the output neuron: the weight $w_1=0$. For the input [10] we obtain for the output of the introduced neuron: $z_1=1$ and thus $\hat{\mathbf{z}}^t=[1, 1]$. The new weight vector becomes $\hat{\mathbf{w}}(2)=\hat{\mathbf{w}}(1)+\hat{\mathbf{z}}(1)=[0, 0]^t+[1, 1]^t=[1, 1]^t$. The new neural net is given in Figure 2.25. Next we apply a counterexample: [11]. The output $y=1$ of the net of Figure 2.25 is not correct. We have to introduce a new neuron that realizes the mask function $\mathbf{x}_{11}$. With this additional neuron the weight vector becomes $\hat{\mathbf{w}}(2)=[1, 1, 0]$. The first-layer extended output vector becomes $\hat{\mathbf{z}}(2)=[1, 1, 1]$. The output $y=1$ of the net is incorrect, so we have to subtract from $\hat{\mathbf{w}}(2)$ the vector $\hat{\mathbf{z}}(2)=[1, 1, 1]^t$. We obtain $\hat{\mathbf{w}}(3)=[0, 0, -1]^t$.

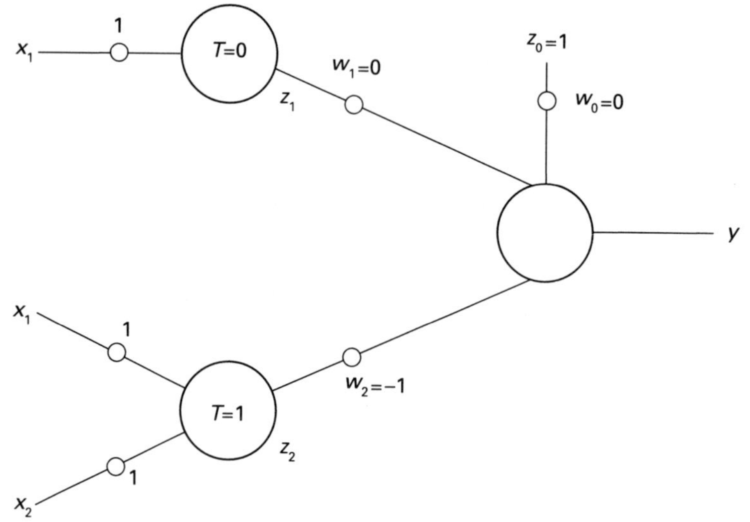

Figure 2.26 The neural net after the second learning step

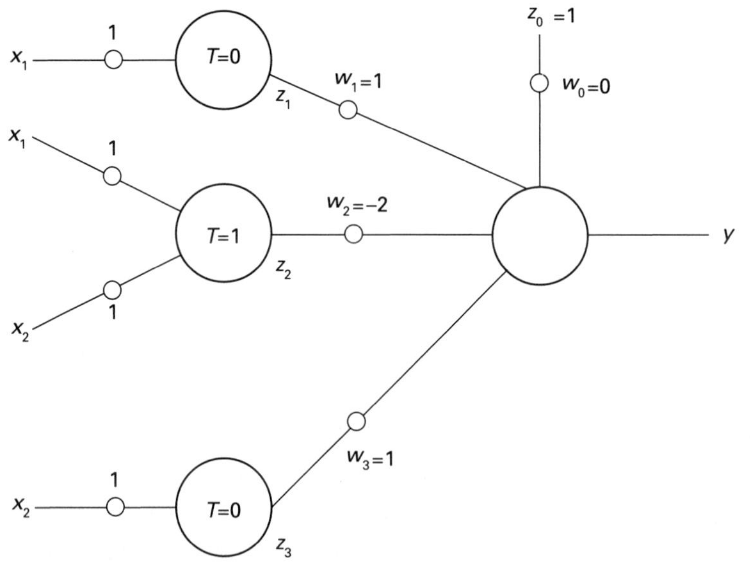

Figure 2.27 The final neural network showing the exclusive-or function

The obtained neural net is given in Figure 2.26. When we proceed in the same way and apply the sequence of inputs [10], [00], [01], [00], [11], [10], [01], [00] we will finally find the network of Figure 2.27 that realizes the exclusive-or function. ■

Practical statement 2.2

In some applications it might occur that the training set contains contradictions. If we apply in that case the recruitment reinforcement rule with a decreasing value of the learning rate $\varepsilon(k)$, the influence of those disruptive elements will be reduced. In that case the output of the final net will become equal to the target values of the elements in the training set that occur with the highest frequency.

Table 2.14

Input vector $x_1 \ldots x_6$	Output	Input vector $x_1 \ldots x_6$	Output
000000	1	110001	1
000001	1	110010	1
000010	1	110100	1
010000	1	000111	0
001000	0	011010	—
000100	0	111000	—
100000	1	101100	0
000011	1	001101	0
001010	1	100101	0
010001	1	001111	—
010010	1	010111	1
010100	1	011101	—
100001	1	011110	1
100010	1	101011	1
101000	1	101101	—
110000	1	110011	1
000101	0	110101	1
001001	0	110110	1
000110	0	111001	1
001100	0	111100	1
100100	0	100111	0
011000	0	011011	—
001011	1	101110	—
001110	—	111010	—
010011	1	011111	1
010101	1	101111	1
010110	1	110111	1
011001	—	111011	1
011100	0	111101	1
100011	1	111110	1
101001	1	111111	1
101010	1	100110	0

2.10 Application of the adaptive recruitment learning rule to switching circuits

As stated before, a binary Perceptron (BPC) can and will realize any logical function, and thus any combinatorial digital circuit, provided that enough examples and counterexamples are given to the net. This behaviour is quite useful if one needs to design a complex switching circuit. Using a BPC can simplify matters considerably. Let us take a look at the next example.

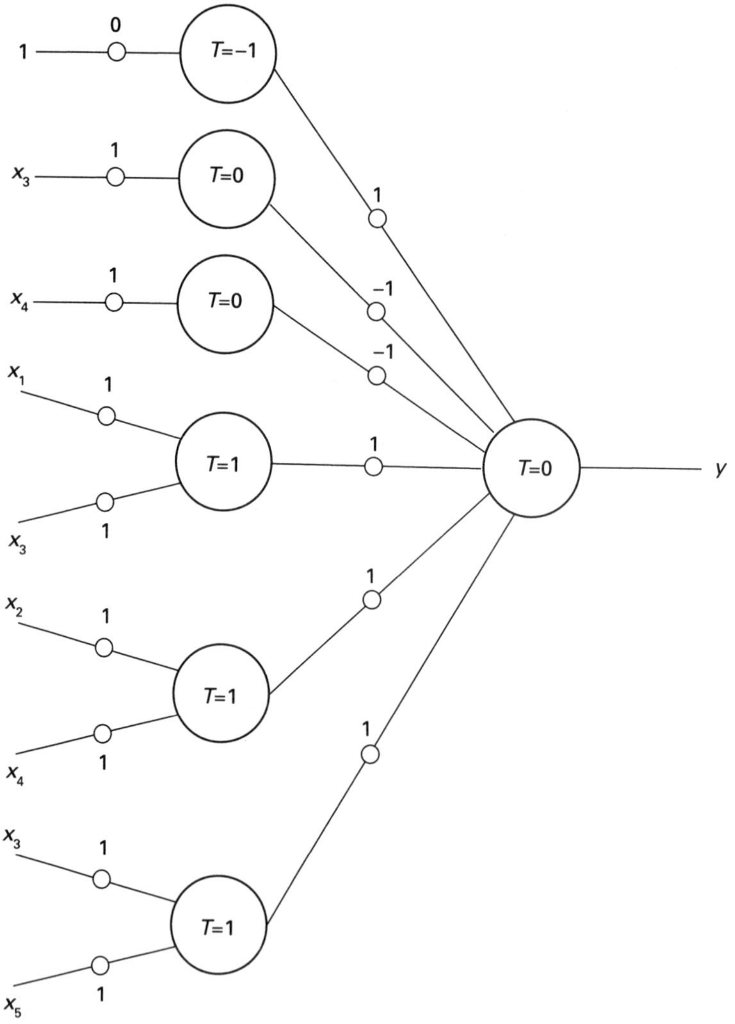

Figure 2.28 The neural network obtained after twenty-two steps illustrating Table 2.14

Table 2.15

Input vector $x_1 \dots x_6$	Output	Input vector $x_1 \dots x_6$	Output
000000	1	110001	1
000001	1	110010	1
000010	1	110100	1
010000	1	000111	0
001000	0	011010	0*
000100	0	111000	0*
100000	1	101100	0
000011	1	001101	0
001010	1	100101	0
010001	1	001111	1*
010010	1	010111	1
010100	1	011101	1*
100001	1	011110	1
100010	1	101011	1
101000	1	101101	1*
110000	1	110011	1
000101	0	110101	1
001001	0	110110	1
000110	0	111001	1
001100	0	111100	1
100100	0	100111	0
011000	0	011011	0*
001011	1	101110	0*
001110	1*	111010	0*
010011	1	011111	1
010101	1	101111	1
010110	1	110111	1
011001	1*	111011	1
011100	1	111101	1
100011	1	111110	1
101001	1	111111	1
101010	1	100110	0

Imagine one has to design a combinatorial circuit that realizes the function specified in Table 2.14 (a dash indicates a 'don't care'). This is a function with six inputs ($x_1 \dots x_6$). The total number of input vectors is sixty-four ($=2^6$).

According to the adaptive recruitment learning rule described in Section 2.7 one has to train the net starting with the zero input vector and then input vectors containing only one 1. Then continue learning using the input vectors containing two 1s, and so on. This assures correct learning.

Table 2.16

$\mathbf{x}_q$	w_q
$\mathbf{x}_{000000}$	1
$\mathbf{x}_{001000}$	-1
$\mathbf{x}_{000100}$	-1
$\mathbf{x}_{001010}$	1
$\mathbf{x}_{010100}$	1
$\mathbf{x}_{101000}$	1
$\mathbf{x}_{011010}$	-1
$\mathbf{x}_{111000}$	-1
$\mathbf{z}_{001110}$	1
$\mathbf{x}_{011001}$	1
$\mathbf{x}_{011100}$	1
$\mathbf{x}_{011011}$	-1
$\mathbf{x}_{101110}$	-2
$\mathbf{x}_{101101}$	1
$\mathbf{x}_{111011}$	1
$\mathbf{x}_{111110}$	1

The net was trained with all vectors containing at most two 1s. The neural net obtained after learning with the twenty-two input vectors is given in Figure 2.28. The output y of the neural net turns out to be consistent with the specification given above.

Although there exists a set of just six sample input vectors that will also yield the same neural net after learning, we may say we are just lucky that we found the correct neural network with twenty-two sample input vectors. If the specification of the logical function to be identified is changed on the positions of the 'don't cares' to the opposite value of the corresponding output of the neural obtained before (see Table 2.15), we obtain the same neural net with the same twenty-two learning input vectors, but the output will be wrong for the input vectors with 'don't cares' in the original table. (Wrongly classified input vectors are marked on Table 2.15 with a '*'.)

The minimal Boolean expression for the logical function specified by Table 2.15 is:

$$y=\bar{x}_1\bar{x}_2x_3x_5+x_1\bar{x}_2\bar{x}_4+x_2\bar{x}_5x_6+x_1x_3x_6+\bar{x}_3\bar{x}_4+x_2x_4$$

This expression was obtained using the McCluskey minimalization algorithm.

In order to guarantee that the neural net will realize some logical function one has to apply all input vectors. If we apply the sixty-four input vectors of Table 2.15 we obtain a neural net as specified in Table 2.16. The first column gives the set of mask functions realized by the sixteen neurons in the first layer. The second column gives the weights connecting the corresponding input neurons to the output neuron. We see that only sixteen first-layer neurons are required to realize the combinatorial function.

2.11 Application of the adaptive recruitment learning rule to hyphenation

During the last decade several software programs have been developed to break words with a hyphen. These programs are in general complicated because they are mostly based on grammatical rules and exceptions to these rules. Furthermore computers are notoriously bad at hyphenation. In the beginning of the automatic hyphenation, errors such as 'the-rapists who pre-ached on wee-knights' could occur. It takes considerable effort and time to develop and implement correct hyphenation algorithms. If we use a neural network, however, we can leave the job of 'developing and implementing the hyphenation algorithm' to a learning process on the basis of examples of hyphenations.

We will show that we can use a binary Perceptron to learn correct hyphenation. In our experiment we used the 70 000 words out of a dictionary of hyphenated words of a Dutch minority language, Frisian.

Because the input of a binary network requires binary vectors we had to code the language alphabet. The code can be arbitrarily chosen but it is profitable to use a code with few '1s' for alphabet symbols that occur frequently in the language and to code 'phonetically similar' symbols with a similar binary code. We used the code shown in Table 2.17.

One method is to present to the binary neural net a binary coded word and require that the target be an output vector of the same length as the uncoded input word with only a '1' on the position of an input symbol after which the word may be broken. In that case one has to limit the length of the words. Moreover the number of inputs and outputs is large.

We took a different approach. We look through a window with a length of seven symbols which moves from left to right across the word. We observe through the

Table 2.17

a	10000000	z	00101110
u	10001000	h	00100100
y	10001100	j	00100110
e	10000100	v	00100111
o	10000110	l	00010000
i	10000111	c	00011000
b	01000000	w	00011100
k	01001000	q	00011110
t	01001100	x	00011111
d	01000100	m	00010100
p	01000110	n	00010110
g	00100000	r	00010111
f	00101000	-	00000000
s	00101100		

Table 2.18

Word-segment	- - P E R - C E P - T R O N - -	
1	- - P E R	0
2	- P E R C	0
3	P E R C E	1
4	E R C E P	0
5	R C E P T	0
6	C E P T R	1
7	E P T R O	0
8	P T R O N	0
9	T R O N -	0
10	R O N - -	0

window a *word-segment* of seven symbols. If the hyphen must be placed after the third symbol we require the target output of the neural net to be '1', and '0' otherwise. Table 2.18 illustrates this method. The word 'Perceptron' is given and with a window of five symbols we move across the word from left to right. We observe successively ten different word-segments. The correct response to each word-segment is given in the last column.

From the 70 000 words in the dictionary we obtained approximately 250 000 different word-segments. It can occur that the same word-segment requires a response '1' as well as the response '0', e.g. the word-segments '-record' and 'record-' from the word 'record'. The word 'record' is supposed to be broken as 'rec-ord' when it is a noun, but as 're-cord' when it is a verb. We eliminated all those contradictory examples from the obtained list of word-segments and then made from them one special lookup table of exceptions. It turned out that 1039 of the 250 000 examples were contradictory. The list of non-contradictory binary coded word-segments with their target values represent the training set D of a binary function f: $\{0, 1\}^{7\times 8} \rightarrow \{0, 1\}$ that can be learned via the adaptive recruitment rule as discussed in Section 2.7. According to our proof in Section 2.7, the learning set D will be classified 100 per cent correctly after learning.

After learning, we obtained a binary neural net of $\pm 20\,000$ neurons in the first layer and one output neuron. In a hardware realization of the obtained neural network, computation time would be negligible because of the parallel structure of the neural net, but in a simulation of the neural network on an ordinary PC it requires several seconds to compute the position of the hyphens in a word. For any realistic application this time is too long. A simple solution to this problem is to divide the learning set D into a large number of disjoint subsets D_i and learn a separate binary Perceptron for each subset. If after learning we have a simple rule to decide to which subset D_i a word-segment belongs, we only have to consider the output of the neural net trained with subset D_i. In an experiment we divided the learning set D with 250 000 elements into $27 \times 27 = 729$ subsets such that all word-segments with the same third and fourth symbol were placed in the same subset. We trained 729 different binary Perceptrons separately with those training sets. After learning, every subnetwork i will correctly classify the elements of the corresponding learning set D_i. The total number of first-layer neurons after learning was 32 522. The mean number of neurons in one subnetwork is 32 522/729 = 45. With an ordinary PC the time to compute the correct positions of the hyphens in a word is reduced in this way to a few milliseconds.

A drawback of the adaptive recruitment rule is the requirement to order the learning set D according to the increasing number of '1s' in the coded elements. A second disadvantage is the requirement that the learning set may not contain contradictions. The recruitment and reinforcement learning rule does not have these disadvantages but will yield a larger neural network and it requires a very long learning time, as may become clear from the next application.

2.12 Application of the recruitment and reinforcement learning rule to contradictory binary data sets

According to the Practical Statement 2.2 the learning set is allowed to contain contradictions if we use the recruitment and reinforcement learning rule as discussed in Section 2.9. If there exists a contradiction in the data set, the final neural net will respond with the target value of the most frequent target for the same input if the learning rate is decreasing with time. In an experiment we used the learning set of Table 2.19. The input vector [010101] occurs twice with target '1' and once with target '0'.

First we performed an experiment using recruitment and reinforcement learning with no contradictions in the data set (we took the target for input 101010 equal to 1). We took a value of 0.5 for the external learning rate ε. Learning was stopped at the moment that all inputs had been correctly classified. After learning, the first layer of the neural network contained at least forty and at most forty-nine neurons depending on the sequence of applied examples in ten experiments. (With the adaptive recruitment rule we found in Section 2.10 a neural network with only sixteen first-layer neurons for the same problem.) If we use another constant value of the learning rate (e.g. $\varepsilon = 100$), the same network is obtained if the sequence of applied examples is the same, only the absolute value of weights is changed in proportion to the change in the value of the learning rate ε. The number of wrongly classified examples that must be applied before correct behaviour was obtained was approximately 126. (With the adaptive recruitment rule sixty-four examples must be applied.)

If we introduce a contradiction in the data set, as mentioned above, and use a linearly decreasing value of the learning rate $\varepsilon(t)$ from 1 to 0.01 in 150 learning steps and constant 0.01 after that, a correct neural network (with output 1 for input 010101) was obtained with the mean of forty-five neurons (in ten experiments) in the first layer. The number of wrongly classified examples that must be applied during learning varied between 110 and 170, depending on the order of applied examples.

Table 2.19

Input vector $x_1 \ldots x_6$	Output	Input vector $x_1 \ldots x_6$	Output
000000	1	110001	1
000001	1	110010	1
000010	1	110100	1
010000	1	000111	0
001000	0	011010	0
000100	0	111000	0
100000	1	101100	0
000011	1	001101	0
001010	1	100101	0
010001	1	001111	1
010010	1	010111	1
010100	1	011101	1
100001	1	011110	1
100010	1	101011	1
101000	1	101101	1
110000	1	110011	1
000101	0	110101	1
001001	0	110110	1
000110	0	111001	1
001100	0	111100	1
100100	0	100111	0
011000	0	011011	0
001011	1	101110	0
001110	1	111010	0
010011	1	011111	1
010101 (2 ×)	1	101111	1
010101 (1 ×)	0		
010110	1	110111	1
011001	1	111011	1
011100	1	111101	1
100011	1	111110	1
101001	1	111111	1
101010	1	100110	0

2.13 Exercises

1. Show that Table 2.2 cannot be realized with a single-neuron binary Perceptron.
2. Show with the use of the consistency property mentioned in Section 2.2 that Table 2.2 cannot be realized with a single-neuron binary Perceptron.
3. Apply the adaptive recruitment learning rule to the data set of Table 2.2.
4. Apply the adaptive recruitment learning rule to the data set of the exclusive-or function.

5. Apply the adaptive recruitment learning rule to the data set of Table 2.2.
6. Explain why the weight vector **w** points in the direction of the positive side of the separating hyperplane defined by: $\hat{\mathbf{w}} \cdot \hat{\mathbf{x}} = 0$.
7. Show with the help of Lemma 2.1 that the linear threshold function $y = S(x_1 + x_2 - x_3)$ is equivalent to $y = S(x_1 + x_2 - x_3 - 0.5)$.
8. Apply the reinforcement learning rule to the data set given in the following table. The initial weights of the single neuron binary Perceptron are: $w_0 = 1$, $w_1 = 1$ and $w_2 = -1$.

x_1	x_2	y
0	0	0
0	1	1
1	0	1
1	1	1

9. Let $\mathbf{x_q}$ $(\mathbf{p}_i)$ be a mask function. Determine the value of $\mathbf{x_q}(\mathbf{p}_i)$ for $\mathbf{q} = 001100$ and $\mathbf{p}_i = 011110$ and $\mathbf{p}_i = 000100$.
10. Determine with two different methods the arithmetical conjunctive normal form of the logical function specified in the following table.

x_1	x_2	x_3	y
0	0	0	0
0	0	1	1
0	1	0	1
0	1	1	0
1	0	0	1
1	0	1	0
1	1	0	0
1	1	1	1

11. Apply the proper adaptive recruitment learning rule to the data set of the table of Exercise 10.
12. Let the set of examples be $E = \{001, 010\}$ and the set of counterexamples be $F = \{011\}$. Apply the proper adaptive recruitment rule. Show that the output of the obtained neural will be equal to 1 for the set L and will be 0 for the set $P - L$. With $P = \{0, 1\}^3$ and L:

$$L = \mathbb{S}\left(\sum_{\mathbf{q}_i \in E} \alpha_i C_{\mathbf{q}_i} - \sum_{\mathbf{q}_j \in F} \alpha_j C_{\mathbf{q}_j} \right) \text{ with } \mathbb{S} \text{ the set step function}$$

with $C_{\mathbf{q}_i}$ the cover of example $\mathbf{q}_i$ and $C_{\mathbf{q}_j}$ the cover of counterexample $\mathbf{q}_j$.
13. Apply the recruitment and reinforcement learning rule to the data set of the table of Exercise 10.

3

THE CONTINUOUS MULTI-LAYER PERCEPTRON

3.1 Introduction

A continuous multi-layer Perceptron is used in a situation where one presupposes the existence of an unknown functional relationship g: $X \to Y$ between two sets X and Y of data with $X \subset \mathbb{R}^n$ and $Y \subset \mathbb{R}^m$, whereas for a given finite *data set* $D = \{[\mathbf{x}_i, \mathbf{t}(\mathbf{x}_i)] | \mathbf{x}_i \in X, t(\mathbf{x}_i) \in Y, 1 \leq i \leq N\}$ it is assumed that $t(\mathbf{x}_i) = g(\mathbf{x}_i)$. In that case one wishes to identify or to approximate the unknown function g on the basis of the set of samples from D.

A multi-layer Perceptron can realize an infinite set of functions $g_{\mathbf{w}}$: $\mathbb{R}^n \to \mathbb{R}^m$ depending on the vector $\mathbf{w}$, composed of all weights in the neural net.

Given a *learning set* L that is a finite subset of the data set D, and a so-called *learning rule* one can change the weight vector $\mathbf{w}$ of the neural net such that $[\mathbf{x}_i, g_{\mathbf{w}}(\mathbf{x}_i)]$ becomes equal or approximately equal to $[\mathbf{x}_i, t(\mathbf{x}_i)]$ for each pair in L. In the learning phase the weight vector $\mathbf{w}$ is changed step by step by presenting the set L to the adaptation algorithm at each step k. After learning, the performance of the net is tested with the *test set* $T = D - L$ (see Figure 3.1).

After the learning phase the neural network will also yield for any input $\mathbf{x}_j$ not in D some output $g_{\mathbf{w}}(\mathbf{x}_j)$. If it is assumed that the *a priori* unknown function g is approximated by $g_{\mathbf{w}}$ if for all $[\mathbf{x}_i, t(\mathbf{x}_i)] \in D$ the value $t(\mathbf{x}_i)$ is approximated 'well enough' by $g_{\mathbf{w}}(\mathbf{x}_i)$, then we call this phenomenon *generalization*.

The multi-layer Perceptron consists of one, two or three layers of neurons. A layer may have an arbitrary number of neurons. In Figure 3.2 a two-layer Perceptron is given with two neurons in the first layer and one neuron in the second layer. Each neuron has a number of inputs and one output. The output of a neuron in some layer constitutes an input for each neuron in the next layer. The output value of a neuron is some function f of the weighted sum $s(\mathbf{x}) = \Sigma\, w_i \cdot x_i + w_0$ of all its input values x_i. The scalars w_i are called the weights of the neuron and $-w_0$ is called the *threshold* of the neuron. The transfer function f is usually the same for all neurons in the neural net. The most common transfer function is the so-called sigmoid function: $f[s(\mathbf{x})] = [1 + \exp - s(\mathbf{x})]^{-1}$ (see Figure 3.3).

In a so-called *learning phase*, input vectors $\mathbf{x}_i = [x_{i1}, x_{i2}, \ldots, x_{in}]^t$ from a given finite

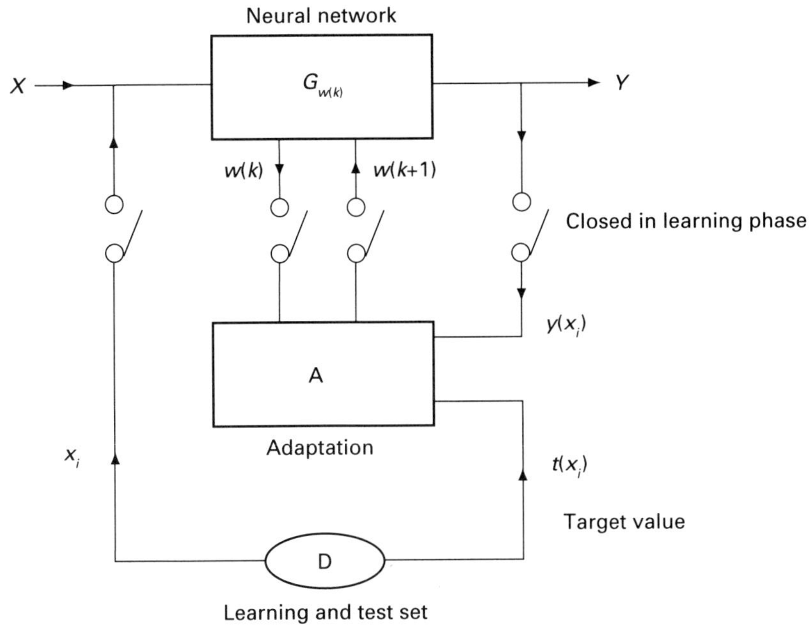

Figure 3.1 Outline for learning and testing

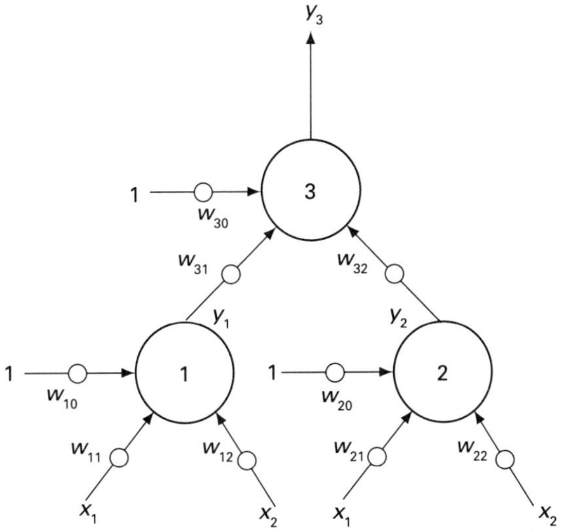

Figure 3.2 A two-layer continuous Perceptron

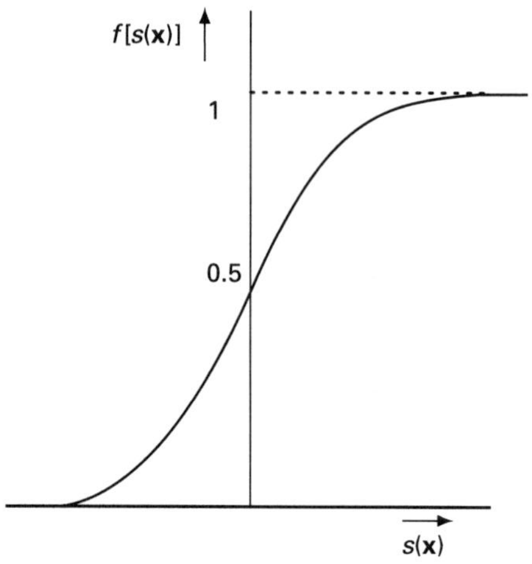

Figure 3.3 The sigmoid transfer function

learning set L are presented to all neurons in the first layer of the neural net. For each input vector $\mathbf{x}_i$ the adaptation values for all the weights in the neural net are computed separately. The computed value of the adaptation $\Delta\mathbf{w}_i$ of all weights in the neural net for the input vector $\mathbf{x}_i$ is such that for each neuron j in the output layer the square of difference between some target value $t_j(\mathbf{x}_i)$ and its actual output value $y_j(\mathbf{x}_i)$ will become smaller if the adaptation were effected. After the computation of the adaptations of weights for all inputs $\mathbf{x}_i$ the weights are actually adapted by $\Delta\mathbf{w} = \Sigma\, \Delta\mathbf{w}_i$.

Multi-layer Perceptrons are frequently used for classification of data into different classes. For instance, one can use the Perceptron of Figure 3.2 to separate a collection of mushrooms into a class A of mushrooms that are very good medicine for some specific illness, and a class B of poisonous mushrooms. Assume the mushrooms look almost the same. We want to have a criterion by which we can decide to what class some particular mushroom belongs. One dangerous way to classify the mushrooms is to eat them, but then the mushrooms are destroyed, so we have to consider another criterion.

Assume the mushrooms of the two classes differ a little bit in the length x_2 and in the thickness x_1 of the stem. Assume we are given a small set D_A of measurements of the length and thickness of mushrooms of class A and another set D_B of measurements of length and thickness of mushrooms of class B. The values of x_1 and x_2 for the set D_A are represented by small circles in Figure 3.4 and for set D_B by small boxes. A solution for the classification (not the best one) is given by the separating lines L_1 and L_2. All points to the right of L_1 and simultaneously to the left of L_2 belong to class A; the points in the remaining area are assumed to represent elements of class B.

We want the Perceptron of Figure 3.2 to learn to find that classification. For now, we will not show how learning is performed but we will demonstrate that for some

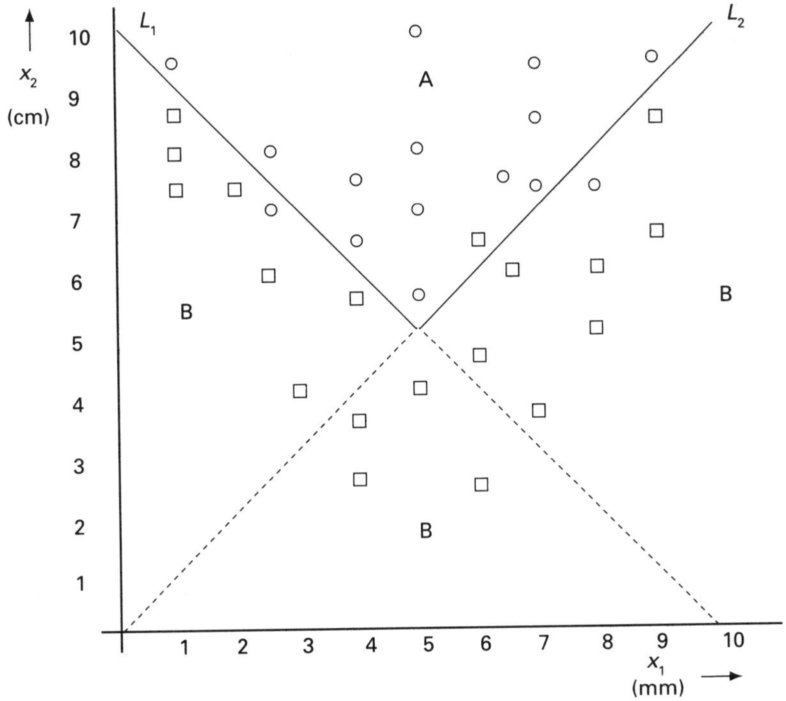

Figure 3.4 Representation of mushrooms by stem thickness (x_1) and stem length (x_2)

of values of the weights w_i in the neural network, the classification is possible by that Perceptron.

We want to have $y_3=1$ for the elements of class A and $y_3=0$ for the elements of class B. Let the weights of neuron 1 be:

$$w_{11}=100 \quad w_{12}=100 \quad w_{10}=-1000$$

The line L_1 is now represented by:

$$w_{11}\cdot x_1+w_{12}\cdot x_2+w_{10}=100x_1+100x_2-1000=0$$

For points on this line we have for the output y_1 of neuron 1 that $y_1=0.5$. Points at the right of this line (and not too close to this line) will give an output $y_1\cong 1$ and points to the left will give $y_1\cong 0$. Let the weights of neuron 2 be:

$$w_{21}=100 \quad w_{22}=-100 \quad w_{20}=0$$

With these weights we have for all points on the line L_2 that

$$w_{21}x_1+w_{22}x_2+w_{20}=0$$

and thus the output y_2 of neuron 2 will be 0.5. Points at the right of this line will give an output $y_2 \cong 1$ and points to the left will give $y_2 \cong 0$.

Thus for the points of class A we have $y_1 \cong 1$ and $y_2 \cong 0$. For elements belonging to class B we obtain:

$$(y_1, y_2) = (0, 0) \text{ or } (0, 1) \text{ or } (1, 1)$$

Let the weights of neuron 3 be:

$$w_{31} = 100 \quad w_{32} = -100 \quad w_{30} = -50$$

Thus the weighted sum of inputs of neuron 3 is:

$$s(\mathbf{x}_i) = 100y_1 - 100y_2 - 50$$

One easily verifies that only for the elements of class A will the output be $y_3 \cong 1$; for the elements of class B we obtain $y_3 \cong 0$.

We see that the classification can be performed by the Perceptron. The set of weights is not unique; there are an infinite number of solutions for the same classification.

An intriguing question is how the Perceptron can learn from the given data sets D_A and D_B a correct set of weights. We will deal with that problem in the next chapter.

3.2 The gradient descent adaptation method

In Figure 3.5 the general structure of a three-layer Perceptron is given. It turns out that any task performed by a multi-layer network can always be solved by a three-layer Perceptron, so we restrict our description of a multi-layer Perceptron to the structure of Figure 3.5.

For any output neuron we can derive the relation between the output vector $\mathbf{y}$ and the input vector $\mathbf{x} = [x_1, x_2, \ldots, x_{n_0}]^t$.

For the value of the ith output in the third layer we have:

$$y_{3i} = f_{3i}(s_{3i})$$

with f_{3i} the transfer function of the particular output neuron and s_{3i} the weighted input of that neuron defined by:

$$s_{3i} = \sum_j w_{3i,j} y_{2j} + w_{3i,0}$$

with $w_{3i,j}$ the weight in the connection between the ith neuron in the third layer and the output y_{2j} of neuron j in the second layer, and $w_{3i,0}$ the threshold weight of the particular output neuron.

For the value of the jth output y_{2j} in the second layer we obtain similarly:

$$y_{2j} = f_{2j}(s_{2j})$$

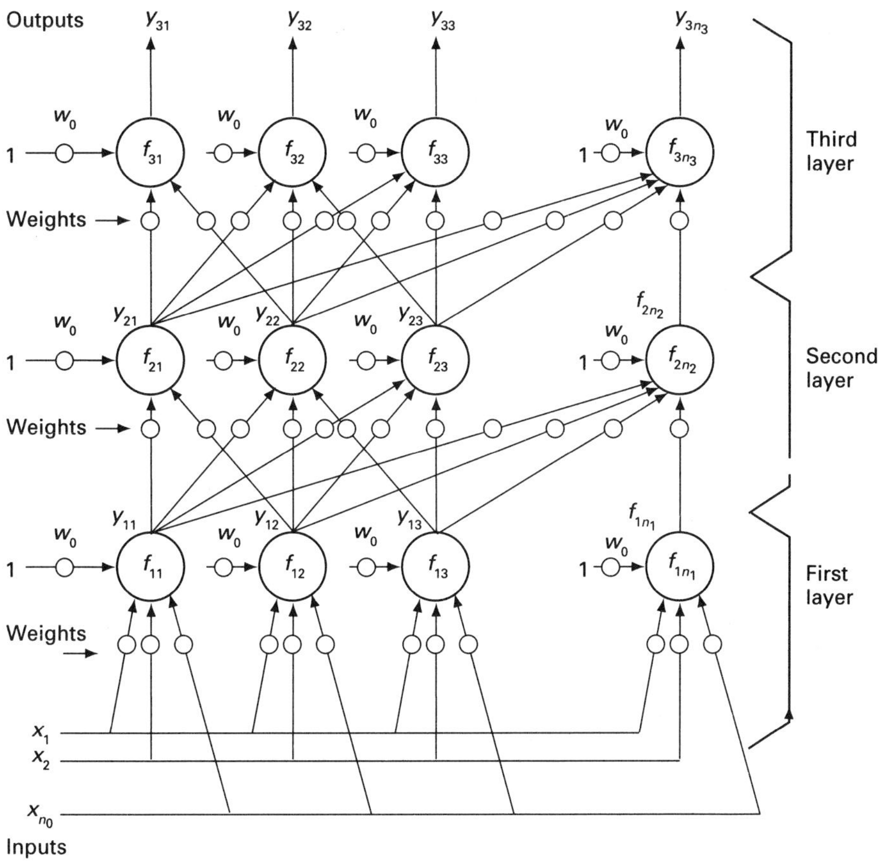

Figure 3.5 General structure of a three-layer continuous Perceptron

with:

$$s_{2j} = \sum_k w_{2j,k} y_{1k} + w_{2j,0}$$

For the value of the kth output y_{1k} in the first layer we have:

$$y_{1k} = f_{1k}(s_{1k})$$

For the value of the weighted input of the kth neuron in the first layer we have:

$$s_{1k} = \sum_m w_{1k,m} x_m + w_{1k,0}$$

with x_m the mth input value.

By substitution we obtain from the equations above for the relation between the

ith output and the inputs $x_1, x_2, \ldots, x_m$ of the network:

$$y_{3i} = f_{3i}\left[\sum_j w_{3i,j} f_{2j}\left\{\sum_k w_{2j,k} f_{1k}\left(\sum_m w_{1k,m} x_m + w_{1k,0}\right) + w_{2j,0}\right\} + w_{3i,0}\right]$$

with the summation over the number of neurons in the particular layer.

By introducing a dummy neuron in layers 2 and 1 with constant transfer functions $f_{20}(\cdot)=1$, $f_{10}(\cdot)=1$ and by adding a constant additional input $x_0=1$ to the net, we can incorporate the thresholds $w_{pq,0}$ of the neurons in the weighted input of the neurons by the terms $w_{3i,0} f_{20}(\cdot)$, $w_{2j,0} f_{10}(\cdot)$ and $w_{1k,0} x_0$.

By this convention we can rewrite the last equation as:

$$y_{3i} = f_{3i}\left[\sum_{j=0}^{n_3} w_{3i,j} f_{2j}\left\{\sum_{k=0}^{n_2} w_{2j,k} f_{1k}\left(\sum_{m=0}^{n_1} w_{1k,m} x_m\right)\right\}\right]$$

The transfer functions f_{3i}, f_{2j} and f_{1k} are in general the same and are non-linear. If the transfer functions are linear then the whole structure collapses to one single neuron as expressed in the following theorem.

Theorem 3.1

A *linear Perceptron* (i.e. all transfer functions are linear: $f(s) = a \cdot s$) can be reduced to one single layer linear Perceptron.

Proof

The relation between the output value of the ith output neuron and the n_0 inputs of the network is:

$$y_{3i} = f_{3i}\left[\sum_{j=1}^{n_2} w_{3i,j} f_{2j}\left\{\sum_{k=1}^{n_1} w_{2j,k} f_{1k}\left(\sum_{m=1}^{n_0} w_{1k,m} x_m + w_{1k,0}\right) + w_{2j,0}\right\} + w_{3i,0}\right]$$

Replacing each transfer function by the scalar factor a, we obtain:

$$\begin{aligned} y_{3i} = a^3 &\left[\sum_{j=1}^{n_2} w_{3i,j}\left\{\sum_{k=1}^{n_1} w_{2j,k}\left(\sum_{m=1}^{n_0} w_{1k,m} x_m\right)\right\}\right] \\ &+ a^3\left[\sum_{j=1}^{n_2} w_{3i,j}\left\{\sum_{k=1}^{n_1} w_{2j,k}(w_{1k,0})\right\}\right] \\ &+ a^2\left[\sum_{j=1}^{n_2} w_{3i,j}\{w_{2j,0}\}\right] \\ &+ a \cdot w_{3i,0} \end{aligned}$$

The three last expressions together represent a constant $\tilde{w}_{3i,0}$.

The triple summation in the first expression:

$$\sum_{j=1}^{n_2} w_{3i,j}\left\{\sum_{k=1}^{n_1} w_{2j,k}\left(\sum_{m=1}^{n_0} w_{1k,m}x_m\right)\right\}$$

can be replaced by a single summation:

$$\sum_{m=1}^{n_0} \tilde{w}_{3i,m}x_m$$

Thus the relation between the output value of the ith output neuron and the n_0 inputs of the network becomes:

$$y_{3i}=\sum_{m=1}^{n_0} \tilde{w}_{3i,m}x_m+\tilde{w}_{3i,0}$$

This equation represents a single linear Perceptron. The same holds for the output values of the other output neurons. This completes the proof. QED

Now we will discuss how we have to train the neural network to behave in a desired way, or in other words how we have to adapt the weights in the neural network such that after the learning period when we present an input vector $\mathbf{x}_i$, the outputs of the net will be equal to (or approximately equal to) some target vector $\mathbf{t}(\mathbf{x}_i)$. This ultimate goal must be represented by some criterion function indicating the performance the neural net. In general this will be a function that is increasing with the error and will only be zero if the error is zero. Thus we have to minimize such a criterion function during the learning period.

Given some finite learning set $L\subseteq\mathbf{D}$ of N pairs $[\mathbf{x}_i, \mathbf{t}(\mathbf{x}_i)]$, let $\mathbf{U}$ be the set of input vectors in L. We want to have for each input vector $\mathbf{x}_i\in\mathbf{U}$ some target output vector $\mathbf{t}(\mathbf{x}_i)$. Let $t_j(\mathbf{x}_i)$ be the target value of the jth output neuron for input vector $\mathbf{x}_i$ and let $y_j(\mathbf{x}_i)$ be the actual output of the jth neuron in the output layer of the neural network. We take the following criterion function as an error measure:

$$\mathbf{E}=\frac{1}{N}\sum_{\mathbf{x}_i\in U} n(\mathbf{x}_i)\sum_{j=1}^{n_3}[t_j(\mathbf{x}_i)-y_j(\mathbf{x}_i)]^2$$

with N the number of samples in the learning set L, $n(\mathbf{x}_i)$ the number of times input $\mathbf{x}_i$ occurs in U and n_3 the number of output neurons in the third layer.

We will call this error measure the *mean squared error* (*MSE*). Because N is a given constant whereas in many applications $n(\mathbf{x}_i)=1$, we can eliminate both from the equation above. To simplify the following expression for the derivative and to conform to the expression for the error usually given in literature, we add a factor 1/2. We obtain in that case for the MSE:

$$\mathbf{E}=\frac{1}{2}\sum_{\mathbf{x}_i\in U}\sum_{j=1}^{n_3}[t_j(\mathbf{x}_i)-y_j(\mathbf{x}_i)]^2$$

This expression for the error is frequently called the *energy* or *error function* of the net.

The error E is a function of all weights in the net. For a given finite learning set L and some initial random distribution of the weights the MSE will in general be large. We want to change, during a so-called *learning period*, the weights step by step, such that E will decrease to its global minimum. Only when the structure of the neural net is able to realize exactly the learning set L, must the minimum of E become zero.

For an infinitesimal increment Δw_i of the weight w_i in the connection of some neuron, somewhere in the net the following holds:

$$\Delta \mathbf{E} = \frac{\partial E}{\partial w_i} \cdot \Delta w_i$$

Thus if we take Δw_i proportional to $-(\partial E/\partial w_i)$, then ΔE will be negative as long as Δw_i is small enough. In Figure 3.6 we have illustrated this adaptation Δw_i graphically. In Figure 3.7 we see that the successive values of E may even increase if Δw_i is not small enough.

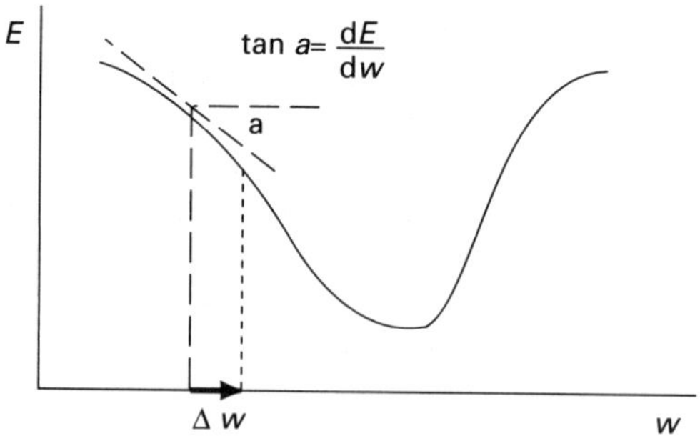

Figure 3.6 The gradient of the error function

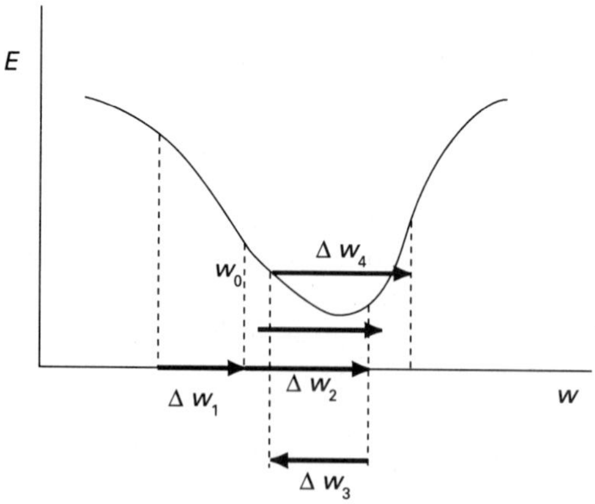

Figure 3.7 The influence of large values of Δw_i

We can recapitulate our theory in the following theorem:

Theorem 3.2a

The MSE of a multi-layer neural network will decrease if a weight w_i is changed according to:

$$\Delta w_i = -\varepsilon \cdot \frac{\partial E}{\partial w_i} \text{ for some } \varepsilon > 0 \text{ sufficiently small}$$

Or in a more general form:

Theorem 3.2b

The MSE of a multi-layer neural network will decrease if the weight vector $\mathbf{w}$ is changed according to:

$$\Delta \mathbf{w} = -\varepsilon \cdot \nabla E(\mathbf{w}) \text{ for some } \varepsilon > 0 \text{ sufficiently small}$$

with $\nabla E(\mathbf{w})$ the gradient vector with respect to $\mathbf{w}$.

The proportionality constant ε is called the *learning rate.* In general it will be profitable to change the learning rate ε during the learning process in a well-defined manner. We will return to this subject later.

The learning rule prescribed by theorem above is frequently called the learning rule based on the *gradient descent method of the MSE.* The theorem does not imply that one will ultimately reach the global minimum of the MSE; it may happen that one ends up in a local minimum of the MSE function (see Figure 3.8).

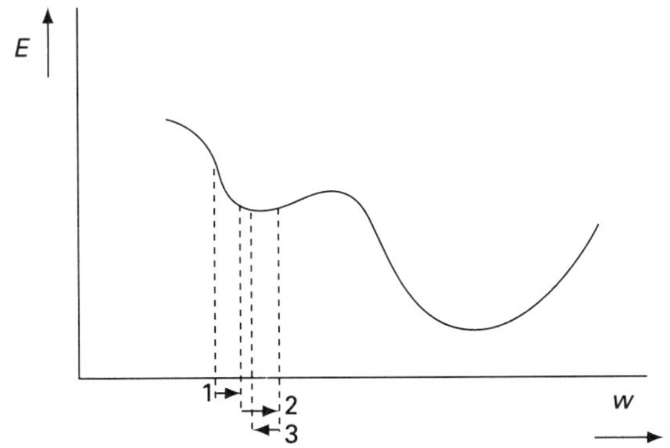

Figure 3.8 A global and local minimum of the error function

Because the expression for the MSE is given by:

$$\mathbf{E} = \frac{1}{2} \sum_{\mathbf{x}_k \in U} \sum_{j=1}^{n_3} [t_j(\mathbf{x}_k) - y_j(\mathbf{x}_k)]^2$$

we can rewrite the learning rule as:

$$\Delta w_i = \sum_{\mathbf{x}_k \in U} \sum_{j}^{n_3} \varepsilon \cdot [t_j(\mathbf{x}_k) - y_j(\mathbf{x}_k)] \cdot \frac{\partial y_j}{\partial w_i}$$

To calculate the adaptation of some weight w_i we have to evaluate the error $t_j(\mathbf{x}_k) - y_j(\mathbf{x}_k)$ and the derivative $\partial y_j / \partial w_i$ for all output neurons and for all $\mathbf{x}_k$ of the learning set L. The calculation of $\partial y_j / \partial w_i$ depends on the structure of the neural network and will be discussed in the following sections.

If we adapt some weight by $\Sigma \Delta w_i$ after the calculation of Δw_i for all $\mathbf{x}_k$ of the learning set L, then we call this procedure *global learning* (or batch training). The number of times the weights are adapted for the whole set L is called the *global learning time C*.

Sometimes the adaptation of a weight w_i is executed (in conflict with our theory) after each element $\mathbf{x}_k$ of L separately. We will call this procedure *local learning* (or incremental training).

Because global learning with a large learning set L can take a considerable amount of time, one frequently uses the so-called *staged training procedure*: One adapts the weights at each step, c_i times (called *cycle time*) for a subset $S_i \subset L$ such that $S_1 \subset S_2 \subset S_3 \subset \cdots \subset S_k = L$. In that case the global learning time is equal to $\mathbf{c} = \Sigma c_i$. Staged training is a valid procedure for minimizing the MSE because it ultimately results in $S_k = L$, as required by the derived learning rule.

3.3 Learning with a single-neuron continuous Perceptron

Although only small-scale problems can be solved with a continuous single-neuron Perceptron, the analysis of its behaviour will provide us with some understanding of the properties of the fundamental building unit of the more general multi-layer Perceptron.

In this section we will determine the adaptation of the weights of the single neuron based on the gradient descent of the MSE as discussed in the previous section.

The output of a single-neuron Perceptron (see Figure 3.9), with some transfer function f, for an input vector $\mathbf{x}_i$ is equal to:

$$g_{\mathbf{w}}(\mathbf{x}_i) = f[s(\mathbf{x}_i)] = f\left(\sum_{j=0}^{n} w_j x_{i,j}\right) \text{ with } x_{i,0} = 1 \text{ and } -w_0 \text{ the threshold}$$

Assume we want to realize with the single-neuron Perceptron for input vector $\mathbf{x}_i$

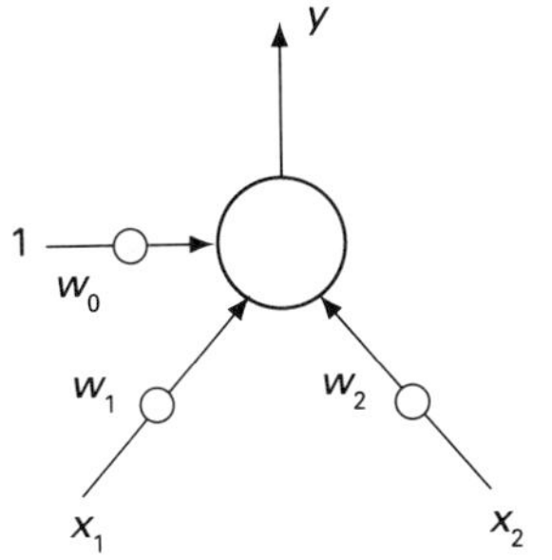

Figure 3.9 A single-neuron continuous Perceptron

a target value $t(\mathbf{x}_i)$ with $\mathbf{x}_i = [x_{i1}, x_{i2}, \ldots, x_{in}]^t \in U$ with U a finite subset of $\mathbb{R}^n$, i.e. $[\mathbf{x}_i, t(\mathbf{x}_i)] \in L$ the learning set.

For the error function we have:

$$E = \frac{1}{2} \sum_{\mathbf{x}_i \in U} [t(\mathbf{x}_i) - g_{\mathbf{w}}(\mathbf{x}_i)]^2$$

For the adaptation of some weight w_j we obtain, according to Theorem 3.2:

$$\Delta w_j = -\varepsilon \cdot \frac{\partial E}{\partial w_j} = \varepsilon \cdot \sum_{\mathbf{x}_i \in U} [t(\mathbf{x}_i) - g_{\mathbf{w}}(\mathbf{x}_i)] \cdot \frac{\mathrm{d}f}{\mathrm{d}s} \cdot x_{i,j}$$

In the following we will distinguish the extended weight vector $\hat{\mathbf{w}} = [w_0, w_1, \ldots, w_n]^t$, from the weight vector $\mathbf{w} = [w_1, w_2, \ldots, w_n]^t$. In the same way we distinguish the extended input vector $\hat{\mathbf{x}} = [1, x_{i1}, x_{i2}, \ldots, x_{in}]^t$ and the input vector $\mathbf{x} = [x_{i1}, x_{i2}, \ldots, x_{in}]^t$.

Because for a given input vector $\mathbf{x}_i$ the value of $(t(\mathbf{x}_i) - g_{\mathbf{w}}(\mathbf{x}_i))$ and of the derivative $\mathrm{d}f/\mathrm{d}s$ is independent of j, we obtain with respect to the incremental vector $\Delta\hat{\mathbf{w}}$ of the weight vector $\hat{\mathbf{w}} = [w_0, w_1, w_2, \ldots, w_n]$ the following theorem:

Theorem 3.3

In order to minimize the MSE for a given set of target values $t(\mathbf{x}_i)$ for a given finite set U, the adaptation of the weight vector $\hat{\mathbf{w}}$ of a single neuron Perceptron must be:

$$\Delta\hat{\mathbf{w}} = \varepsilon \cdot \sum_{\mathbf{x}_i \in U} [t(\mathbf{x}_i) - g(\mathbf{x}_i)] \frac{\mathrm{d}f}{\mathrm{d}s} \hat{\mathbf{x}}_i$$

We see that, for a single input vector $\mathbf{x}_i$, $|\Delta\hat{\mathbf{w}}|$ is proportional to $|\hat{\mathbf{x}}_i|$ and in the direction of the extended input vector $\hat{\mathbf{x}}_i$ if $t(\mathbf{x}_i) > g(\mathbf{x}_i)$ and in the opposite direction of $\hat{\mathbf{x}}_i$ if $t(\mathbf{x}_i) < g(\mathbf{x}_i)$. The same holds for the relation between $\Delta\mathbf{w}$ and $\mathbf{x}_i$ in the non-extended input space. For the case of three-dimensional extended input space, see Figure 3.10.

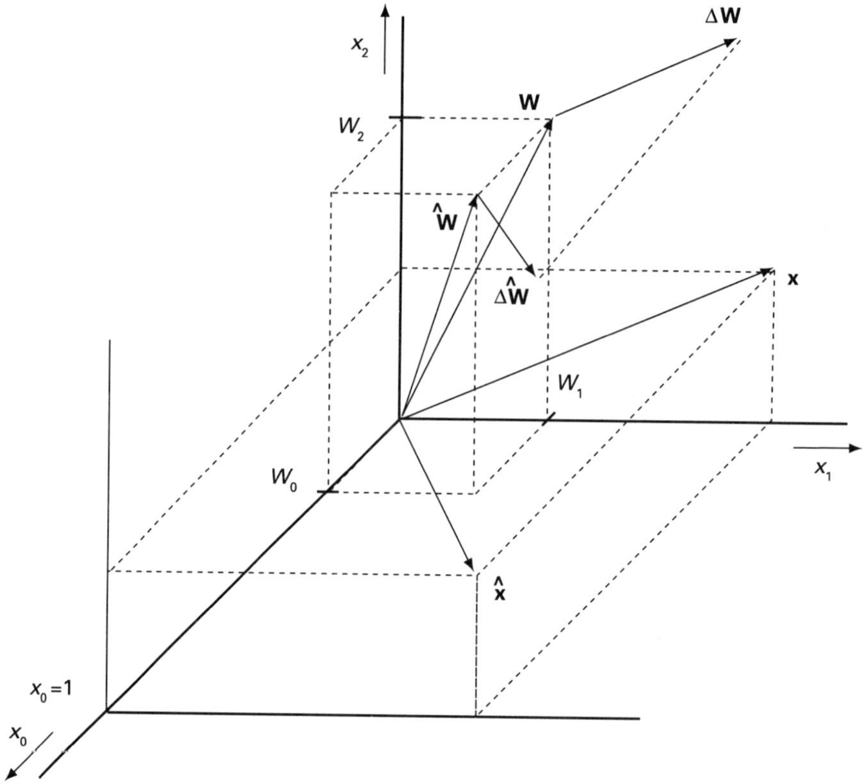

Figure 3.10 The extended input/weight space. The weight vector adaptation $\Delta\hat{\mathbf{w}}$ is proportional to the input vector $\hat{\mathbf{x}}$

Example 3.1

Take a single-neuron Perceptron with two inputs x_1 and x_2 and a threshold w_0. For the sake of simplicity we take a linear transfer function for the neuron, defined by $f(s)=s$, with $s=w_0+w_1x_1+w_2x_2$. Assume $U=\{[1, 1]^t, [1, -1]^t, [-1, 1]^t\}$.

The target functions are $t([1, 1]^t)=0.9$, $t([1, -1]^t)=0.1$ and $t([-1, 1]^t)=0.1$. Assume the initial weights are: $w_1=0.5$, $w_2=0.5$ and $w_0=0$.

If we present the elements of U to the net we obtain:

$$\mathbf{x}=[1, 1]^t \Rightarrow s(\mathbf{x})= 0.5+0.5=1 \Rightarrow g(\mathbf{x})=1 \Rightarrow t(\mathbf{x})-g(\mathbf{x})=0.9-1=-0.1$$
$$\mathbf{x}=[1, -1]^t \Rightarrow s(\mathbf{x})= 0.5-0.5=0 \Rightarrow g(\mathbf{x})=0 \Rightarrow t(\mathbf{x})-g(\mathbf{x})=0.1-0=+0.1$$
$$\mathbf{x}=[-1, 1]^t \Rightarrow s(\mathbf{x})= -0.5+0.5=0 \Rightarrow g(\mathbf{x})=0 \Rightarrow t(\mathbf{x})-g(\mathbf{x})=0.1-0=+0.1$$

The adaptation of weights is prescribed by:

$$\Delta\hat{\mathbf{w}}=\varepsilon\sum_i\,[t(\mathbf{x}_i)-g(\mathbf{x}_i)]\frac{\mathrm{d}f}{\mathrm{d}s}\hat{\mathbf{x}}_i$$

Because for all $s(\mathbf{x}_i)$: $\mathrm{d}f/\mathrm{d}s=1$ we have:

$$\begin{bmatrix}\Delta w_0\\ \Delta w_1\\ \Delta w_2\end{bmatrix}=\varepsilon(-0.1)\begin{bmatrix}1\\1\\1\end{bmatrix}+\varepsilon(0.1)\begin{bmatrix}1\\1\\-1\end{bmatrix}+\varepsilon(0.1)\begin{bmatrix}1\\-1\\1\end{bmatrix}=\varepsilon\begin{bmatrix}0.1\\-0.1\\-0.1\end{bmatrix}$$

We have stipulated that ε must be small enough (e.g. $\varepsilon=0.01$) to guarantee the decline in E. Incidentally in this simple linear case we can take $\varepsilon=1$ and after adaptation we will obtain the values of the weights for which the MSE is zero.

For this simple case we can easily calculate without learning the values of the three weights such that the MSE is zero. For a zero MSE we have:

$$\begin{aligned}&\text{for } \mathbf{x}=[1,\,1]^t && t(\mathbf{x})=g(\mathbf{x})=0.9=s(\mathbf{x})=+w_1+w_2+w_0\\ &\text{for } \mathbf{x}=[1,\,-1]^t && t(\mathbf{x})=g(\mathbf{x})=0.1=s(\mathbf{x})=+w_1-w_2+w_0\\ &\text{for } \mathbf{x}=[-1,\,1]^t && t(\mathbf{x})=g(\mathbf{x})=0.1=s(\mathbf{x})=-w_1+w_2+w_0\end{aligned}$$

It follows that the error is zero for $w_1=0.4$, $w_2=0.4$ and $w_0=0.1$.

In Figure 3.11, a three-dimensional plot for the error E as a function of w_1 and w_2 is given with $w_0=0.1$. Figure 3.12 shows the contour lines of constant error levels. ■

According to the learning rule, we adapt the weights after calculating the error for the whole set U of input vectors of the learning set L. We have called this procedure *global learning*.

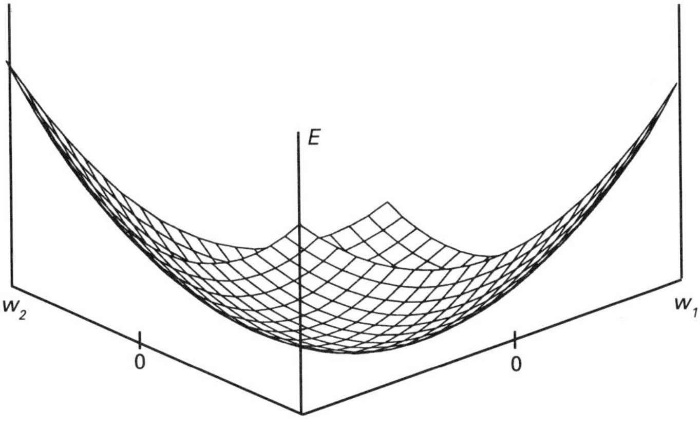

Figure 3.11 The error function with $w_0=0.1$ of Example 3.1

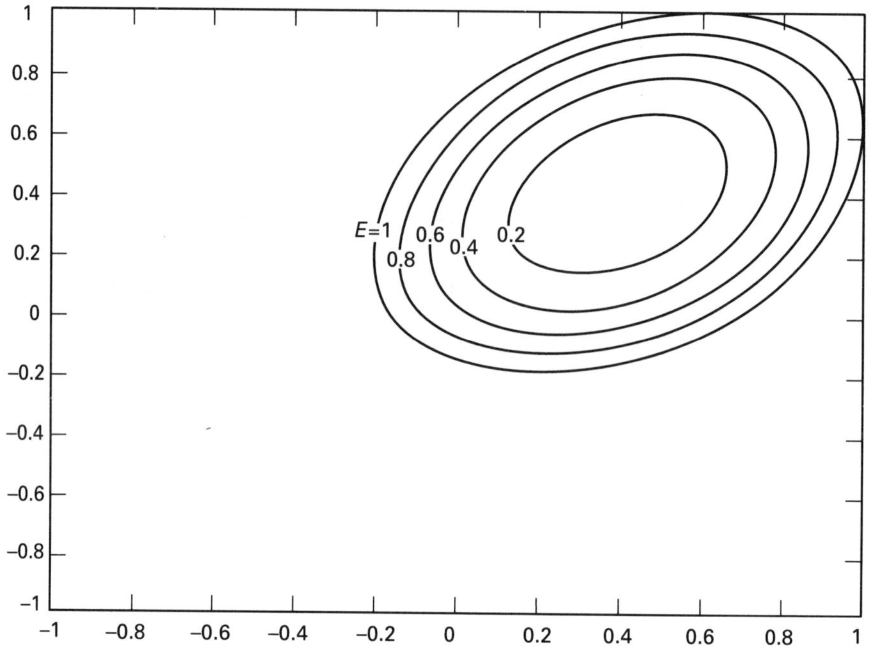

Figure 3.12 The constant-level contour lines of the error function of Figure 3.11

What will happen if we adapt the weights after the calculation of the error for each element of U separately? This method of learning is called *local learning*, and the following example shows that, in general, global and local learning differ after a finite number of adaptations (globally one step and locally four steps). In general, after a finite number of steps in case of local learning, the MSE will depend on the particular sequence of elements of U.

Example 3.2

Take the previous example with the same initial distribution of weights: $w_1 = 0.5$, $w_2 = 0.5$ and $w_0 = 0$.

We calculate first the error for $\mathbf{x} = (1, 1)$:

$$\mathbf{x} = [1, 1]^t \quad \Rightarrow s(\mathbf{x}) = 0.5 + 0.5 = 1 \quad \Rightarrow g(\mathbf{x}) = 1 \quad \Rightarrow t(\mathbf{x}) - g(\mathbf{x}) = 0.9 - 1 = -0.1$$

We take $\varepsilon = 1$.

According to the learning rule the weights become $w_1 = 0.4$, $w_2 = 0.4$ and $w_0 = -0.1$.

For the new set of weights we calculate the error for $\mathbf{x}=[1, -1]^t$:

$$\mathbf{x}=[1, -1]^t \quad \Rightarrow s(\mathbf{x})=0.4-0.4-0.1=-0.1 \quad \Rightarrow g(\mathbf{x})=-0.1$$
$$\Rightarrow t(\mathbf{x})-g(\mathbf{x})=0.1-(-0.1)=0.2$$

After adaptation the weights will be $w_1=0.6$, $w_2=0.2$ and $w_0=0.1$.

We calculate next the error for $\mathbf{x}=[-1, 1]^t$:

$$\mathbf{x}=[-1, 1]^t \quad \Rightarrow s(\mathbf{x})=-0.6+0.2+0.1=-0.3 \quad \Rightarrow g(\mathbf{x})=-0.3$$
$$\Rightarrow t(\mathbf{x})-g(\mathbf{x})=0.1-(-0.3)=0.4$$

After adaptation the weights will be $w_1=0.2$, $w_2=0.6$ and $w_0=0.5$.

Thus the final weights are $w_1=0.2$, $w_2=0.6$ and $w_0=0.5$. The MSE in this case is not zero. ■

In general the MSE will have several minima, and depending on the initial distribution of weights, one may end up in some local minimum instead of the global minimum. It turns out that with local learning one can frequently avoid local minima.

3.4 The exact fitting of the data set with a single-neuron Perceptron

In this section we investigate whether the function realized after learning by a single continuous Perceptron can go exactly through the samples of the given data set $D=\{(\mathbf{x}_i, t(\mathbf{x}_i)\}$.

The following theorem will show that only under very restricted conditions can a function $g_{\mathbf{w}}(\mathbf{x})$ that is realizable by a single-neuron Perceptron, go exactly through the samples of the data set.

Theorem 3.4

A function $g_{\mathbf{w}}$: $\mathbb{R}^n \rightarrow \mathbb{R}$ that can be realized by a single-neuron Perceptron with transfer function f can go exactly (the MSE is zero) through the samples of the data set $[\mathbf{x}_i, t(\mathbf{x}_i)] \in D$, iff for every element $[\mathbf{x}_i, t(\mathbf{x}_i)]$ from D the vector:

$$\tilde{\mathbf{x}}_i=\{\hat{\mathbf{x}}_i, f^{-1}[t(\mathbf{x}_i)]\} \text{ with } \hat{\mathbf{x}}_i=[1, x_{i1}, x_{i2}, \ldots, x_{in}]$$

is a linear combination of some unique set of $n+1$ linear independent vectors:

$$\tilde{\mathbf{x}}_j=\{\hat{\mathbf{x}}_j, f^{-1}[t(\mathbf{x}_j)]\}$$

obtained from $n+1$ elements $[\mathbf{x}_j, t(\mathbf{x}_j)]$ from the data set D.

Proof

Let the data set D contain a subset S with $n+1$ pairs $[\mathbf{x}_j, t(\mathbf{x}_j)]$ such that we can form $n+1$ independent equations:

$$\sum_{k=0}^{n} w_k \cdot x_{j,k} = f^{-1}[t(\mathbf{x}_j)] \quad \text{for each element in } S \text{ (NB } x_{j,0}=1)$$

From these equations all $n+1$ weights and thus $g_{\mathbf{w}}$, which exactly go through the samples of the subset S, can be determined.

If in addition for every element $[\mathbf{x}_i, t(\mathbf{x}_i)]$ of the data set D we can write:

$$\{\hat{\mathbf{x}}_i, f^{-1}[t(\mathbf{x}_i)]\} = \left\{\sum_{j=1}^{n+1} \alpha_j \hat{\mathbf{x}}_j, \sum_{j=1}^{n+1} \alpha_j f^{-1}[t(\mathbf{x}_j)]\right\} \quad \text{with } [\mathbf{x}_j, t(\mathbf{x}_j)] \in S$$

and not all $\alpha_j = 0$, then for every element $(\mathbf{x}_i, t(\mathbf{x}_i)) \in D$ the following relation should hold:

$$g_{\mathbf{w}}(\mathbf{x}_i) = f\left(\sum_{k=0}^{n} w_k \cdot x_{i,k}\right) = f\left(\sum_{k=0}^{n} w_k \sum_{j=1}^{n+1} \alpha_j \cdot x_{j,k}\right) = f\left(\sum_{j=1}^{n+1} \alpha_j \sum_{k=0}^{n} w_k \cdot x_{j,k}\right)$$

$$= f\left\{\sum_{j=1}^{n+1} \alpha_j f^{-1}[t(\mathbf{x}_j)]\right\} = f\{f^{-1}[t(\mathbf{x}_i)]\} = t(\mathbf{x}_i)$$

Thus the function $g_{\mathbf{w}}$ also goes exactly through all the samples of D. If, on the other hand, the data set should contain more than one subset S with the property mentioned above we would find a different function $g_{\mathbf{w}}$ for each such a set and hence no unique solution for $g_{\mathbf{w}}$ exists. QED

Example 3.3

Take a single-neuron Perceptron with two inputs x_1 and x_2 and a threshold w_0. The transfer function of the neuron is defined by:

$$f(s) = s \quad \text{with } s = w_0 + w_1 x_1 + w_2 x_2$$

Assume $\mathbf{U} = \{\mathbf{x}_1, \mathbf{x}_2, \mathbf{x}_3, \mathbf{x}_4\} = \{(1, 1), (1, -1), (-1, 1), (-1, 3)\}$.

The target values are: $t(1, 1) = 0.9$, $t(1, -1) = 0.1$, $t(-1, 1) = 0.1$ and $t(-1, 3) = 0.9$.

The first three extended input vectors $\hat{\mathbf{x}}_i$ are linear independent, thus we can calculate the required values of the three weights.

For a zero MSE we must have:

$$\text{for } \mathbf{x} = (1, 1) \quad t(\mathbf{x}) = g(\mathbf{x}) = 0.9 = s(\mathbf{x}) = w_0 + w_1 + w_2$$

$$\text{for } \mathbf{x} = (1, -1) \quad t(\mathbf{x}) = g(\mathbf{x}) = 0.1 = s(\mathbf{x}) = w_0 + w_1 - w_2$$

$$\text{for } \mathbf{x} = (-1, 1) \quad t(\mathbf{x}) = g(\mathbf{x}) = 0.1 = s(\mathbf{x}) = w_0 - w_1 + w_2$$

Thus $w_0 = 0.1$, $w_1 = 0.4$ and $w_2 = 0.4$.

For the fourth sample we have $[\hat{\mathbf{x}}_4, f^{-1}(t(\mathbf{x}_4))]$ is a linear combination of

$[\hat{\mathbf{x}}_1, f^{-1}(t(\mathbf{x}_1))], [\hat{\mathbf{x}}_2, f^{-1}(t(\mathbf{x}_2))]$ and $[\hat{\mathbf{x}}_3, f^{-1}(t(\mathbf{x}_3))] = [1, -1, 3, 0.9] = [1, 1, 1, 0.9] - [1, 1, -1, 0.1] + [1, -1, 1, 0.1]$, thus the function $g_{\mathbf{w}}(\mathbf{x})$ also goes through the fourth sample. ■

Example 3.4

It will be clear that we can exactly fit the data points when the data points belong to the function $g_{\mathbf{w}}(\mathbf{x})$ realizable by a single-neuron Perceptron. For example, given three data points $D = \{[x_1, t(\mathbf{x}_1)], [x_2, t(x_3)], [x_3, t(x_3)]\}$ with $t(x_i) = g_{\mathbf{w}}(x_i) = f(w_0 + w_1 x_i)$ for some w_0 and some w_1, we can then determine from two examples w_0 and w_1 and the third example will satisfy the condition mentioned in Theorem 3.3. ■

From Theorem 3.3 and its proof we can conclude that we can calculate the weights if there exists a function $g_{\mathbf{w}}(\mathbf{x})$ that goes exactly through all samples of the data set D, by selecting $n+1$ independent input vectors.

Practical statement 3.1

If a function realizable by a single-neuron Perceptron can go exactly through the samples of the data set, then we do not need to learn that Perceptron, because we can calculate beforehand the required weights from the data set *D*.

Another consequence of Theorem 3.3 is the following statement:

Practical statement 3.2

If x in $[\mathbf{x}, t(\mathbf{x})] \in D$ is of dimension n and the data set D contains $p \leqslant n+1$ independent extended input vectors $\hat{\mathbf{x}}_i$, then the function $g_{\mathbf{w}}$: $\mathbb{R}^n \to \mathbb{R}$ realized by the single-neural Perceptron can go exactly through all the samples of the data set *D*.

If $p = n+1$, the function $g_{\mathbf{w}}(\mathbf{x})$ is unique. If $p < n+1$, then there is an infinite number of solutions for $g_{\mathbf{w}}(\mathbf{x})$.

The last sentence in the statement above reveals that generalization from a data set with a neural net can be dangerous.

Because any 'non-academic' data set D does not fulfil the requirements of Theorem 3.3 we also have the following statement:

Practical statement 3.3

If x in $[\mathbf{x}, t(\mathbf{x})] \in D$ is of dimension n, and the data set D contains more than $n+1$ elements, then the function $g_{\mathbf{w}}(\mathbf{x})$ realized by the single-neuron Perceptron cannot go exactly through all the samples of the data set *D*.

3.5 The approximate fitting of the data set with a single-neuron Perceptron

In the previous chapter we found that in most practical cases that any function realizable with a single-neuron Perceptron cannot go exactly through the samples of the data set. However, we can always find with a single-neuron Perceptron a function $g_\mathbf{w}$ such that $g_\mathbf{w}(\mathbf{x})$ is 'approximately' equal to the targets $t(\mathbf{x})$ for all elements in the data set D. The quality of the 'approximation' is measured by the MSE.

Example 3.5

Assume the transfer function of the neuron is given by $f(s)=[1+\exp(-s)]^{-1}$ and $s=w_0x_0+w_1x_1$, with $x_0=1$. Thus there is one external input x_1. Let $D=\{(x_i, t(x_i))\}=\{(1, 0.1), (-1, 0.1), (2, 0.4), (-2, 0.4), (3, 0.9), (-3, 0.9),\}$ be the data set. [NB: ten times $t(x_i)$ gives the square of x_i.] The MSE depends on the value of w_0 and w_1. See Figure 3.13.

The minimum of the MSE occurs at $w_0=-0.13$ and $w_1=0$. Thus the function to be realized after learning by the single-neuron Perceptron is the constant: $g_\mathbf{w}(\mathbf{x})=[1+\exp(0.13)]^{-1}=0.47$ (see Figure 3.14); MSE$=0.327$.

One might say that the function $g_\mathbf{w}(\mathbf{x})=0.47$, which will be learned by the single-neuron Perceptron, is a bad approximation of the function $g(\mathbf{x})=0.1x^2$, which one might presume to be the function underlying the data set D. However, if the function $h(x)$ drawn in Figure 3.13 was the generator of the data set D, then $g_\mathbf{w}(\mathbf{x})=0.47$ would not be a bad approximation at all. So without any additional information,

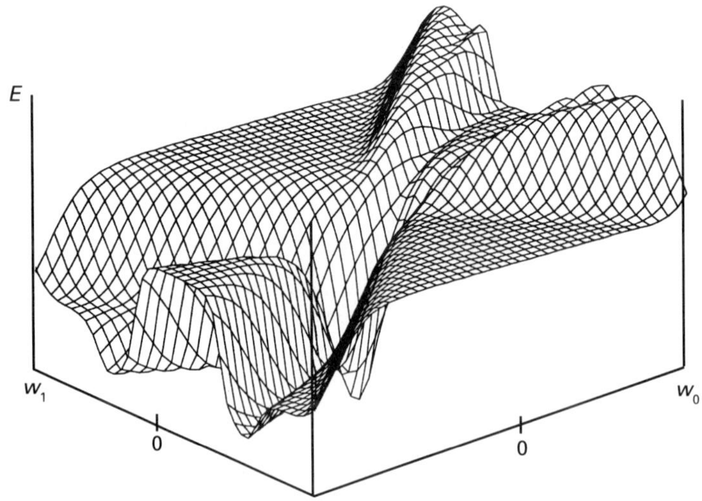

Figure 3.13 The error function of Example 3.5

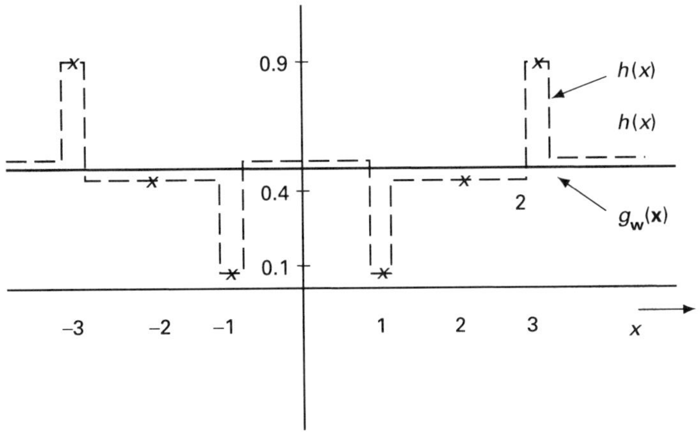

Figure 3.14 The constant function $g_{\mathbf{w}}(\mathbf{x})$ with a minimal error for Example 3.5

apart from the value of the MSE, one cannot qualify the correctness of the function learned by the neural net. ■

After learning, we only know that the function realized by the single-neuron Perceptron will approximate the given data set as closely as possible (measured by the MSE), as long as we do not end up in a local minimum of the MSE.

In many cases, however, we do not have to use the learning algorithm to find the desired function $g_{\mathbf{w}}$ approximating the data set D, because we can determine the pertinent weight vector $\hat{\mathbf{w}}$ beforehand.

Theorem 3.5

Given a single-neuron Perceptron that can realize functions $g_{\mathbf{w}}$: $\mathbb{R}^n \rightarrow \mathbb{R}$, then the weight vector $\mathbf{w}$ of the function $g_{\mathbf{w}}$: $\mathbb{R}^n \rightarrow \mathbb{R}$ with a minimal MSE for a given data set D can be determined from a set of linear equations if:

1. The 'error' $\Delta g_{\mathbf{w}}(\mathbf{x}_i) = g_{\mathbf{w}}(\mathbf{x}_i) - t(\mathbf{x}_i)$ for all $\mathbf{x}_i$ of D is 'small'.
2. The derivative $\mathrm{d}E/\mathrm{d}s = 0$ only for its global minimum.
3. The transfer function f of the neuron has an inverse f^{-1}.

Proof

Let

$$s(\mathbf{x}_i) = \sum_{k=0}^{n} w_k x_{ik}$$

be the weighted input of the neuron. In case of the minimal MSE we have for each $[\mathbf{x}_i, t(\mathbf{x}_i)] \in D$: $g_{\mathbf{w}}(\mathbf{x}_i) = f[s(\mathbf{x}_i)]$, and the 'error' $\Delta g_{\mathbf{w}}(\mathbf{x}_i) = g_{\mathbf{w}}(\mathbf{x}_i) - t(\mathbf{x}_i)$ in the output of

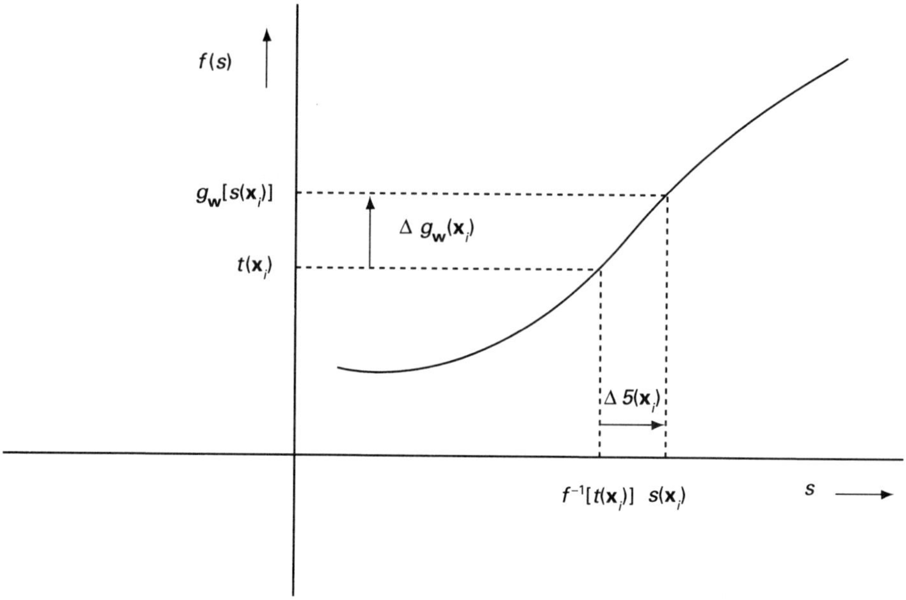

Figure 3.15 The graphical representation of the relation between $\Delta s(\mathbf{x}_i)$ and $\Delta g_{\mathbf{w}}(\mathbf{x}_i)$

the neuron for input $\mathbf{x}_i$ is assumed to be 'small'. (For simplicity we will from now on drop the subscript $\mathbf{w}$ of $g_{\mathbf{w}}$.)

If there were no error, the weighted input for the input vector $\mathbf{x}_i$ would be $s(\mathbf{x}_i) = f^{-1}\{t(\mathbf{x}_i)\}$; however, $t(\mathbf{x}_i) \neq g(\mathbf{x}_i)$, thus there will be an error $\Delta s(\mathbf{x}_i)$ in the weighted input. Thus we can write:

$$s(\mathbf{x}_i) = f^{-1}\{t(\mathbf{x}_i)\} + \Delta s(\mathbf{x}_i)$$

If $\Delta s(\mathbf{x}_i)$ is small (or higher derivatives of $f(s)$ are negligible) we may replace the equation above by (see also Figure 3.15):

$$f[s(\mathbf{x}_i)] = t(\mathbf{x}_i) + \frac{\mathrm{d}f}{\mathrm{d}s}\Delta s(\mathbf{x}_i)$$

On the other hand we have:

$$f[s(\mathbf{x}_i)] = t(\mathbf{x}_i) + \Delta g(\mathbf{x}_i)$$

Thus:

$$\Delta s(\mathbf{x}_i) = \frac{\Delta g(\mathbf{x}_i)}{\dfrac{\mathrm{d}f}{\mathrm{d}s}}$$

For each of the elements $(\mathbf{x}_i, t(\mathbf{x}_i)) \in D$ we now have a linear equation:

$$\sum_{k=0}^{n} w_k x_{ik} = f^{-1}\{t(\mathbf{x}_i)\} + \Delta s(\mathbf{x}_i) = f^{-1}\{t(\mathbf{x}_i)\} + \left(\frac{\mathrm{d}f}{\mathrm{d}s}\right)^{-1} \Delta g(\mathbf{x}_i)$$

with the derivative $\mathrm{d}f/\mathrm{d}s$ evaluated for $f^{-1}\{t(\mathbf{x}_i)\}$.

For the N elements of D we can find N such independent equations, thus we also have N unknown values of $\Delta g(\mathbf{x}_i)$ and $n+1$ unknown values of weights w_k. However, because the MSE must be minimal we also have $\partial E/\partial w_j = 0$ for each weight w_j.

With:

$$E = \sum_{\mathbf{x}_i \in D} \frac{1}{2}\{t(\mathbf{x}_i) - g(\mathbf{x}_i)\}^2$$

$$g(\mathbf{x}_i) = f(s(\mathbf{x}_i)) = f\left(\sum_{k=0}^{n} w_k x_{ik}\right)$$

we obtain for each weight w_j an additional equation:

$$\frac{\partial E}{\partial w_j} = \sum_{\mathbf{x}_i \in D} \Delta g(\mathbf{x}_i) \frac{\mathrm{d}f}{\mathrm{d}s} x_{ij} = 0$$

The derivative $\mathrm{d}f/\mathrm{d}s$ should be evaluated at $s(\mathbf{x}_i) = f^{-1}(g(\mathbf{x}_i))$; for small values of $\Delta g(\mathbf{x}_i)$, however, we may evaluate the derivative at $f^{-1}[t(\mathbf{x}_i)]$. We now have $N+n+1$ unknowns and the same number of linear equations, and so we can determine the weight vector $\mathbf{w}$ and hence $g_{\mathbf{w}}$ and also the minimum MSE. QED

Example 3.6

According to Theorem 3.3 there is no function that is realizable by a single-neuron Perceptron exactly going through the samples of the following data set:

$$D = \{(\mathbf{x}_i, t(\mathbf{x}_i))\} = \{([1, 1]^t, 0.9), ([1, -1]^t, 0.1), ([-1, 1]^t, 0.1), ([-1, -1]^t, 0.1)\}$$

However, the data set can be approximated by a function $g_{\mathbf{w}}$ to be learned by the neural net and we can calculate the desired weights.

For simplicity we take a linear transfer function $f(s) = s$. Because in this case $f^{-1}(t(\mathbf{x}_i)) = t(\mathbf{x}_i)$ and $\mathrm{d}f/\mathrm{d}s = 1$, we have $\Delta s(\mathbf{x}_i) = \Delta g(\mathbf{x}_i)$. Now the following relations must hold:

$$\mathbf{x}_1 = [1, 1]^t \quad \Rightarrow s(\mathbf{x}_1) = w_0 + w_1 + w_2 = 0.9 + \Delta g(\mathbf{x}_1)$$

$$\mathbf{x}_2 = [1, -1]^t \quad \Rightarrow s(\mathbf{x}_2) = w_0 + w_1 - w_2 = 0.1 + \Delta g(\mathbf{x}_2)$$

$$\mathbf{x}_3 = [-1, 1]^t \quad \Rightarrow s(\mathbf{x}_3) = w_0 - w_1 + w_2 = 0.1 + \Delta g(\mathbf{x}_3)$$

$$\mathbf{x}_4 = [-1, -1]^t \quad \Rightarrow s(\mathbf{x}_4) = w_0 - w_1 - w_2 = 0.1 + \Delta g(\mathbf{x}_4)$$

For the minimal MSE we must have $\partial E/\partial w_j = 0$ for each weight w_j. Thus for each

$j \in \{0, 1, 2\}$:

$$\sum_i \Delta g(\mathbf{x}_i) \frac{\mathrm{d}f}{\mathrm{d}s} x_{ij} = 0 \quad \left(\frac{\mathrm{d}f}{\mathrm{d}s} = 1 \text{ in our case and } x_{i0} = 1 \right)$$

we obtain the following relations:

$$\begin{aligned}
\Delta g(\mathbf{x}_1)(1) + \Delta g(\mathbf{x}_2)(1) + \Delta g(\mathbf{x}_3)(1) + \Delta g(\mathbf{x}_4)(1) \quad &= 0 \\
\Delta g(\mathbf{x}_1)(1) + \Delta g(\mathbf{x}_2)(1) + \Delta g(\mathbf{x}_3)(-1) + \Delta g(\mathbf{x}_4)(-1) &= 0 \\
\Delta g(\mathbf{x}_1)(1) + \Delta g(\mathbf{x}_2)(-1) + \Delta g(\mathbf{x}_3)(1) + \Delta g(\mathbf{x}_4)(-1) &= 0
\end{aligned}$$

From this set of seven equations we conclude:

$$\Delta g(\mathbf{x}_1) = -0.2 \quad \Delta g(\mathbf{x}_2) = 0.2 \quad \Delta g(\mathbf{x}_3) = 0.2 \quad \Delta g(\mathbf{x}_4) = -0.2$$

$$w_1 = 0.2 \quad w_2 = 0.2 \quad w_0 = 0.3$$

$$\text{MSE} = 0.08$$

■

The previous example was a demonstration of Theorem 3.5 for the simple case of a linear transfer function for the neuron. The following example shows the application of Theorem 3.5 in case of a non-linear transfer function. If the reader is convinced that the theorem holds for a non-linear transfer function he or she may skip this example.

Example 3.7

Assume the transfer function of the neuron is given by:

$$f(s) = 1 - \mathrm{e}^{-s^2}$$

and $s = w_0 x_0 + w_1 x_1$, with $x_0 = 1$. Thus there is one external input x_1. Let $D = \{(x_i, t(x_i)\} = \{(1, 0.1), (-1, 0.1), (2, 0.4), (-2, 0.4), (3, 0.9), (-3, 0.9)\}$ be the data set. [NB: ten times $t(x_i)$ yields the square of x_i.] For each extended input $\hat{\mathbf{x}}_i = (x_{i0}, x_{i1}) = (1, x_{i1})$ we have an equation of the form:

$$\sum_{k=0}^{n} w_k x_{ik} = f^{-1}[t(\mathbf{x}_i)] + \left(\frac{\mathrm{d}f}{\mathrm{d}s} \right)^{-1} \Delta g(\mathbf{x}_i)$$

With $f^{-1}(t) = \pm[\ln(1-t)^{-1}]^{1/2}$ and $\mathrm{d}f/\mathrm{d}s = 2s[1 - f(s)]$ we obtain for each of the elements $[\mathbf{x}_i, t(\mathbf{x}_i)] \in D$ a linear equation:

$$\sum_{k=0}^{n} w_k x_{ik} = f^{-1}\{t(\mathbf{x}_i)\} + \Delta s(\mathbf{x}_i) = f^{-1}\{t(\mathbf{x}_i)\} + \left(\frac{\mathrm{d}f}{\mathrm{d}s} \right)^{-1} \Delta g(\mathbf{x}_i)$$

We take first $f^{-1}\{t(\mathbf{x}_i)\} = +[\ln(1-t)^{-1}]^{1/2}$ and then $-[\ln(1-t)^{-1}]^{1/2}$.

$$\begin{aligned}
w_0 + w_1 &= 0.32 + 1.71\Delta g(\mathbf{x}_1) \\
w_0 - w_1 &= -0.32 - 1.71\Delta g(\mathbf{x}_2) \\
w_0 + 2w_1 &= 0.71 + 1.17\Delta g(\mathbf{x}_3) \\
w_0 - 2w_1 &= -0.71 - 1.17\Delta g(\mathbf{x}_4) \\
w_0 + 3w_1 &= 1.52 + 3.30\Delta g(\mathbf{x}_5) \\
w_0 - 3w_1 &= -1.52 - 3.30\Delta g(\mathbf{x}_6)
\end{aligned}$$

For each weight w_j we obtain an additional equation:

$$\frac{\partial E}{\partial w_j} = \sum_{\mathbf{x}_i \in \mathbf{D}} \Delta g(\mathbf{x}_i) \frac{\mathrm{d}f}{\mathrm{d}s} x_{ij} = 0$$

Thus:

$$0.58\Delta g(\mathbf{x}_1) - 0.58\Delta g(\mathbf{x}_2) + 0.85\Delta g(\mathbf{x}_3) - 0.85\Delta g(\mathbf{x}_4) + 0.3\Delta g(\mathbf{x}_5) - 0.3\Delta g(\mathbf{x}_6) = 0$$

$$0.58\Delta g(\mathbf{x}_1) + 0.58\Delta g(\mathbf{x}_2) + 1.70\Delta g(\mathbf{x}_3) + 1.70\Delta g(\mathbf{x}_4) + 0.9\Delta g(\mathbf{x}_5) + 0.9\Delta g(\mathbf{x}_6) = 0$$

The solution for this set of equations is:

$$w_0 = 0 \quad w_1 = 0.39$$

$$\Delta g(\mathbf{x}_1) = 0.042 \quad \Delta g(\mathbf{x}_2) = 0.042 \quad \Delta g(\mathbf{x}_3) = 0.041$$

$$\Delta g(\mathbf{x}_4) = 0.041 \quad \Delta g(\mathbf{x}_5) = -0.10 \quad \Delta g(\mathbf{x}_6) = -0.10$$

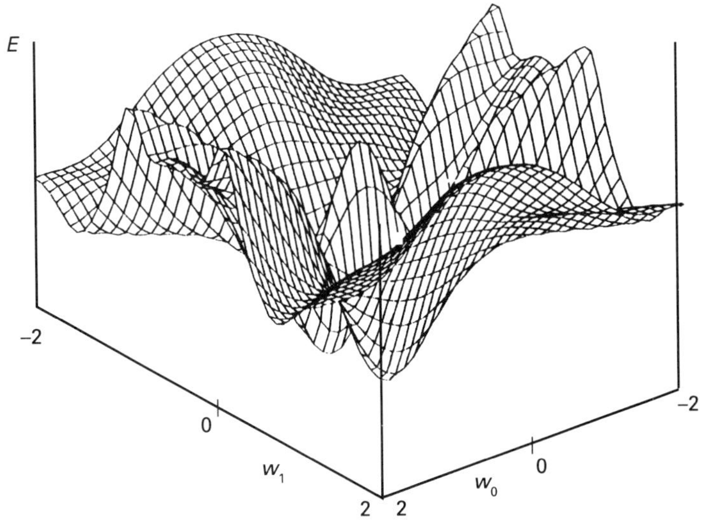

Figure 3.16 The MSE for Example 3.7

If we take first $f(s)^{-1} = -[\ln(1-t)^{-1}]^{-1/2}$, we obtain $w_0 = 0$ and $w_1 = -0.39$ with the same values for $\Delta g(\mathbf{x}_i)$.

The MSE as a function of w_0 and w_1 is given in Figure 3.16. ■

From the discussion above we can conclude the following:

Practical statement 3.4

If a data set D can be approximated well enough by a function $g_{\mathbf{w}}$ realized by a single-neuron Perceptron, then we do not have to learn that function because we can calculate the weights beforehand.

3.6 Generalizing with a single-neuron continuous Perceptron

In the literature on artificial neural networks one frequently encounters statements about the generalization capability of neural networks, such as: 'Neural networks are capable of adapting themselves with the aid of a learning rule and a set of examples to model relationships among the data without any *a priori* assumptions about the nature of the relationships.' A similar statement is: 'After learning, neural networks may be used to predict characteristics of new samples or to derive empirical models from examples in situations in which no theoretically based model is known.' Although to a certain extent these types of statements are true, one must be careful with the substatements that no model, or no *a priori* information about the nature of the relationship between examples is assumed.

Generalization is the process of inductive inference of general relationships from a finite number of samples. For example, one is inclined to assume that the sun will rise tomorrow because we have observed a finite number of times, without any counterexample, that the sun has so far risen on every day in our lifetime. Another example, already discussed in the preface of the book, is the inference of a new number in a finite sequence of numbers: one is inclined to say that the next example in sequence 1, 4, 9, 16,... will be the number 25, because we observe a simple regularity in the sequence: the kth element in the sequence is k^2. However, without being prejudiced in favour of some type of 'model' any number may follow the given sequence. For example, one might as well say that the next number is 29 because one is in favour of the regularity where the nth number $y(n)$ in the sequence is defined by $y(n)$ = 'sum of first n primes'. If a large number of people (with some knowledge of elementary calculus!) are asked to guess the next number in the sequence given above, almost all will say that the next number is 25. This phenomenon reveals the human attitude to select always the most 'simple' model to explain a sequence of experimental observations.

Practical statement 3.5

'Modelling' can be defined as the process of formulating a finite set of interrelated rules (or the construction of a finite set of interconnected mechanisms) by which one can generate the (potentially infinite) set of observed data. The simplicity of the model is very subjective because it depends on the domain of knowledge of the person who is doing the modelling.

If one is using complex neural networks to 'model' relationships behind a given set of data, it is hard to demonstrate that one is using (or assuming) *a priori* information about the kind of relationship. In the case of a single-neuron Perceptron one can easily show that one needs (or is assuming) *a priori* information about the relationship between the given data in order to be justified in accepting the outcome of the learning process.

If one is using, for example, a sigmoid transfer function $g_{\mathbf{w}}(\mathbf{x})=\{1+\exp[-s(\mathbf{x})]\}^{-1}$ and a single-neuron Perceptron with one input, then whatever the data set may be, the types of relationships behind the data that one will find with the neural network will be restricted to a function from the set of sigmoid functions. All functions realized will be increasing ($w_1>0$) or decreasing ($w_1<0$) sigmoid functions of the input x (see Figure 3.17). One extreme for $w_1=0$ will be the constant $g_{\mathbf{w}}(\mathbf{x})=[1+\exp(-w_0)]^{-1}$ and the other extreme for $w_1\to\infty$ with w_0/w_1 finite will be the threshold function $y=1$ for $x>-w_0/w_1$ and $y=0$ for $x<-w_0/w_1$ (see Figure 3.18).

If, for example, the data set contains pairs $[x_i, t(x_i)]$ from the set $\{(x_i, y_i)|y_i=0.01x_i^2\}$ uniformly distributed over a symmetrical x-domain, then the neural net will find after learning that the relation between x_i and y_i is some constant $g_{\mathbf{w}}(x_i)=[1+\exp(-w_0)]^{-1}$ (see Example 3.5).

Furthermore, the outcome of the neural net depends on the type of transfer function used, and the result will also depend on the data set. If, for example, the data again contains pairs $(x_i, t(x_i))$ from the set $\{(x_i, y_i)|y_i=0.01x_i^2\}$ but now uniformly distributed over the x-interval $[0, x_{max}]$, then the neural net will find after learning that the relation between x_i and y_i can be approximated by an increasing sigmoid

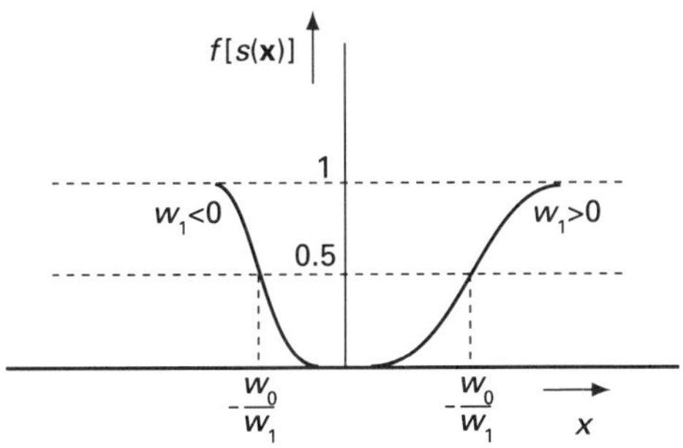

Figure 3.17 Decreasing and increasing sigmoid functions

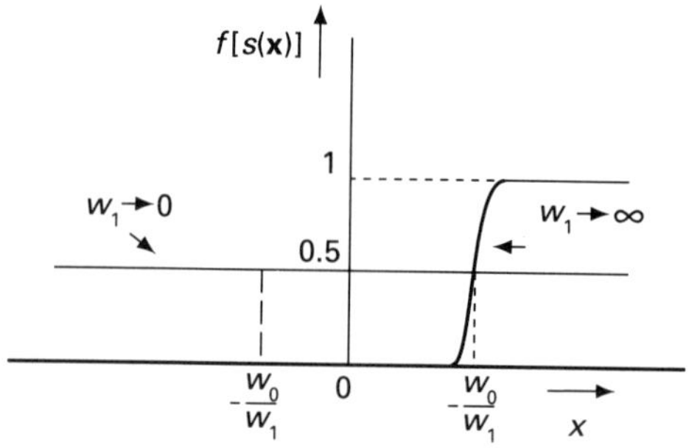

Figure 3.18 The extremes of the sigmoid functions

function $g_{\mathbf{w}}(x_i) = [1 + \exp - (w_0 + w_1 x)]^{-1}$ with $w_0 < 0$ and $w_1 > 0$. For example, if $x_i \in \{0, 1, 2, 3, 4, 8\}$ we will find after learning with the single-neuron Perceptron: $g_{\mathbf{w}}(x_i) = [1 + \exp(-4.04 + 0.58x_i)^{-1}]$. From our discussion we may conclude the following statement:

Practical statement 3.6

Using a neural network to find the relation behind the data in the data set, one is assuming *a priori* **that the relation can be modelled by a function of the class of functions realizable with the neural net, and secondly one is assuming that the data set is representative of the relation.**

In general, the sigmoid function $g_{\mathbf{w}} = [1 + \exp - s(\mathbf{w})]^{-1}$ with $s(\mathbf{x}) = \Sigma\, w_i x_i$, is used as the transfer function of a neuron in a continuous multi-layer Perceptron. Sometimes other transfer functions are used, such as the following:

The $y = \text{tang h}(s(\mathbf{x}))$.
The piecewise linear function.
The radial base function $y = \exp(-s(\mathbf{x}) - \mu)^2/\sigma^2$.

The requirements for a transfer function are:

1. The function is continuous differentiable (because the adaptation $\Delta\mathbf{w}$ is proportional to $\mathrm{d}f/\mathrm{d}s$).
2. The derivative $\mathrm{d}f/\mathrm{d}s$ is finite, preferably going to zero for large $|s|$. (The adaptation of weights $\Delta\mathbf{w}$ must be small but is proportional to $\mathrm{d}f/\mathrm{d}s$.)
3. The function is non-linear (see Theorem 3.1).

3.7 The classification of data with a single-neuron Perceptron

Perceptrons are frequently used for classification of data into two or more disjoint classes. As an example we refer to the classification of mushrooms into two separate classes discussed in Section 3.1. We now consider again the case of a two-class classification problem. It turns out that in many situations we can obtain very good classification results for two-class problems with just one single-neuron Perceptron. In that situation one has two sets A and B of objects. Objects are presented by the values of n measurements of n different attributes of the objects: $\mathbf{x}=(x_1, x_2, \ldots, x_n)$. The *classifier* realizes a so-called *discriminator function* $D(\mathbf{x})$ which assigns to the *measurement vector* $\mathbf{x}$ the label l_A of class A or the label l_B of class B. In general a penalty is associated with a misclassification. Let $c(l_A, B)$ be the cost of assigning label l_A to the measurement vector $\mathbf{x}$ if the corresponding object is an element of class B, and $c(l_B, A)$ the cost for the complementary misclassification. If $p(A)$ and $p(B)$ are the probabilities of respectively class A and class B, and $p(l_A|B)$ and $p(l_B|A)$ the conditional probabilities of misclassification, then the expected value of the cost of a classification, the *risk*, will be:

$$R=p(A)p(l_B|A)c(l_B, A)+p(B)p(l_A|B)c(l_A, B)$$
$$\text{with } p(A)+p(B)=1$$

For an optimal classifier we want to have the risk as small as possible.

In general we only have a finite set D_A of examples of class A and a finite set D_B of examples of class B. How do we have to design the classifier if we only have the finite set D_A of examples of class A and the finite set D_B of examples of class B? Given the sets of examples it is likely that the optimal classifier will divide the input space X into two regions X_A and X_B such that most examples of D_A are in region X_A and most of the examples of D_B are in X_B. One might, for instance, assume that a $n-1$-dimensional hyperplane will divide the n-dimensional input space $X=\mathbb{R}^n$ into regions X_A and X_B. Figure 3.19 shows the two-dimensional case with:

$$D_A=\{(-3, 3), (1, 3), (-3, 7), (-7, 3), (-3, -1), (0, 0), (-1, 1)\}$$
$$D_B=\{(2, -2), (2, 2), (6, -2), (2, -6), (-2, -2), (0, 0), (-1, 1)\}$$

One might just as well assume, however, that the input space can be divided into two regions by means of several hyperplanes (see, for example, Figure 3.20 for the same sets D_A and D_B). One might even assume that the elements in data set D_A are

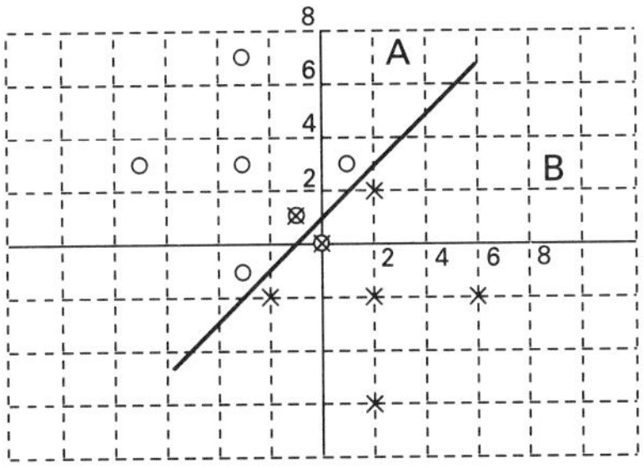

Figure 3.19 A two-dimensional classification problem

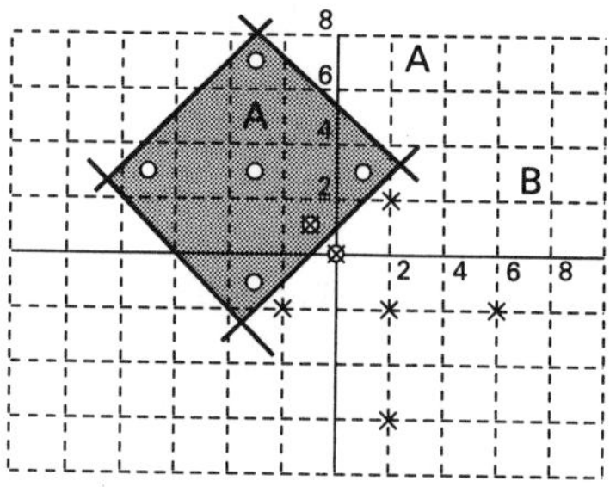

Figure 3.20 Classification with four hyperplanes

the only elements of class A, and class B consists of all remaining elements of the input space $X = \mathbb{R}^n$.

Without any additional information the less-biased assumption is that any vector $\mathbf{x} \in \mathbb{R}^n$ according to some probability ratio may belong to class A as well as to class B. Accepting this assumption, one has to have recourse to the assumption that given a measurement vector $\mathbf{x}$, there is a conditional probability $p(\mathrm{A}|\mathbf{x})$ for each $\mathbf{x}$ that $\mathbf{x}$ belongs to an object of class A and a conditional probability $p(\mathrm{B}|\mathbf{x}) = 1 - p(\mathrm{A}|\mathbf{x})$ that $\mathbf{x}$ belongs to an object of class B.

The discrimination function of the optimal classifier must be:

$$D(\mathbf{x}) = l_{\mathrm{A}} \quad \text{if } p(\mathrm{A}|\mathbf{x})/p(\mathrm{B}|\mathbf{x}) > T$$

$$D(\mathbf{x}) = l_{\mathrm{B}} \quad \text{if } p(\mathrm{A}|\mathbf{x})/p(\mathrm{B}|\mathbf{x}) < T$$

For a minimal risk the threshold T will depend on the costs $c(l_{\mathrm{A}}|\mathrm{B})$ and $c(l_{\mathrm{B}}|\mathrm{A})$ of misclassification.

For the continuous case we have:

$$p(\mathrm{A}|\mathbf{x}) = \frac{f_{\mathrm{A}}(\mathbf{x}) \cdot p(\mathrm{A})}{f(\mathbf{x})} \quad p(\mathrm{B}|\mathbf{x}) = \frac{f_{\mathrm{B}}(\mathbf{x}) \cdot p(\mathrm{B})}{f(\mathbf{x})} \quad \text{(Bayes's rule)}$$

with $f_{\mathrm{A}}(\mathbf{x})$ and $f_{\mathrm{B}}(\mathbf{x})$ the class-conditional probability density functions.

With the use of these expression for the *a posteriori* probabilities, we can rewrite the discriminator function as follows:

$$D(\mathbf{x}) = l_{\mathrm{A}} \quad \text{if } f_{\mathrm{A}}(\mathbf{x})/f_{\mathrm{B}}(\mathbf{x}) > T \cdot p(\mathrm{B})/p(\mathrm{A})$$

$$D(\mathbf{x}) = l_{\mathrm{B}} \quad \text{if } f_{\mathrm{A}}(\mathbf{x})/f_{\mathrm{B}}(\mathbf{x}) < T \cdot p(\mathrm{B})/p(\mathrm{A})$$

The quotient $f_{\mathrm{A}}(\mathbf{x})/f_{\mathrm{B}}(\mathbf{x})$ is called the *likelihood ratio* L for class A. The quantity $K = T \cdot p(\mathrm{B})/p(\mathrm{A})$ is called the *likelihood ratio threshold.* In cases where the costs $c(l_{\mathrm{A}}|\mathrm{B})$ and $c(l_{\mathrm{B}}|\mathrm{A})$ are equal, we have to select the threshold T (and thus K) such that the risk R:

$$R = p(\mathrm{A})p(l_{\mathrm{B}}|\mathrm{A}) + p(\mathrm{B})p(l_{\mathrm{A}}|\mathrm{B})$$

$$= \int_{X_{\mathrm{B}}} p(\mathrm{A}) f_{\mathrm{A}}(\mathbf{x})\, d\mathbf{x} + \int_{X_{\mathrm{A}}} p(\mathrm{B}) f_{\mathrm{B}}(\mathbf{x})\, d\mathbf{x}$$

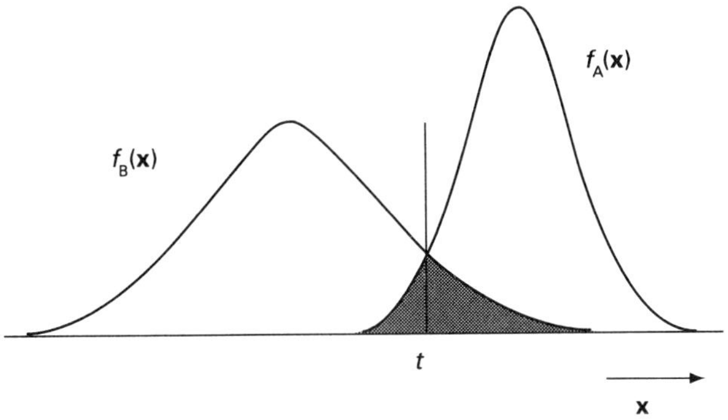

Figure 3.21 The risk (shaded area) for a one-dimensional classification problem

is minimal. The risk for a one-dimensional input space with $p(\mathrm{A})=p(\mathrm{B})$ is given by the shaded area in Figure 3.21.

In order to design an optimal 'Bayes' classifier, one has to assume (or to know) the class-conditional probability density functions $f_A(\mathbf{x})$ and $f_B(\mathbf{x})$, and in addition the class probabilities $p(\mathrm{A})$ and $p(\mathrm{B})$. The first condition for the derivation of the optimal 'Bayes' classifier implies that one has to assume the functional form of the density functions and to estimate the parameters (mean and variance) of the density function from the sets of examples D_A and D_B.

In the following sections we will see that if we use a continuous Perceptron as a classifier we do not have to make such severe assumptions about the functional form of the density functions, and also that we do not have to make estimations of the parameters of the density functions. We will also show in the subsequent sections that under certain conditions a single-neuron Perceptron will be an optimal classifier.

A single-neuron Perceptron assigns to every measurement vector $\mathbf{x}$ an output value $g_{\mathbf{w}}(\mathbf{x})$ depending on the weight vector $\mathbf{w}$. If we use the sigmoid transfer function $f(s)=[1+\exp(-s)]^{-1}$ with $s=\Sigma\, w_i x_i$, then the output $g_{\mathbf{w}}(\mathbf{x})$ will vary between 0 and 1.

We can use the output $g_{\mathbf{w}}(\mathbf{x})$ as a label for the class to which $\mathbf{x}$ belongs. If we use the sigmoid function as the transfer function of the neuron, then there are three different ways of learning to use the output $g_{\mathbf{w}}(\mathbf{x})$ as a label for class A and class B:

1. *Hyperplane boundary classification by learning one and zero labelling.* In this case the n-dimensional input space $X=\mathbb{R}^n$ is divided by a $n-1$-dimensional hyperplane defined by $\Sigma\, w_i x_i=0$ into the regions X_A and X_B. For $\mathbf{x}\in X_A$ we have $g_{\mathbf{w}}(\mathbf{x})>0.5$ and for $\mathbf{x}\in X_B$ we have $g_{\mathbf{w}}(\mathbf{x})\leqslant 0.5$. In the learning phase for $\mathbf{x}\in D_A$ the target value $t(\mathbf{x})=1$ and for $\mathbf{x}\in D_B$ the target value $t(\mathbf{x})=0$. So we have for the discriminator function $D(\mathbf{x})=l_A$ if $g_{\mathbf{w}}(\mathbf{x})>0.5$ and $D(\mathbf{x})=l_B$ if $g_{\mathbf{w}}(\mathbf{x})\leqslant 0.5$ if $p(\mathrm{A})=p(\mathrm{B})$.
2. *Hyperplane boundary classification by learning double threshold labelling.* In this case too the n-dimensional input space $X=\mathbb{R}^n$ is divided by a $n-1$-dimensional hyperplane defined by $\Sigma\, w_i x_i=0$ into X_A and X_B. For $\mathbf{x}\in X_A$ we have $g_{\mathbf{w}}(\mathbf{x})>0.5$ and for $\mathbf{x}\in X_B$ we have $g_{\mathbf{w}}(\mathbf{x})\leqslant 0.5$. In the learning phase, however, for an input

$\mathbf{x} \in D_A$ the target value is $t(\mathbf{x}) \geqslant a$ with $0.5 < a < 1$ (e.g. $a = 0.9$) and for an input $\mathbf{x} \in D_B$ the target value is $t(\mathbf{x}) \leqslant b$ with $0 < b < 0.5$ (e.g. $b = 0.1$) given: $p(A) = p(B)$.

3. *Hyperplane boundary classification by single threshold labelling.* Again the n-dimensional input space $X = \mathbb{R}^n$ is divided by a $n-1$-dimensional hyperplane defined by $\Sigma w_i x_i = 0$ into X_A and X_B. For $\mathbf{x} \in X_A$ we have $g_\mathbf{w}(\mathbf{x}) > 0.5$ and for $\mathbf{x} \in X_B$ we have $g_\mathbf{w}(\mathbf{x}) \leqslant 0.5$. In the learning phase, however, for an input $\mathbf{x} \in D_A$ the target value is $t(\mathbf{x}) > 0.5$, and for an input $\mathbf{x} \in D_B$ the target value is $t(\mathbf{x}) \leqslant 0.5$. We will see that in the last case we have to modify the learning rule as discussed in previous sections.

In the next section we will discuss the different classification methods.

3.8 Hyperplane boundary classification by one–zero labelling

In case of hyperplane boundary classification by one–zero labelling with a single-neuron Perceptron with sigmoid transfer function, the n-dimensional input space $X = \mathbb{R}^n$ is divided by a hyperplane defined by the dot product $\hat{\mathbf{w}} \cdot \hat{\mathbf{x}} = 0$ into the regions X_A and X_B. In the learning phase for $\mathbf{x} \in D_A$ the target value $t(\mathbf{x}) = 1$, and for $\mathbf{x} \in D_B$ the target value $t(\mathbf{x}) = 0$. After learning, we use $g_\mathbf{w}(\mathbf{x})$ as a label for the class to be identified: $\mathbf{x} \in A$ if $g_\mathbf{w}(\mathbf{x}) > 0.5$ and $\mathbf{x} \in B$ if $g_\mathbf{w}(\mathbf{x}) \leqslant 0.5$ if $p(A) = p(B)$.

We observe that during learning we have to learn for each $\mathbf{x}$ the target value $t(\mathbf{x})$, thus the learning goal is identical with the approximate fitting of a data set $D = D_A \cup D_B$ as discussed in Section 3.5. Therefore we can use the same learning rule as discussed in the previous sections.

If the data set is linear separable by a hyperplane defined by the dot product $\hat{\mathbf{w}} \cdot \hat{\mathbf{x}} = 0$, then the final weight vector will be such that the functions $g_\mathbf{w}(\mathbf{x})$ becomes (almost) a threshold function and the final MSE becomes zero. Figure 3.22 shows the one-dimensional case. The 'hyperplane' $w_0 + w_1 x_1 = 0$ will be in this case a point $t = -w_0/w_1$.

One may note that if the data set is linear separable we also have the same problem as discussed in Section 2.4 on classification with a single-neuron binary Perceptron. The desired output is also zero or one, and only the input vectors are now real valued instead of binary valued, but the convergence theorem is independent of the type of input vector. Thus to solve the given classification problem we can also use the reinforcement learning rule of Section 2.4.

We say that the data set D is *separable* if the data sets D_A and D_B do not have input vectors in common. Most data sets are separable but are not linear separable, i.e. we cannot divide the input space X by one single hyperplane into regions X_A and X_B such that $D_A \subseteq X_A$ and $D_B \subseteq X_B$. Although most data sets are not linear separable we can frequently obtain very good classification results with one single-neuron continuous Perceptron.

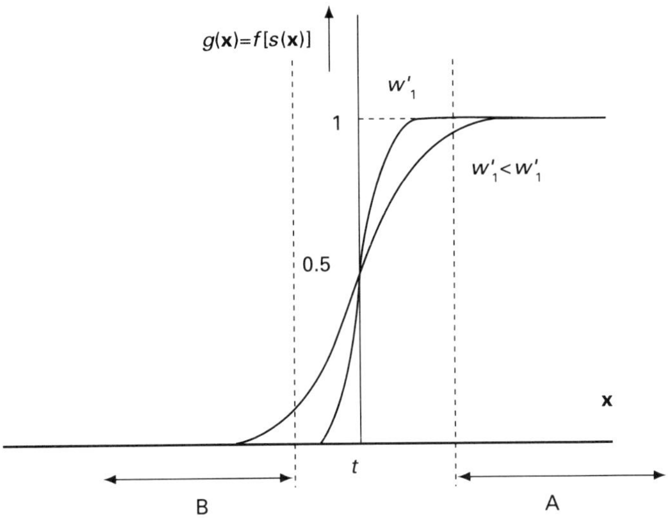

Figure 3.22 The sigmoid function becoming a threshold function

Example 3.8

Assume we have a two-dimensional data set $D = D_A \cup D_B$ depicted by the points in Figure 3.23.

The data set is D_A generated by a Gaussian distribution function with the following probability density function:

$$f_A(x, y) = \frac{1}{\sqrt{2\pi\sigma_x}} \exp \frac{-(x-\mu_x)^2}{2\sigma_x^2} + \frac{1}{\sqrt{2\pi\sigma_y}} \exp \frac{-(y-\mu_y)^2}{2\sigma_y^2}$$

with $\mu_x = 0$, $\sigma_x = 0.2$, $\mu_y = 0.465$ and $\sigma_y = 0.4$. Similarly we have the data set D_B generated by the same type of Gaussian distribution function but now with $\mu_x = 0$, $\sigma_x = 0.4$, $\mu_y = -0.465$ and $\sigma_y = 0.2$.

For an optimal minimal risk classifier one can prove that the boundary between the two regions if X_A and X_B is defined by the condition that $f_A(x, y) = f_B(x, y)$, if $p(A) = p(B)$ and $c(l_A|B) = c(l_B|A)$. The optimal classification boundary, or discrimination line is given in Figure 3.22 by a curved line. One can calculate that the probability of error in that case is 5.14 per cent.

We can also divide the input space by one straight line (the boundary hyperplane is in this case a straight line given in Figure 3.22); the probability of error will then be slightly larger. The optimal position of the line is $y = -0.11$. It turns out that the probability of error in that case is 5.15 per cent. If a single-neuron Perceptron could find this boundary we have a very good result. In Section 3.11 we will show that with a single-neuron Perceptron we can learn to find this boundary. ■

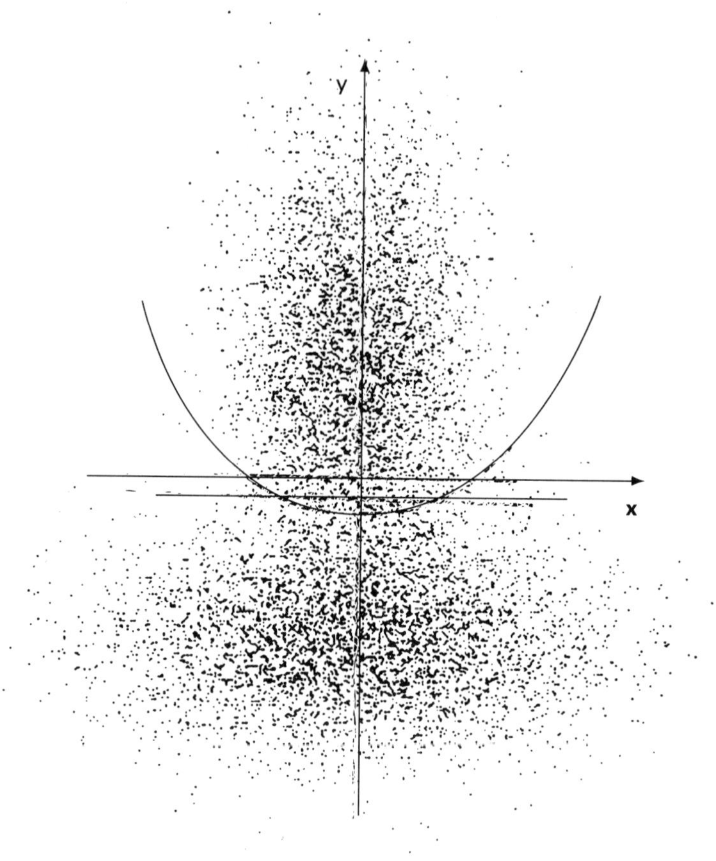

Figure 3.23 Two-dimensional classification problem with Gaussian distributions

The preceding example illustrates the following practical statement:

Practical statement 3.7

Many two-class classification problems (in which the optimal classification boundary is an open, non-linear, convex boundary and the intersection area of both classes is not too large) can be solved reasonably well with one single-neuron continuous Perceptron.

Classifiers that divide the n-dimensional input space by a $n-1$-dimensional hyperplane will be called *hyperplane boundary classifiers.* The single-neuron

Perceptron is an optimal hyperplane boundary classifier if a certain condition is satisfied:

Theorem 3.6

If the data sets D_A and D_B are representative for the underlying distribution functions of class A and class B, then the single-neuron Perceptron will divide after learning with one–zero labelling the n-dimensional input space $X=\mathbb{R}^n$ by an optimally located $n-1$-dimensional hyperplane if for the final weight vector: $|\mathbf{w}|\rightarrow\infty$.

Proof

In the previous section we found that for an optimal classifier the risk (for equal costs of misclassification):

$$R=p(\mathrm{A})p(l_b|\mathrm{A})+p(\mathrm{B})p(l_a|\mathrm{B})$$
$$=\int_{X_B} p(\mathrm{A})f_A(\mathbf{x})\,d\mathbf{x}+\int_{X_A} p(\mathrm{B})f_B(\mathbf{x})\,d\mathbf{x}$$

must be minimal. For the one-dimensional case with $p(\mathrm{A})=p(\mathrm{B})$ the shaded area in Figure 3.24 must be minimal for a minimal risk.

By training a single-neuron Perceptron to fit the data set $D=D_A\cup D_B$ by one–zero labelling, the MSE will be minimized. We have to prove that by minimizing the MSE, the risk will also be minimized.

In Section 3.2 we have defined the MSE as:

$$\mathbf{E}=\frac{1}{N}\sum_{\mathbf{x}_i\in D} n(\mathbf{x}_i)[t(\mathbf{x}_i)-g_{\mathbf{w}}(\mathbf{x}_i)]^2$$

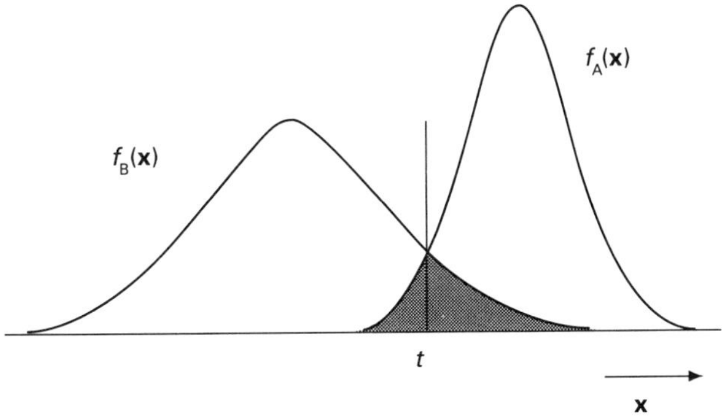

Figure 3.24 The risk (shaded area) for a one-dimensional classification problem

with N the number of samples in the data set D, and $n(\mathbf{x}_i)$ the number of times input $\mathbf{x}_i$ occurs in the data set.

For input vectors $\mathbf{x}_i \in D_A$ the target is $t(\mathbf{x}_i)=1$, and for input vectors of D_B the target is $t(\mathbf{x}_i)=0$. Thus:

$$\mathbf{E}=\frac{1}{N}\sum_{\mathbf{x}_i\in D_A} n_A(\mathbf{x}_i)[1-g_{\mathbf{w}}(\mathbf{x}_i)]^2+\frac{1}{N}\sum_{\mathbf{x}_i\in D_B} n_B(\mathbf{x}_i)[-g_{\mathbf{w}}(\mathbf{x}_i)]^2$$

with N the number of samples in the data set $D=D_A \cup D_B$ and $n_A(\mathbf{x}_i)$ (respectively n_B) the number of times input $\mathbf{x}_i$ occurs in D_A (respectively D_B). We can rewrite the above equation as:

$$\mathbf{E}=\frac{N_A}{N}\sum_{\mathbf{x}_i\in D_A} \frac{n_A(\mathbf{x}_i)}{N_A}[1-g_{\mathbf{w}}(\mathbf{x}_i)]^2+\frac{N_B}{N}\sum_{\mathbf{x}_i\in D_B} \frac{n_B(\mathbf{x}_i)}{N_B}[g_{\mathbf{w}}(\mathbf{x}_i)]^2$$

with N_A (respectively N_B) the total number of samples in D_A (respectively D_B).

If the data set is representative of the underlying distributions, then N_A/N is an estimate of the class probability $p(A)$ and $n_A(\mathbf{x}_i)/N_A$ is an estimate of the probability $f_A(\mathbf{x}_i)\,\mathbf{dx}$ of finding an input vector in the infinitesimal n-dimensional cube $\mathbf{dx}$ surrounding $\mathbf{x}_i$. Thus the discrete sum above is a discrete approximation of:

$$\mathbf{E}=\int_X p(A) f_A(\mathbf{x})[1-g_{\mathbf{w}}(\mathbf{x})]^2\,\mathbf{dx}+\int_X p(B) f_B(\mathbf{x})[g_{\mathbf{w}}(\mathbf{x})]^2\,\mathbf{dx}$$

After learning, the input space $X=\mathbb{R}^n$ will be divided by the hyperplane defined by $\hat{\mathbf{w}}\cdot\hat{\mathbf{x}}=0$ into the regions X_A and X_B. Thus we can replace the equation above by:

$$\mathbf{E}=\int_{X_A} p(A) f_A(\mathbf{x})[1-g_{\mathbf{w}}(\mathbf{x})]^2\,\mathbf{dx}+\int_{X_B} p(A) f_A(\mathbf{x})[1-g_{\mathbf{w}}(\mathbf{x})]^2\,\mathbf{dx}$$

$$+\int_{X_A} p(B) f_B(\mathbf{x})[g_{\mathbf{w}}(\mathbf{x})]^2\,\mathbf{dx}+\int_{X_B} p(B) f_B(\mathbf{x})[g_{\mathbf{w}}(\mathbf{x})]^2\,\mathbf{dx}$$

If, after learning, $|\mathbf{w}|$ is large ($|\mathbf{w}|\to\infty$), then $g_{\mathbf{w}}(\mathbf{x})$ will become a threshold function with $g_{\mathbf{w}}(\mathbf{x})\cong 1$ if $\mathbf{x}\in X_A$ and $g_{\mathbf{w}}(\mathbf{x})\cong 0$ if $\mathbf{x}\in X_B$. The equation for the MSE reduces to:

$$E=\int_{X_B} p(A) f_A(\mathbf{x})\,\mathbf{dx}+\int_{X_A} p(B) f_B(\mathbf{x})\,\mathbf{dx}$$

This expression is identical to the expression for the risk where there is an equal cost of misclassification. We know that the gradient descent learning rule will minimize the MSE and thus it will also minimize the risk.

One easily verifies that if $g_{\mathbf{w}}(\mathbf{x})=p(\mathbf{A})f_A(\mathbf{x})/(p(\mathbf{A})f_A(\mathbf{x})+p(\mathbf{B})f_B(\mathbf{x}))$ then the MSE will be minimal.

QED

To understand what is happening during learning, we will analyze some simple classification situations.

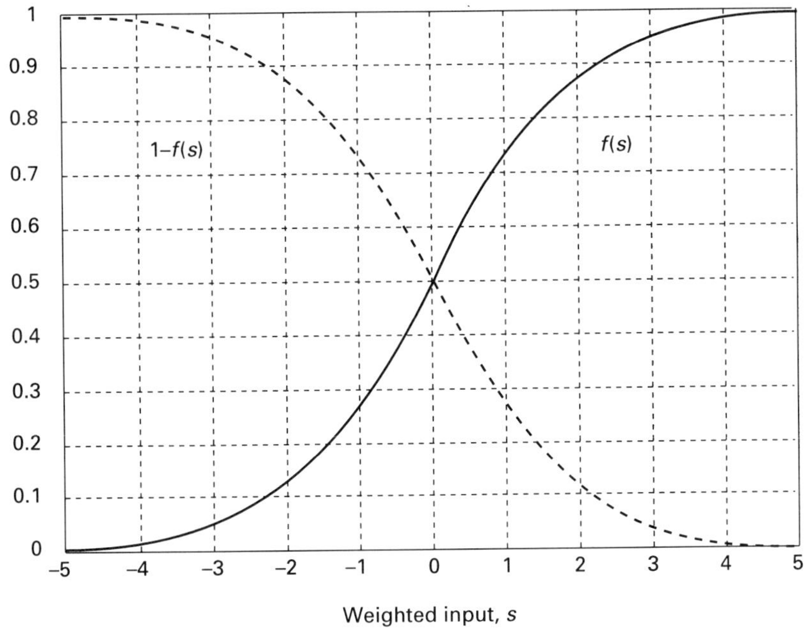

Figure 3.25 The sigmoid transfer function $f(s)$

According to Theorem 3.3 the adaptation of the weight vector **w** must be:

$$\Delta\hat{\mathbf{w}} = \varepsilon \cdot \sum_{\mathbf{x}_i \in U} [t(\mathbf{x}_i) - g_{\mathbf{w}}(\mathbf{x}_i)] \frac{\mathrm{d}f}{\mathrm{d}s} \hat{\mathbf{x}}_i$$

with $g_{\mathbf{w}}(\mathbf{x}_i) = f[s(\mathbf{x}_i)]$, $\hat{\mathbf{w}} = [w_0, w_1, \ldots, w_n]^t$ the extended weight vector and $\hat{\mathbf{x}}_i = [1, x_{i1}, x_{i2}, \ldots, x_{in}]^t$ the extended input vector. (Note that we have to multiply the equation above by $n(\mathbf{x}_i)/N$ in the most complete form.)

The contribution of different input vectors to the adaptation of the weight vector will be different. For an element of D_A the target value is 1. The value of $t(\mathbf{x}_i) - f[s(\mathbf{x}_i)]$ is large for large negative values of $s(\mathbf{x}_i)$ and will become zero for large positive values of the weighted input (see Figure 3.25). The factor $\mathrm{d}f/\mathrm{d}s = f(s)[1 - f(s)]$ has a maximum of 0.25 for $s = 0$ (see Figure 3.26). The product:

$$\gamma(\mathbf{x}_i) = \{t(\mathbf{x}_i) - f[s(\mathbf{x}_i)]\} \frac{\mathrm{d}f}{\mathrm{d}s}$$

will be called the *internal learning rate* for $\mathbf{x}_i$.

For $t(\mathbf{x}_i) = 1$ the internal learning rate as a function of the weighted input $s(\mathbf{x}_i)$ will be:

$$\gamma(s) = [1 - f(s)] f(s)[1 - f(s)]$$

and is given in Figure 3.27 by curve A.

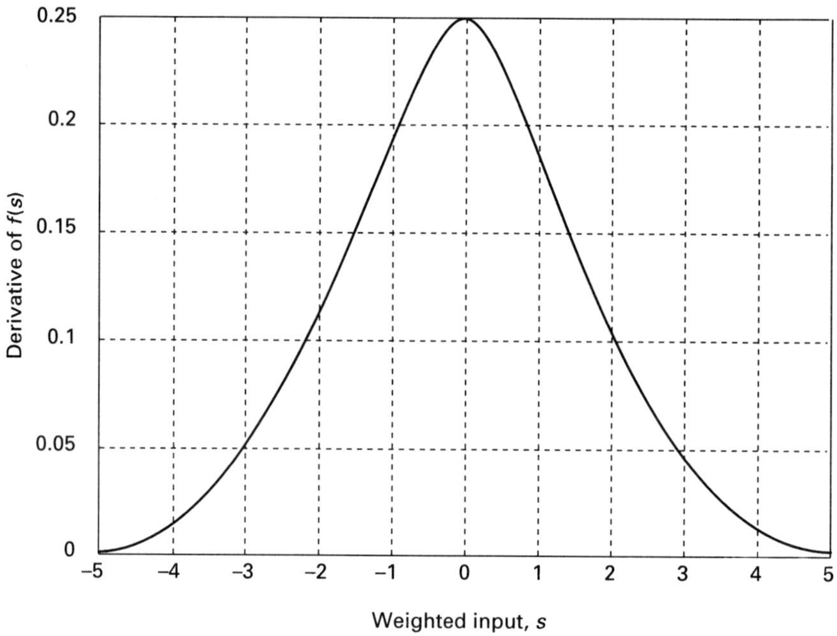

Figure 3.26 The derivative $\mathrm{d}f/\mathrm{d}s$ of the sigmoid transfer function

For $t(\mathbf{x}_i)=0$ the internal learning rate is:

$$\gamma(s) = -f(s)f(s)[1-f(s)]$$

represented by curve B in Figure 3.27.

The learning rule can now be written as follows:

$$\Delta\hat{\mathbf{w}} = \sum_i \varepsilon\gamma(\mathbf{x}_i)\hat{\mathbf{x}}_i$$

The learning rule implies:

$$\Delta w_0 = \sum_i \varepsilon\gamma(\mathbf{x}_i)$$

$$\Delta\mathbf{w} = \sum_i \varepsilon\gamma(\mathbf{x}_i)\mathbf{x}_i$$

We now consider the simple situation that we have to classify two points: $D_A=\{\mathbf{x}_a\}$ and $D_B=\{\mathbf{x}_b\}$ (see Figure 3.28). Let the initial weight vector of a single-neuron Perceptron with two external inputs x_1 and x_2 be such that the separating hyperplane $w_0+w_1x_1+w_2x_2=0$ is perpendicular to $\mathbf{x}_a-\mathbf{x}_b$ with $\mathbf{x}_a$ on the positive side ($w_0+w_1x_1+w_2x_2>0$) of the hyperplane and $\mathbf{x}_b$ on the negative side ($w_0+w_1x_1+w_2x_2<0$). We also assume that the distance from $\mathbf{x}_a$ to the hyperplane

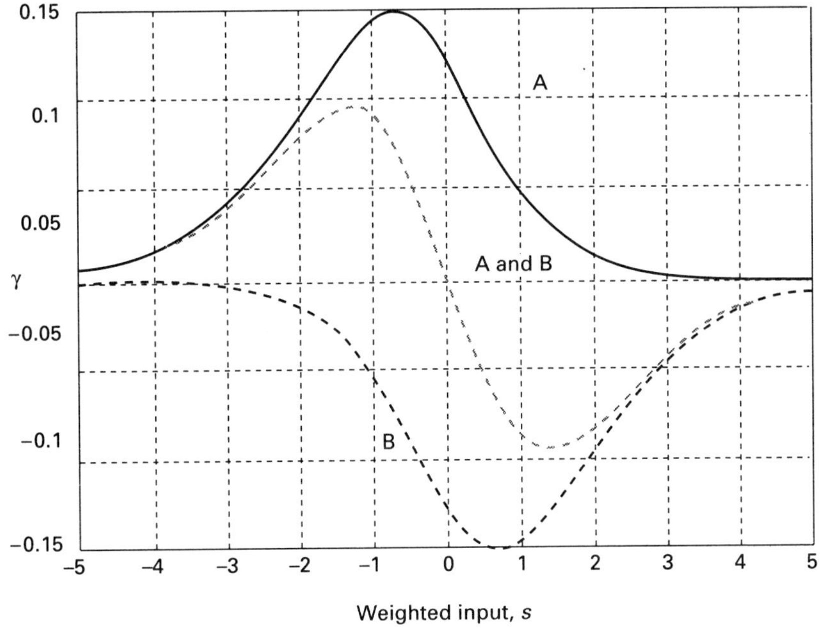

Figure 3.27 The internal learning rate $\gamma(s)$ for $t(\mathbf{x}_i)=1$ (curve A) and for $t(\mathbf{x}_i)=0$ (curve B)

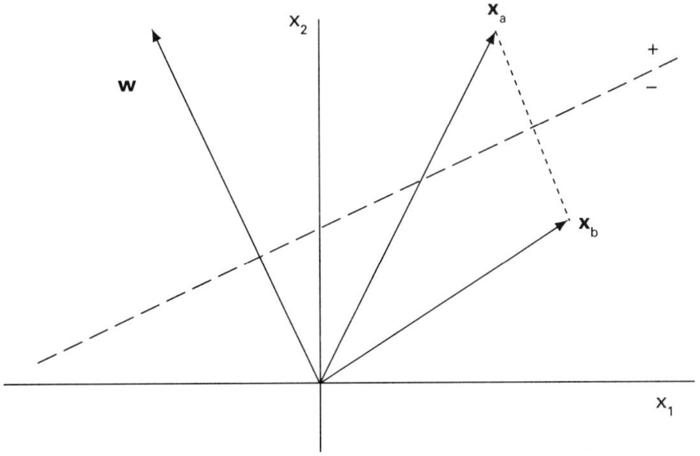

Figure 3.28 A two-point, two-dimensional classification problem

is the same as the distance from $\mathbf{x}_b$ to the hyperplane. At first glance one might assume that learning will not change the weight vector because the initial hyperplane separates in an ideal way the input space, and the two inputs will thus be classified correctly [$g_{\mathbf{w}}(\mathbf{x}_a) > 0.5$ and $g_{\mathbf{w}}(\mathbf{x}_b) < 0$]. However, at the first learning step the targets $t(\mathbf{x}_a) = 1$ and $t(\mathbf{x}_b) = 0$ might not be satisfied and the MSE will not be zero. During learning, the weights will be adapted such that the MSE decreases while keeping the classification correct.

Because $\mathbf{x}_a$ and $\mathbf{x}_b$ are on the opposite sides of the hyperplane we obtain for the weighted inputs: $s(\mathbf{x}_a) > 0$ and $s(\mathbf{x}_b) < 0$. Now we recall from Section 2.2 that the weight vector $\mathbf{w} = [w_1, w_2]^t$ is perpendicular to the separating plane and is pointing into the direction of the region X_A where $s(\mathbf{x}) > 0$. From Section 2.2 we recall the relation between the distance $\delta(\mathbf{x})$ (in the direction of $\mathbf{w}$) from the hyperplane to a point $\mathbf{x}$, and the weighted input $s(\mathbf{x})$:

$$\delta(\mathbf{x}) = \frac{\mathbf{w}^t\mathbf{x} + w_0}{|\mathbf{w}|} = \frac{s(\mathbf{x})}{|\mathbf{w}|}$$

Because $\delta(\mathbf{x}_a) = -\delta(\mathbf{x}_b)$ we have $s(\mathbf{x}_a) = -s(\mathbf{x}_b)$ and thus $\gamma(\mathbf{x}_a) = -\gamma(\mathbf{x}_b)$ (see Figure 3.28). The adaptation of the weight vector:

$$\Delta\mathbf{w} = \varepsilon\gamma(\mathbf{x}_a)\mathbf{x}_a + \varepsilon\gamma(\mathbf{x}_b)\mathbf{x}_b = \varepsilon\gamma(\mathbf{x}_a)(\mathbf{x}_a - \mathbf{x}_b)$$

is in the direction of $\mathbf{x}_a - \mathbf{x}_b$ and thus in the direction of the weight vector $\mathbf{w}$. Thus the weight vector is multiplied by some scalar. Because the weight vector $\mathbf{w}$ before and after adaptation is in the same direction, the orientation of the separating hyperplane will also be the same.

We recall from Section 2.2 that the distance along $\mathbf{w}$ from the origin to the hyperplane is given by:

$$d = \frac{-w_0}{|\mathbf{w}|}$$

Because $\gamma(\mathbf{x}_a) = -\gamma(\mathbf{x}_b)$ we have $\Delta w_0 = \varepsilon\gamma(\mathbf{x}_a) + \varepsilon\gamma(\mathbf{x}_b) = 0$. Thus before and after adaptation w_0 will be the same. We found that $|\mathbf{w}|$ will be different before and after adaptation and thus the distance d from the origin to the hyperplane will change. The hyperplane comes closer to $\mathbf{x}_b$.

As we continue learning we observe that the hyperplane twists around the initial hyperplane but still between $\mathbf{x}_a$ and $\mathbf{x}_b$ (if ε is small enough) while $|\mathbf{w}|$ increases, i.e. until $g_{\mathbf{w}}(\mathbf{x})$ becomes a threshold function and the separating hyperplane will be the same as the initial one but now with zero MSE.

Example 3.9

Consider the simple one-dimensional classification problem: $D_A = \{\mathbf{x}_a\} = \{3\}$ and $D_B = \{\mathbf{x}_b\} = \{1\}$. The initial extended weight vector is $\hat{\mathbf{w}} = [w_0, w_1] = [-4, 2]$ (see Figure 3.29).

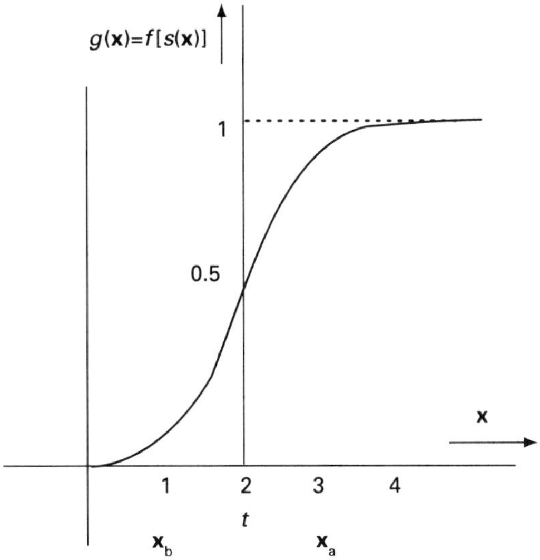

Figure 3.29 The initial realized sigmoid function for Example 3.9

The initial separating hyperplane (a point) is defined by $-4+2x_1=0$. For the weighted inputs we find $s(\mathbf{x}_a)=-4+2\cdot3=2$, $s(\mathbf{x}_b)=-4+2\cdot1=-2$. From Figure 3.27 we obtain for the internal learning rates $\gamma(\mathbf{x}_a)\cong0.01$ and $\gamma(\mathbf{x}_b)\cong-0.01$. We find for the value of the adaptation of the weight vector:

$$\begin{bmatrix}\Delta w_0\\ \Delta w_1\end{bmatrix}=\varepsilon\gamma(\mathbf{x}_a)\begin{bmatrix}x_{a0}\\ x_{a1}\end{bmatrix}+\varepsilon\gamma(\mathbf{x}_b)\begin{bmatrix}x_{b0}\\ x_{b1}\end{bmatrix}=\varepsilon(0.01)\begin{bmatrix}1\\ 3\end{bmatrix}+\varepsilon(-0.01)\begin{bmatrix}1\\ 1\end{bmatrix}=\varepsilon\begin{bmatrix}0.00\\ 0.02\end{bmatrix}$$

With $\varepsilon=10$ the new weight vector becomes:

$$\begin{bmatrix}w_0\\ w_1\end{bmatrix}=\begin{bmatrix}4.0\\ 2.2\end{bmatrix}$$

The separating point $-w_0/w_1\cong1.9$ is moving to the left and w_1 is increased. An additional adaptation will show that the next separating point will be located to the right of the original separating point $x=2$. Finally we will end up with $|\hat{\mathbf{w}}|=\infty$ and $-w_0/w_1=2$. During learning we will jump in the cutter of the error landscape (see Figure 3.30) from one slope to the other until w_0 and w_1 approach infinity and the MSE becomes zero. ■

If for $\mathbf{x}_a\in D_A$ and $\mathbf{x}_b\in D_B$ we have $\mathbf{x}_a=\mathbf{x}_b=\mathbf{x}$, the contribution of those points to the adaptation will be:

$$\Delta\mathbf{w}=\varepsilon\gamma(\mathbf{x}_a)\mathbf{x}_a+\varepsilon\gamma(\mathbf{x}_b)\mathbf{x}_b=\varepsilon[\gamma(\mathbf{x}_a)-\gamma(\mathbf{x}_b)]\mathbf{x}$$

The value of $\gamma(\mathbf{x}_a)-\gamma(\mathbf{x}_b)$ is given by curve 'A and B' in Figure 3.26. We see from

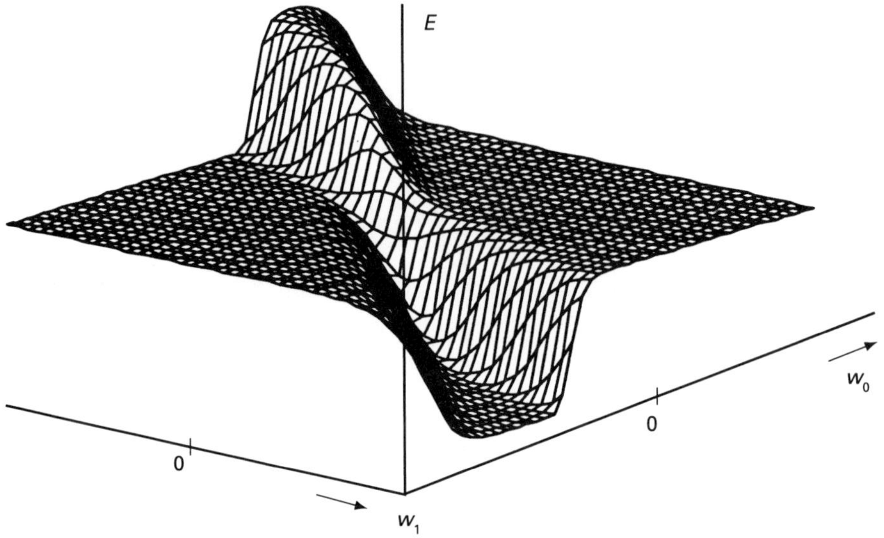

Figure 3.30 The error function of Example 3.9

Figure 3.26 that if the two points coincide on the actual separating hyperplane, then they do not contribute to the adaptation.

If the two coinciding points are on the positive side of the hyperplane ($s(\mathbf{x}_a)=s(\mathbf{x}_b)>0$), they will be treated as a point of D_B. If the two coinciding points are on the negative side of the hyperplane ($s(\mathbf{x}_a)=s(\mathbf{x}_b)<0$), they will be treated as a point of D_A.

We conclude that the contribution of a data point $\mathbf{x}_i$ to the adaptation of the weight vector depends on the internal learning rate $\gamma(\mathbf{x}_i)$, and the internal learning rate depends on the weighted input $s(\mathbf{x}_i)$, while $s(\mathbf{x}_i)=\delta|\mathbf{w}|$ with δ the distance from the hyperplane to the point $\mathbf{x}_i$. From Figure 3.27 we conclude that the internal learning rate for $\mathbf{x}_i$ is relatively small for $|s(\mathbf{x}_i)|>5$. This implies that data points with an actual output $g_{\mathbf{w}}(\mathbf{x}_i)\cong 1$ or $g_{\mathbf{w}}(\mathbf{x}_i)\cong 0$ hardly contribute to the adaptation, while on the other hand they consume learning time. In the next section we will give a method that eliminates those points automatically from the learning set.

3.9 Hyperplane boundary classification by double threshold labelling

In the case of hyperplane boundary classification by double threshold labelling with a single-neuron Perceptron with sigmoid transfer function, the n-dimensional input space $X=\mathbb{R}^n$ is also divided by a $n-1$-dimensional hyperplane, defined by the dot product $\hat{\mathbf{w}} \cdot \hat{\mathbf{x}} = 0$, into the regions X_A and X_B. For $\mathbf{x} \in X_A$ we require $g_{\mathbf{w}}(\mathbf{x}) \geqslant 0.5$ and for $\mathbf{x} \in X_B$, $g_{\mathbf{w}}(\mathbf{x}) < 0.5$ if $p(A) = p(B)$.

In the learning phase, however, we try to label each element of D_A with a target value $t(\mathbf{x}_i) \geqslant a$ with $0.5 < a < 1$ (e.g. $a = 0.9$) and we try to label each element of D_B with a target value $t(\mathbf{x}) \leqslant b$ with $0 < b < 0.5$ (e.g. $b = 0.1$). Thus if during learning $g_{\mathbf{w}}(\mathbf{x}_i) \geqslant a$ for $\mathbf{x}_i \in D_A$, then the squared error is zero, and if $g_{\mathbf{w}}(\mathbf{x}_i) < a$, then the squared error is $\{a - g(\mathbf{x}_i)\}^2$.

If $\mathbf{x}_i \in D_B$ we have for the target value $t(\mathbf{x}_i) < b$, thus if during the learning phase $g_{\mathbf{w}}(\mathbf{x}_i) \leqslant b$, then the squared error is zero, and if $g_{\mathbf{w}}(\mathbf{x}_i) > b$, then the squared error is $\{b - g(\mathbf{x}_i)\}^2$.

This learning rule implies that there might be elements of D for which we do not have to adapt the weights because the error is zero. This is of great advantage for the speed of the learning process (especially for networks with many neurons) because in general the data set will be large and the adaptation of weights is time consuming if every element of the data set D is contributing to the adaptation of weights, as with the classification procedure of the previous section. In cases where the data set is linear separable (see Figure 3.31 for the one-dimensional case), classification with the double threshold boundary with a final MSE$=0$ is optimal.

Classification by learning with double threshold labelling will not always be optimal (but will still be very good) if the data set is not linear separable because in that case the final weights will not become very large and thus $g_{\mathbf{w}}(\mathbf{x})$ will not become a threshold function as required for optimal classification (see previous section).

These observations and Practical Statement 3.5 lead to the following practical statement.

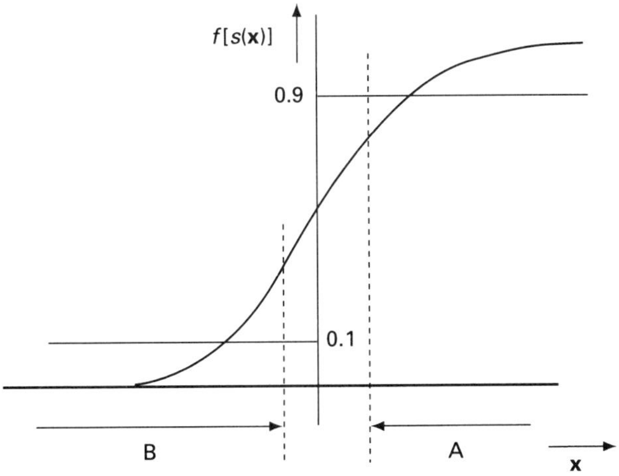

Figure 3.31 Classification of linearly separable one-dimensional data

Practical statement 3.8

Many two-class classification problems can be learned quickly and be solved reasonably well (if not always optimally) with one single-neuron continuous Perceptron if we use the classification method of double threshold labelling.

With the method of classification of double threshold labelling, only a subset $D_{\mathrm{Aw}} \subseteq D_{\mathrm{A}}$ and a subset $D_{\mathrm{Bw}} \subseteq D_{\mathrm{B}}$ (depending on the actual weight vector $\mathbf{w}$) of the given data sets $D_{\mathrm{A}} \subseteq X_{\mathrm{A}}$ and $D_{\mathrm{B}} \subseteq X_{\mathrm{B}}$ are wrongly classified, and only these inputs will contribute to the adaptation of weights.

Given some weight vector $\mathbf{w}$ we will minimize the MSE $\mathbf{E}$ for elements in the set $D_{\mathrm{Aw}} \cup D_{\mathrm{Bw}}$:

$$\mathbf{E} = \frac{1}{N}\left\{ \sum_{\mathbf{x}_i \in D_{\mathrm{Aw}}} n(\mathbf{x}_i)[a - g_{\mathbf{w}}(\mathbf{x}_i)]^2 + \sum_{\mathbf{x}_i \in D_{\mathrm{Bw}}} n(\mathbf{x}_i)[b - g_{\mathbf{w}}(\mathbf{x}_i)]^2 \right\}$$

with N the number of elements in $D_{\mathrm{Aw}} \cup D_{\mathrm{Bw}}$ and $n(\mathbf{x}_i)$ the number of times the elements $\mathbf{x}_i$ occur in D_{Aw} respectively in D_{Bw}. Because N is a constant and assuming $n(\mathbf{x}_i) = 1$ or $n(\mathbf{x}_i) = 0$, we can simplify the expression to:

$$\mathbf{E} = \sum_{\mathbf{x}_i \in D_{\mathrm{Aw}}} [a - g_{\mathbf{w}}(\mathbf{x}_i)]^2 + \sum_{\mathbf{x}_i \in D_{\mathrm{Bw}}} [b - g_{\mathbf{w}}(\mathbf{x}_i)]^2$$

To minimize $\mathbf{E}$, according to Theorem 3.3, the adaptation of the weight vector $\mathbf{w}$ must be:

$$\Delta \hat{\mathbf{w}} = \varepsilon \left\{ \sum_{\mathbf{x}_i \in D_{\mathrm{Aw}}} [a - g_{\mathbf{w}}(\mathbf{x}_i)] \frac{\mathrm{d}f}{\mathrm{d}s} \hat{\mathbf{x}}_i + \sum_{\mathbf{x}_i \in D_{\mathrm{Bw}}} [b - g_{\mathbf{w}}(\mathbf{x}_i)] \frac{\mathrm{d}f}{\mathrm{d}s} \hat{\mathbf{x}}_i \right\}$$

with $g_{\mathbf{w}}(\mathbf{x}_i) = f(s(\mathbf{x}_i))$, $\hat{\mathbf{w}} = [w_0, w_1, \ldots, w_n]^{\mathrm{t}}$ the extended weight vector and $\hat{\mathbf{x}}_i = [1, x_{i1}, x_{i2}, \ldots, x_{in}]^{\mathrm{t}}$ the extended input vector.

The given learning rule is correct as long as the set $D_{\mathrm{Aw}} \cup D_{\mathrm{Bw}}$ is constant. After adaptation of the weight vector $\mathbf{w}$ the set $D_{\mathrm{Aw}} \cup D_{\mathrm{Bw}}$ may be changed into a new set $D_{\mathrm{Aw'}} \cup D_{\mathrm{Bw'}}$ because $\mathbf{w}$ is changed into $\mathbf{w}'$. Thus we subsequently minimize the MSE for a sequence of wrongly classified sets. While in the case of classification with one–zero labelling the MSE will gradually decrease during learning (if the learning external rate ε is small enough), it may happen that with this classification method jumps in the decreasing curve of the MSE occur during learning, because the sets of wrongly classified examples may change abruptly after the adaptation of the weight vector $\mathbf{w}$.

The product:

$$\gamma(\mathbf{x}_i) = [t(\mathbf{x}_i) - f(s(\mathbf{x}_i))] \frac{\mathrm{d}f}{\mathrm{d}s}$$

is called the *internal learning rate* for $\mathbf{x}_i$. The contribution of different elements of the data set D to the adaptation of the weight vector can be quite different, as

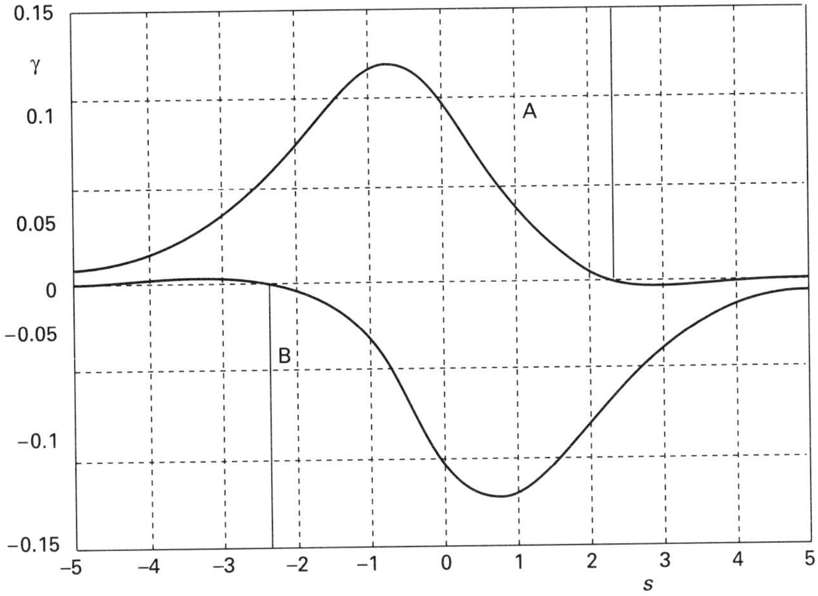

Figure 3.32 Internal learning rate for $t(\mathbf{x}_i)=0.9$ (curve A) and $t(\mathbf{x}_i)=0.1$ (curve B)

expressed by the internal learning rate. For a wrongly classified element of D_A the target value is, for example, 0.9. Because $df/ds=f(s)[1-f(s)]$ the internal learning rate as a function of the weighted input $s(\mathbf{x}_i)$ for the elements of D_{Aw} will be:

$$\gamma(s)=[0.9-f(s)]f(s)[1-f(s)]$$

and is given in Figure 3.32 by curve A. The curve is only used for values of s to the left of the solid vertical line because at the right side $f(s)>0.9$ and no adaptation will occur.

For a wrongly classified element of D_B the target value is, for example, 0.1. The internal learning rate is:

$$\gamma(s)=[0.1-f(s)]f(s)[1-f(s)]$$

represented by curve B in Figure 3.32. The curve is only used for values of s to the right of the solid vertical line because at the left side $f(s)<0.1$ and no adaptation will occur.

The learning rule can now be written as:

$$\Delta\hat{\mathbf{w}}=\sum_{\mathbf{x}_i\in D_{Aw}}\varepsilon\gamma(\hat{\mathbf{x}}_i)\hat{\mathbf{x}}_i+\sum_{\mathbf{x}_i\in D_{Bw}}\varepsilon\gamma(\hat{\mathbf{x}}_i)\hat{\mathbf{x}}_i$$

The first sum represents a weighted sum of wrongly labelled vectors from D_{Aw}. The second sum represents a weighted sum of wrongly labelled vectors from D_{Bw}. Let us denote the first sum by $\hat{\mathbf{x}}_A$ and the second sum by $\hat{\mathbf{x}}_B$; then we can write the adaptation as follows:

$$\Delta\hat{\mathbf{w}} = \hat{\mathbf{x}}_A + \hat{\mathbf{x}}_B$$

If it happens that $\hat{\mathbf{x}}_A + \hat{\mathbf{x}}_B = \lambda\hat{\mathbf{w}}$, then after adaptation the new value of the weight vector becomes $\hat{\mathbf{w}} + \Delta\hat{\mathbf{w}} = (1+\lambda)\hat{\mathbf{w}}$ and the separating hyperplane, defined by $s(\mathbf{x})=0$, will be the same before and after adaptation but the MSE will be reduced.

Example 3.9

Let $D_A = \{\mathbf{x}_1, \mathbf{x}_2, \mathbf{x}_3\}$ with $\mathbf{x}_1 = (1, 2)$, $\mathbf{x}_2 = (-2, -1)$ and $\mathbf{x}_3 = (-1, 2)$. Let $D_B = \{\mathbf{x}_4, \mathbf{x}_5, \mathbf{x}_6\}$ with $\mathbf{x}_4 = (2, 1)$, $\mathbf{x}_5 = (-1, -2)$ and $\mathbf{x}_6 = (2, -2)$. Given is $D_A \subseteq X_A$ and $D_B \subseteq X_B$.

With a single-neuron Perceptron we want to separate the two-dimensional input space with a one-dimensional hyperplane defined by $s(\mathbf{x}) = w_0 + w_1x_1 + w_2x_2 = 0$ (a line) into the region X_A and X_B such that $g_\mathbf{w}(\mathbf{x}) > 0.5$ for $\mathbf{x} \in X_A$ and $g_\mathbf{w}(\mathbf{x}) < 0.5$ for $x \in X_B$.

Assume the initial weights are $w_0 = 0$, $w_1 = -1$ and $w_2 = 1$. The initial separating line defined by $s(\mathbf{x}) = -x_1 + x_2 = 0$ is given by the solid line in Figure 3.33. Although all points of D_A and D_B are already correctly classified, the output targets for learning are not satisfied.

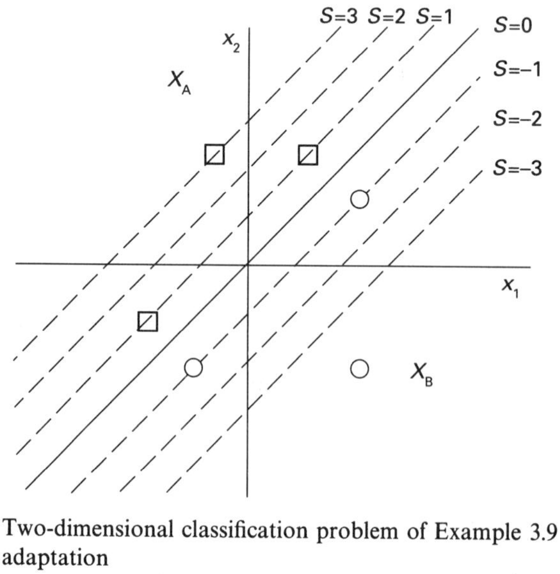

Figure 3.33 Two-dimensional classification problem of Example 3.9 before adaptation

We calculate the weighted inputs as follows:

$$s(\mathbf{x}_1)=(-1)(1)+(1)(2) \quad =1$$
$$s(\mathbf{x}_2)=(-1)(-2)+(1)(-1)=1$$
$$s(\mathbf{x}_3)=(-1)(-1)+(1)(2) \quad =3$$
$$s(\mathbf{x}_4)=(-1)(2)+(1)(1) \quad =-1$$
$$s(\mathbf{x}_5)=(-1)(-1)+(1)(-2)=-1$$
$$s(\mathbf{x}_6)=(-1)(2)+(1)(-2) \quad =-4$$

With the help of Figure 3.25 we find:

$$g_{\mathbf{w}}(\mathbf{x}_1)\cong 0.73$$
$$g_{\mathbf{w}}(\mathbf{x}_2)\cong 0.73$$
$$g_{\mathbf{w}}(\mathbf{x}_3)\cong 0.95$$
$$g_{\mathbf{w}}(\mathbf{x}_4)\cong -0.27$$
$$g_{\mathbf{w}}(\mathbf{x}_5)\cong -0.27$$
$$g_{\mathbf{w}}(\mathbf{x}_6)\cong 0.01$$

We see that only for $\mathbf{x}_3$ and for $\mathbf{x}_6$ are the learning targets satisfied. We thus have to adapt the weights. From Figure 3.32 we obtain:

$$\gamma(\mathbf{x}_1)=\gamma(\mathbf{x}_2)\cong 0.03$$
$$\gamma(\mathbf{x}_4)=\gamma(\mathbf{x}_5)\cong -0.03$$

With the learning rule we obtain:

$$\Delta\hat{\mathbf{w}}=\varepsilon\gamma(\mathbf{x}_1)\hat{\mathbf{x}}_1+\varepsilon\gamma(\mathbf{x}_2)\hat{\mathbf{x}}_2+\varepsilon\gamma(\mathbf{x}_4)\hat{\mathbf{x}}_4+\varepsilon\gamma(\mathbf{x}_5)\hat{\mathbf{x}}_5$$

With $\varepsilon=10$ we obtain:

$$\Delta\hat{\mathbf{w}}=0.3\begin{bmatrix}1\\1\\2\end{bmatrix}+0.3\begin{bmatrix}1\\-2\\-1\end{bmatrix}-0.3\begin{bmatrix}1\\2\\1\end{bmatrix}-0.3\begin{bmatrix}1\\-1\\-2\end{bmatrix}-\begin{bmatrix}0\\-1.2\\1.2\end{bmatrix}$$

The new weight vector becomes:

$$\mathbf{w}=\begin{bmatrix}0\\-2.2\\2.2\end{bmatrix}$$

This new weight vector defines the same separating line $s(\mathbf{x})=-2.2x_1+2.2x_2=0$ but now the target values are satisfied. If we compare Figure 3.32 (before adaptation)

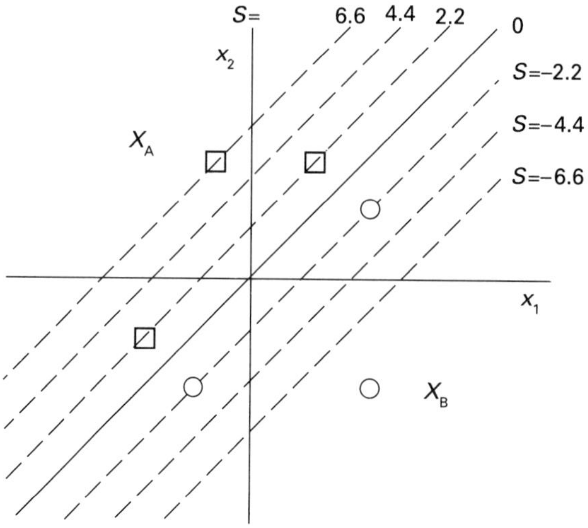

Figure 3.34 Two-dimensional classification problem of Example 3.9 after adaptation

with Figure 3.34 (after adaptation) we observe that the lines with constant values of s are moved towards the separating line. For $s=2.2$ we have $f(s)=0.9$ and for $s=-2.2$, $f(s)=0.1$: the realized function becomes almost a threshold function. ■

3.10 Hyperplane boundary classification by single threshold labelling

In the previous section on classification with one–zero labelling as well as with double threshold labelling we used the output of a single-neuron Perceptron to classify the inputs: if $g_{\mathbf{w}}(\mathbf{x})>0.5$ then $\mathbf{x}\in X_A$, and if $g_{\mathbf{w}}(\mathbf{x})\leqslant 0.5$ then $\mathbf{x}\in X_B$. However, during learning with the given learning sets the targets were not equal to $t(\mathbf{x})>0.5$ for $\mathbf{x}\in D_A\subseteq X_A$ and $t(\mathbf{x})\leqslant 0.5$ for $\mathbf{x}\in D_B\subseteq X_B$. The question arises as to why we did not use the same conditions for learning as for classification. The learning rule is implied by our goal to minimize the MSE:

$$E=\sum_{\mathbf{x}_i\in D}[t(\mathbf{x}_i)-g_{\mathbf{w}}(\mathbf{x}_i)]^2$$

with the learning rule:

$$\Delta\hat{\mathbf{w}}=\varepsilon\sum_{\mathbf{x}_i\in D}[t(\mathbf{x}_i)-g_{\mathbf{w}}(\mathbf{x}_i)]\frac{\mathrm{d}f}{\mathrm{d}s}\hat{\mathbf{x}}_i$$

the weights are modified such that the MSE becomes as small as possible. Suppose we require during learning that $t(\mathbf{x})>0.5$ for $\mathbf{x}\in D_A$ and $t(\mathbf{x})\leqslant 0.5$ for $\mathbf{x}\in D_B$. If at some stage during learning the output $g_{\mathbf{w}}(\mathbf{x})$ would become equal to 0.5 for all $\mathbf{x}\in D_A\cup D_B$, then the MSE would be zero and our goal would be reached. An output $g_{\mathbf{w}}(\mathbf{x})=0.5$ for all $\mathbf{x}\in D_A\cup D_B$ can be realized if all weights are zero because then $s(\mathbf{x})=0$. Thus during learning the weights will become zero and the only global minimum of $E=0$ is reached. In that case all elements of D_A are misclassified and all elements of D_B are correctly classified. Thus we cannot use the single threshold classification method without additional precautionary measures.

We must realize that the learning rule will always minimize the MSE but that criterion is not always identical with minimizing the number of wrongly classified elements of the data set D.

From this analysis we can conclude as follows:

Practical statement 3.9

A small value of the MSE does not always imply that the classification with a single-neuron Perceptron is correct.

Although classification of data with the unmodified single threshold labelling is dangerous because of the small values of the final weights, we can still get reasonable results because the separating final hyperplane depends on the quotients of the (small) values of weights. We will give an illustration in the following example.

Example 3.10

Assume we are given a two-dimensional data set $D=D_A\cup D_B$ depicted in Figure 3.35. With

$$D_A=\{(-3,3),(1,3),(-3,7),(-7,3),(-3,-1),(0,0),(-1,1)\}$$

$$D_B=\{(2,-2),(2,2),(6,-2),(2,-6),(-2,-2),(0,0),(-1,1)\}$$

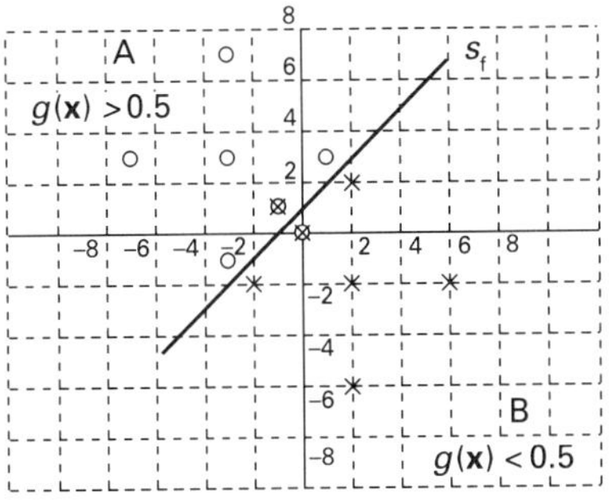

Figure 3.35 Two-dimensional classification problem of Example 3.10 with final desired separating hyperplane S_f

Table 3.1

$\mathbf{x}_i$	$s = w_0 + w_1x_1 + w_2x_2$	$g(\mathbf{x}_i)$	$t(\mathbf{x}_i) - g(\mathbf{x}_i)$	$\partial f/\partial s$
−3, 3	−0.3 + 0.6 = 0.3	0.57	0.00	0.25
1, 3	0.1 + 0.6 = 0.7	0.66	0.00	0.23
−3, 7	−0.3 + 1.4 = 1.1	0.75	0.00	0.19
−7, 3	−0.7 + 0.6 = −0.1	0.48	0.02	0.25
−3, −1	−0.3 − 0.2 = −0.5	0.38	0.12	0.24
0, 0	= 0.0	0.50	0.00	0.25
−1, 1	−0.1 + 0.2 = 0.1	0.52	0.00	0.25

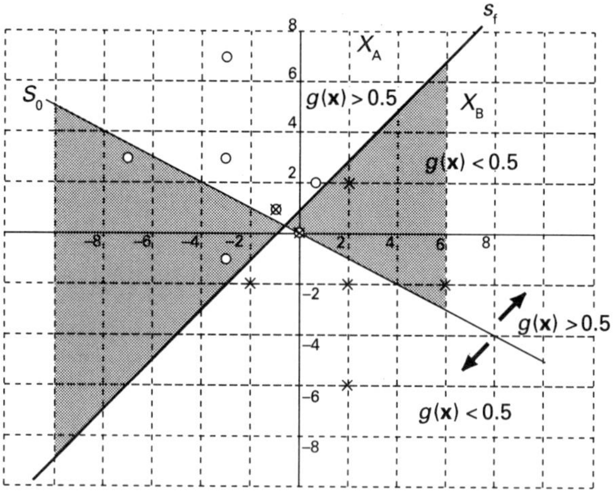

Figure 3.36 Two-dimensional classification problem of Example 3.10 with initial S_0 and final desired separating hyperplane S_f

We have the target values: $t(\mathbf{x}_i) > 0.5$ if $\mathbf{x}_i \in D_A$ and $t(\mathbf{x}_i) \leqslant 0.5$ if $\mathbf{x}_i \in D_B$. We take a single-neuron Perceptron with the sigmoidal transfer function:

$$f[s(\mathbf{x}_i)] = \frac{1}{1 + e^{-s(\mathbf{x}_i)}} \text{ and } s(\mathbf{x}_i) = \sum_{j=0}^{n} w_j \cdot x_{ij}$$

Let:

$$g(\mathbf{x}_i) = f[s(\mathbf{x}_i)]$$

then

$$\frac{\mathrm{d}f}{\mathrm{d}s} = f(s) \cdot [1 - f(s)]$$

In Table 3.1 and Table 3.2 we have calculated for the sets D_a and D_b the values

Table 3.2

$\mathbf{x}_i$	$s=w_0+w_1x_1+w_2x_2$	$g(\mathbf{x}_i)$	$t(\mathbf{x}_i)-g(\mathbf{x}_i)$	$\partial g/\partial s$
2, −2	0.2−0.4=−0.2	0.45	0.00	0.25
2, 2	0.2+0.4=0.6	0.64	−0.14	0.23
6, −2	0.6−0.4=0.2	0.55	−0.05	0.25
2, −6	0.2−1.2=−1.0	0.27	0.00	0.20
−2, −2	−0.2−0.4=−0.6	0.35	0.00	0.23
0, 0	=0.0	0.50	0.00	0.25
−1, 1	−0.1+0.2=0.1	0.52	−0.02	0.25

of $s(\mathbf{x}_i)$, $g_\mathbf{w}(\mathbf{x}_i)$, $t(\mathbf{x}_i)-g_\mathbf{w}(\mathbf{x}_i)$ and $\mathrm{d}f/\mathrm{d}s$ for initial weights $w_0=0$, $w_1=0.1$ and $w_2=0.2$ corresponding to the initial separating line S_0 given in Figure 3.36. Table 3.1 shows values for D_A with $t(\mathbf{x}_i)>0.5$, and Table 3.2 for D_B with $t(\mathbf{x}_i)\leqslant 0.5$.

The misclassified extended input vectors from D_A are $D_{\mathrm{Aw}}=\{[1,-7,3],[1,-3,-1]\}$. The misclassified extended vectors from D_B are $D_{\mathrm{Bw}}=\{[1,2,2],[1,6,-2],[1,-1,1]\}$ (see the shaded area in Figure 3.36).

The adaptation of the weight vector must be:

$$\Delta\hat{\mathbf{w}}=\varepsilon\sum_{\mathbf{x}_i\in D_{\mathrm{Aw}}}\{0.5-g_\mathbf{w}(\mathbf{x}_i)\}\frac{\mathrm{d}f}{\mathrm{d}s}\hat{\mathbf{x}}_i+\varepsilon\sum_{\mathbf{x}_i\in D_{\mathrm{Bw}}}\{0.5-g_\mathbf{w}(\mathbf{x}_i)\}\frac{\mathrm{d}f}{\mathrm{d}s}\hat{\mathbf{x}}_i$$

Thus:

$$\Delta\hat{\mathbf{w}}=\varepsilon\cdot 0.02\cdot 0.25\cdot\begin{bmatrix}1\\-7\\3\end{bmatrix}+\varepsilon\cdot 0.12\cdot 0.24\cdot\begin{bmatrix}1\\-3\\-1\end{bmatrix}$$

$$-\varepsilon\cdot 0.14\cdot 0.13\cdot\begin{bmatrix}1\\2\\2\end{bmatrix}-\varepsilon\cdot 0.05\cdot 0.25\cdot\begin{bmatrix}1\\6\\-2\end{bmatrix}-\varepsilon\cdot 0.02\cdot 0.25\cdot\begin{bmatrix}1\\-1\\1\end{bmatrix}$$

With $\varepsilon=0.5$ we obtain:

$$\Delta\hat{\mathbf{w}}=\begin{bmatrix}-0.01\\-0.13\\-0.03\end{bmatrix}$$

Thus we obtain for the weight vector after adaptation:

$$\hat{\mathbf{w}}_1=\begin{bmatrix}0.0\\0.1\\0.2\end{bmatrix}+\begin{bmatrix}-0.01\\-0.13\\-0.03\end{bmatrix}=\begin{bmatrix}-0.01\\-0.03\\0.17\end{bmatrix}$$

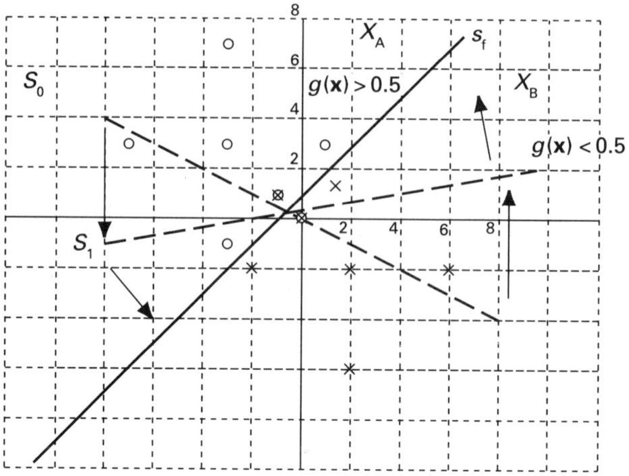

Figure 3.37 Two-dimensional classification problem of Example 3.10 with initial separating hyperplane S_0, the separating hyperplane after one learning step S_1 and the final desired separating hyperplane S_f

The equation for the separating line (see line S_1 in Figure 3.37) will be $-0.01-0.03x_1+0.17x_2=0$. For simulation results see Figure 3.38 (global learning) and Figure 3.39 (local learning). ■

In Figure 3.39 the results for local learning of a simulation experiment for the data given at the beginning of this example are given. After the presentation of a single example of the data set the weights are adapted. An element of the data set is called example k if that element is the kth element in the enumeration of the tables given before. We see that given the initial separating line, the first three examples are classified correctly and thus will not contribute to the adaptation of weights at that stage of learning. ■

One way to circumvent the problem of bad classification with the single threshold labelling method is using a target value for $\mathbf{x} \in D_A$ slightly higher than 0.5 (e.g. 0.51) and a target value for $\mathbf{x} \in D_B$ slightly smaller than 0.5 (e.g. 0.49). However, we are then in fact using the double threshold method as described in the previous section.

Another way to prevent the weights becoming zero and still use the target $t(\mathbf{x})>0.5$ for an element of X_A and $t(\mathbf{x})\leqslant 0.5$ for $\mathbf{x} \in X_B$ is to multiply after each adaptation step the weights with a scalar such that $|\hat{\mathbf{w}}|$ remains constant. Multiplying all weights with the same scalar does not change the separating hyperplane defined by the dot product $\hat{\mathbf{w}} \cdot \hat{\mathbf{x}}=0$. By keeping $|\hat{\mathbf{w}}|$ constant during learning we are searching for a minimum of the MSE for values of $\hat{\mathbf{w}}$ on a hypersphere in the solution space W. We know, however, that the value of $|\hat{\mathbf{w}}|$ must become infinite to reach an optimal solution (see Section 3.7). If we divide the learning process in a sequence of time intervals such

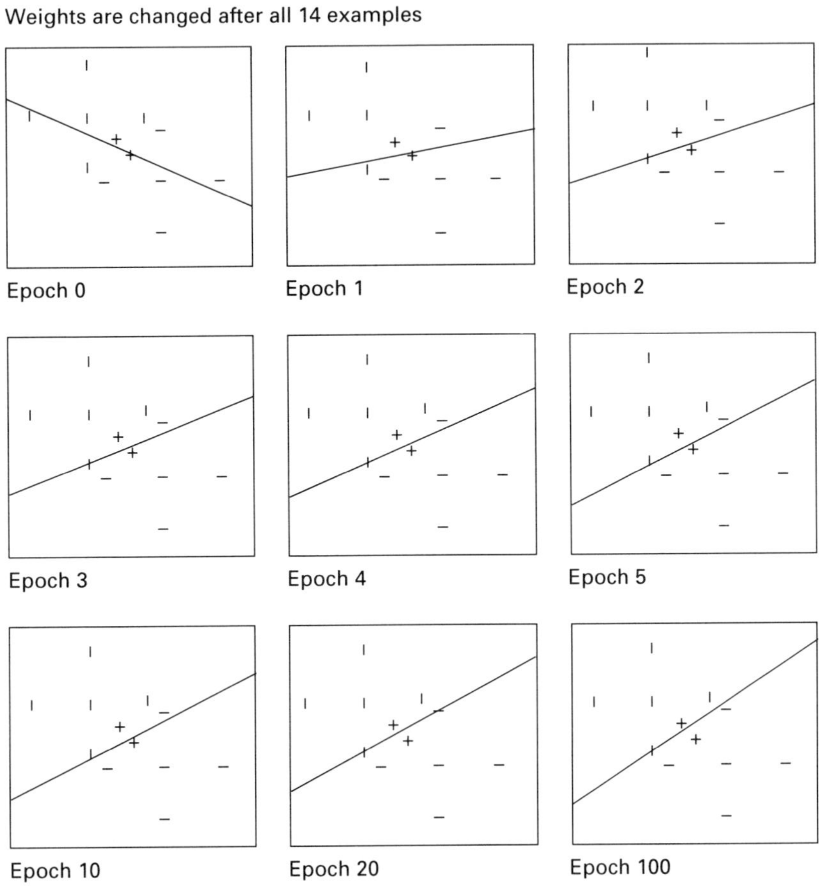

Figure 3.38 Simulation result after several epochs (=number of times the total training set of fourteen examples is supplied) of global learning

that $|\hat{\mathbf{w}}|$ is kept constant in each interval until a minimum for the MSE is reached and then change $|\mathbf{w}|$ subsequently to a larger constant value for the next sequence, we systematically search through the complete solution space W and end up with the required large value of $|\hat{\mathbf{w}}|$.

Example 3.11

Suppose we have one neuron with one input x_1 and we have the following data sets $D_A = \{-2.0, -1.9, -1.8, -1.7, 1.95\}$ and $D_B = \{0.5, 0.6, 0.7, 0.8\}$. Note that the two sets are not linear separable. After learning with single threshold labelling we want to have for the output of the neuron $g_{\mathbf{w}}(\mathbf{x}) > 0.5$ for $\mathbf{x} \in D_A$ and $g_{\mathbf{w}}(\mathbf{x}) \leqslant 0.5$ for $\mathbf{x} \in D_B$. An optimal solution for the separating hyperplane $w_0 + w_1 x_1 = 0$ (a point in this case)

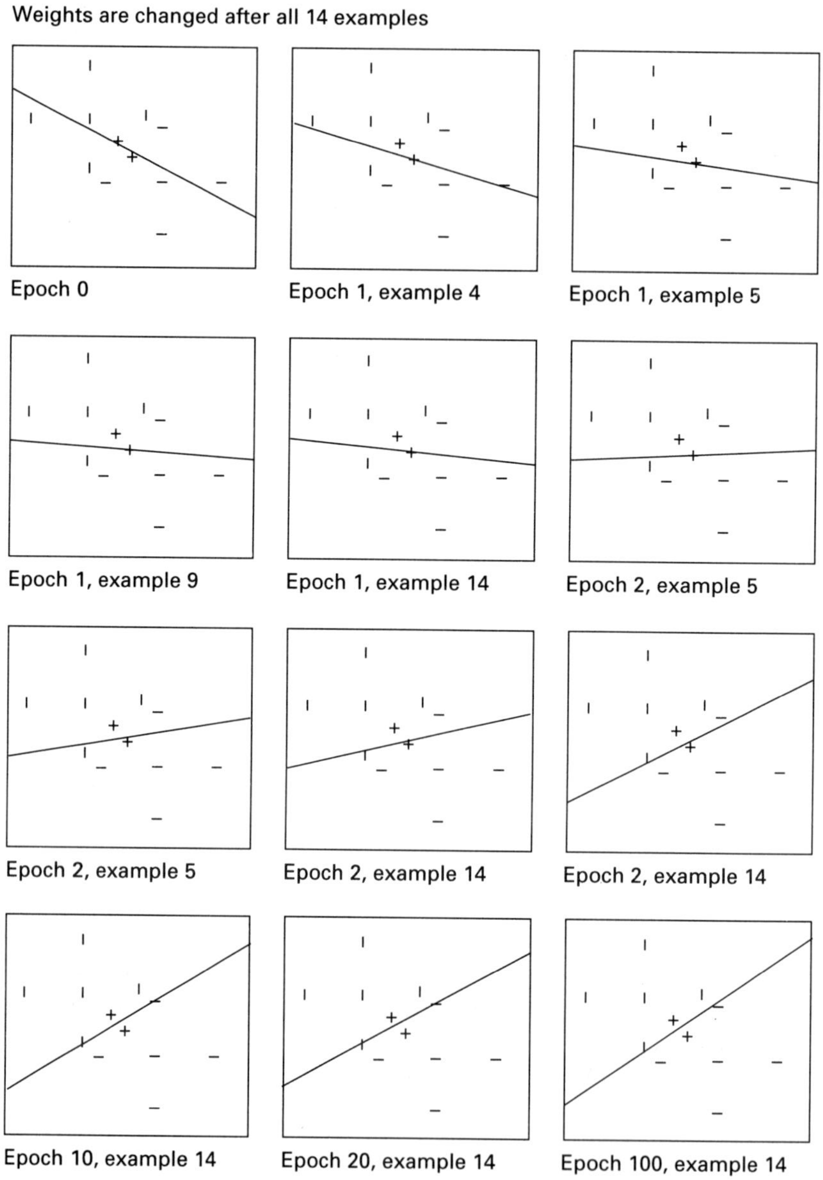

Figure 3.39 Simulation result after several epochs (=number of times fourteen randomly chosen examples of the total training set are supplied) of local learning

is obtained when $-w_0/w_1$ has a value between -1.7 and 0.5; in that case eight elements will be classified correctly and one element is misclassified.

If we use the unmodified single threshold labelling method, the weights will indeed go to zero but the separation point $-w_0/w_1$ goes to a constant value. There are two separation points to which the network converges; which one is reached depends on the initial weights. The values of $-w_0/w_1$ will be either 1.06 or -2.67. In both cases five inputs are misclassified.

In order to see why we obtain these solutions we calculate the error measure E with $\hat{\mathbf{w}}$ on a circle around the origin in the w_0–w_1 plane for a small value of $|\hat{\mathbf{w}}|$. We take $w_0=|\mathbf{w}|\cos\phi$ and $w_1=|\mathbf{w}|\sin\phi$ and vary ϕ from 0 to 2π rad. In Figure 3.40 we see that for $|\hat{\mathbf{w}}|=1$ we obtain two minima for the MSE – one at $\phi=5.53$ rad (corresponding to the separating point $-w_0/w_1=|\mathbf{w}|\cos\phi/|\mathbf{w}|\sin\phi=-0.073/-0.068=1.06$), and a second at $\phi=3.50$ rad (corresponding to a separating point $-w_0/w_1=0.094/-0.035=-2.67$).

When we use constant $|\mathbf{w}|$ during subsequent time intervals of learning as described above and start the first learning interval with $|\mathbf{w}|=1$, we will find one of the solutions mentioned before. If we use $|\mathbf{w}|=2$ in the second learning interval, then we will again find two possible solutions: one larger than -2.67 and the other smaller than 1.06. Which solution is found depends on the solution found in the previous time interval.

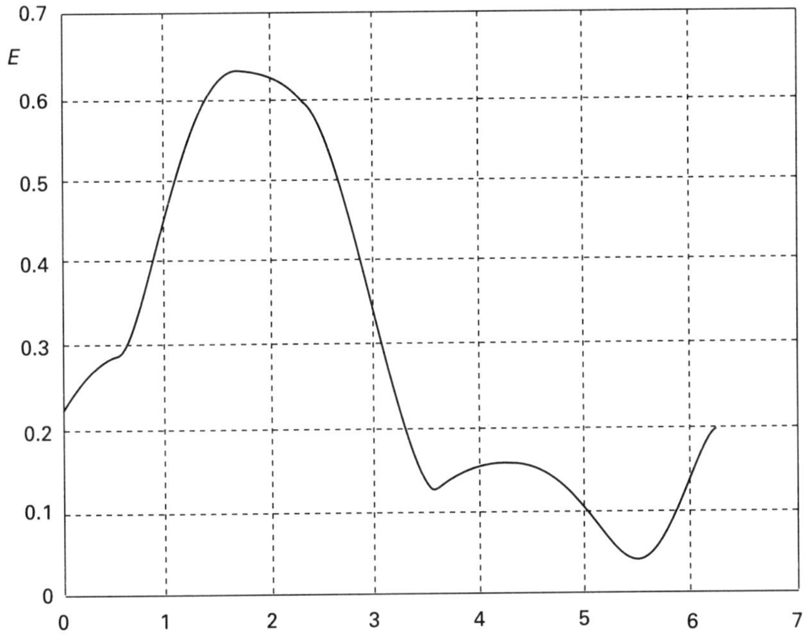

Figure 3.40 The MSE for example 3.11 with $|\mathbf{w}|=1$, $w_0=|\mathbf{w}|\cos\phi$, $w_1=|\mathbf{w}|\sin\phi$ and $0\leqslant\phi\leqslant2\pi$.

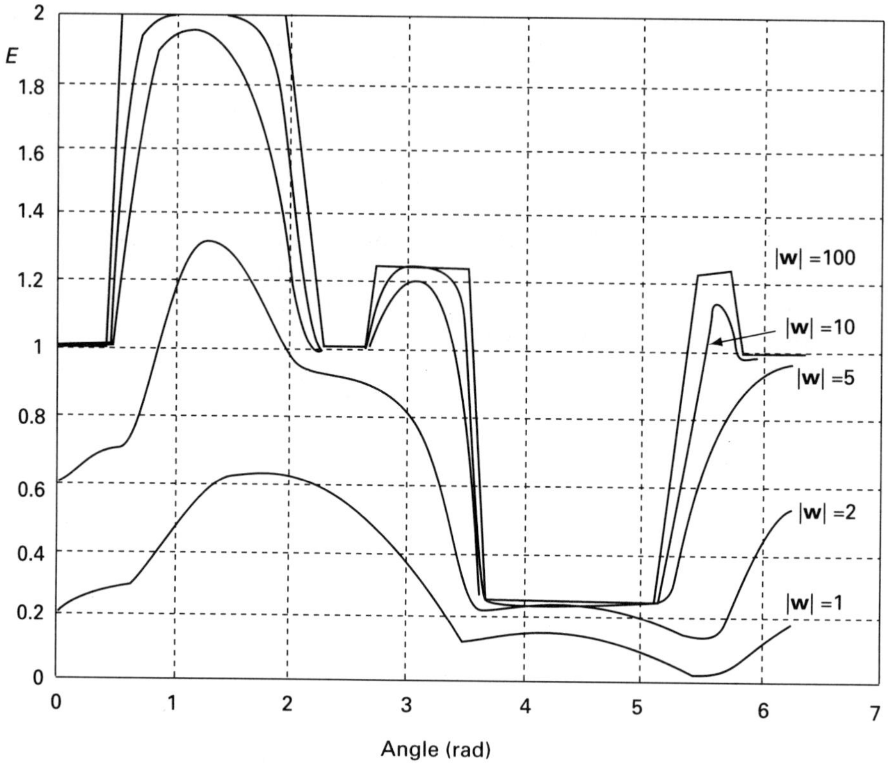

Figure 3.41 The MSE for Example 3.11 with respectively $|\mathbf{w}|=1$, $|\mathbf{w}|=2$, $|\mathbf{w}|=5$ and $|\mathbf{w}|=100$. With $w_0=|\mathbf{w}|\cos\phi$, $w_1=|\mathbf{w}|\sin\phi$ and $0\leqslant\phi\leqslant 2\pi$

This means that both solutions move in the direction of the correct interval between -1.7 and 0.5. This phenomenon becomes clear from Figure 3.41 where we plotted the MSE for several values of $|\hat{\mathbf{w}}|$. When $|\hat{\mathbf{w}}|$ becomes infinite we see that the minimum of the MSE will be in the interval for ϕ between 3.67 and 5.18 rad, corresponding to the separating points -1.7 and 0.5. A closer look reveals that there will be one minimum just to the left of $\phi=5.18$ rad (corresponding to the point 0.5). ■

3.11 Application to the classification of normally distributed classes

In Section 3.8 we presented a classification problem with a two-dimensional data set $D=D_A\cup D_B$ depicted by the points in Figure 3.42. The data set is D_A generated by

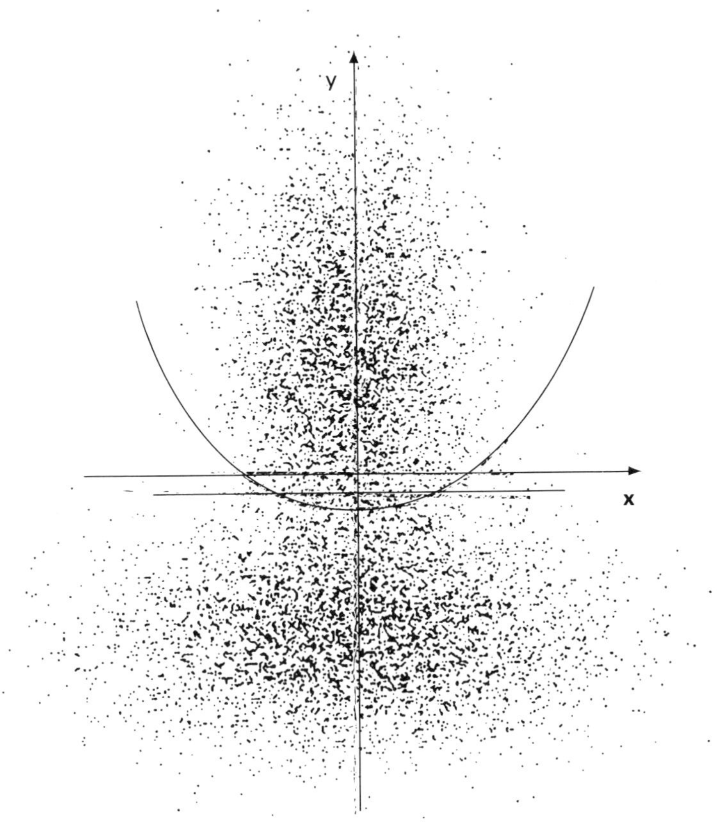

Figure 3.42 Two-dimensional classification problem with two Gaussian distributed data sets

a Gaussian distribution function with probability density function:

$$f_A(x, y) = \frac{1}{\sqrt{2\pi\sigma_x}} \exp \frac{-(x-\mu_x)^2}{2\sigma_x^2} + \frac{1}{\sqrt{2\pi\sigma_y}} \exp \frac{-(y-\mu_y)^2}{2\sigma_y^2}$$

with $\mu_x = 0$, $\sigma_x = 0.2$, $\mu_y = 0.465$ and $\sigma_y = 0.4$.

Similarly we have the data set D_B generated by the same type of Gaussian distribution function but now with $\mu_x = 0$, $\sigma_x = 0.4$, $\mu_y = -0.465$ and $\sigma_y = 0.2$.

Although the optimal classification boundary, or discrimination line, is a curved line, with an error of 5.14 per cent, we can divide the input space by one straight line such that the probability of error will then be slightly larger: 5.15 per cent. The optimal position of the line is $y = -0.11$. If we use a single-neuron Perceptron and use the one–zero labelling method with 100 000 examples generated by the distribution we will find an almost horizontal separation hyperplane located at $y = -0.11$ with an error of 5.15 per cent.

When we use the method of double threshold labelling with targets 0.9 and 0.1 we will again find an almost horizontal separating hyperplane located at $y = -0.11$ with an error of 5.18 per cent. A typical example of the final weights is $w_0 = -0.7280$, $w_1 = -0.0971$ and $w_2 = -6.3304$, corresponding with a separating line $y = 0.015x - 0.115$.

When we use the method of double threshold labelling with targets 0.51 and 0.49 we finally again find an almost horizontal separating hyperplane located at $y = -0.15$ with an error of 5.39 per cent.

When we use the unmodified single threshold labelling method the position of the final, almost horizontal separating hyperplane will be located at $y = -0.176$ and we obtain an error of 6.66 per cent. As a typical example we will find for the values of (small) final weights: $w_0 = 0.0022$, $w_1 = 0.0042$ and $w_2 = 0.0131$.

3.12 Learning rule for a two-layer continuous Perceptron

We consider first the situation with two neurons in the first layer and one neuron in the second layer (see Figure 3.43). The adaptation rule for this simple case is almost the same as for a general two-layer network with n_1 neurons in the first and n_2 neurons in the second layer. We will come back to it later.

For the output y_3 for the neuron in the second layer we have:

$$y_3 = f_3(s_3)$$

with $s_3 = w_{31}y_1 + w_{32}y_2 + w_{30}$ the weighted input of neuron 3.

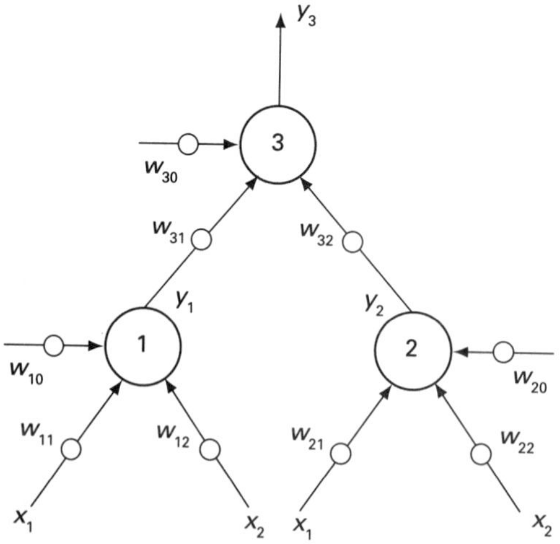

Figure 3.43 A simple two-layer continuous Perceptron

For the output y_1 and y_2 we have respectively:

$$y_1 = f_1(s_1) \text{ and } y_2 = f_2(s_2)$$

with $s_1 = w_{11}x_1 + w_{12}x_2 + w_{10}$ and $s_2 = w_{21}x_1 + w_{22}x_2 + w_{20}$.

The function $g_{\mathbf{w}}(\mathbf{x}) = y_3$ realized by the neural net is:

$$g_{\mathbf{w}}(\mathbf{x}) = y_3 = f_3[w_{31}f_1\{w_{11}x_1 + w_{12}x_2 + w_{10}\} + w_{32}f_2\{w_{21}x_1 + w_{22}x_2 + w_{20}\} + w_{30}]$$

Assume we want to realize with the two-layer Perceptron of Figure 3.42 for input vector $\mathbf{x}_i$ a target value $t_3(\mathbf{x}_i)$ for the output of neuron 3 with $\mathbf{x}_i = [x_{i1}, x_{i2}] \in \mathbf{U}$, with $\mathbf{U}$ a finite subset of $\mathbb{R}^2$, i.e. $[\mathbf{x}_i, t(\mathbf{x}_i)]$ is an element of the given data set D.

For the error function we have:

$$E = \frac{1}{2} \sum_{\mathbf{x}_i \in \mathbf{U}} [t_3(\mathbf{x}_i) - y_3(\mathbf{x}_i)]^2$$

For the adaptation of some weight w_j we obtain, according to Theorem 3.2:

$$\Delta w_j = -\varepsilon \frac{\partial E}{\partial w_j}$$

Thus the adaptation of w_{30} becomes:

$$\Delta w_{30} = -\varepsilon \frac{\partial E}{\partial w_{30}} = \varepsilon \sum_{\mathbf{x}_i \in \mathbf{U}} [t_3(\mathbf{x}_i) - y_3(\mathbf{x}_i)] \frac{\mathrm{d}f_3}{\mathrm{d}s_3}$$

In the same way we obtain for the adaptation of w_{31} and w_{32}:

$$\Delta w_{31} = -\varepsilon \frac{\partial E}{\partial w_{31}} = \varepsilon \sum_{\mathbf{x}_i \in \mathbf{U}} [t_3(\mathbf{x}_i) - y_3(\mathbf{x}_i)] \frac{\mathrm{d}f_3}{\mathrm{d}s_3} y_1(\mathbf{x}_i)$$

$$\Delta w_{32} = -\varepsilon \frac{\partial E}{\partial w_{32}} = \varepsilon \sum_{\mathbf{x}_i \in \mathbf{U}} [t_3(\mathbf{x}_i) - y_3(\mathbf{x}_i)] \frac{\mathrm{d}f_3}{\mathrm{d}s_3} y_2(\mathbf{x}_i)$$

The adaptation of w_{10} will be:

$$\Delta w_{10} = -\varepsilon \frac{\partial E}{\partial w_{10}} = \varepsilon \sum_{\mathbf{x}_i \in \mathbf{U}} [t_3(\mathbf{x}_i) - y_3(\mathbf{x}_i)] \frac{\mathrm{d}f_3}{\mathrm{d}s_3} w_{31} \frac{\mathrm{d}f_1}{\mathrm{d}s_1}$$

In the same way we obtain for the adaptation of w_{11} and w_{12}:

$$\Delta w_{11} = -\varepsilon \frac{\partial E}{\partial w_{11}} = \varepsilon \sum_{\mathbf{x}_i \in \mathbf{U}} [t_3(\mathbf{x}_i) - y_3(\mathbf{x}_i)] \frac{\mathrm{d}f_3}{\mathrm{d}s_3} w_{31} \frac{\mathrm{d}f_1}{\mathrm{d}s_1} x_{i1}$$

$$\Delta w_{12} = -\varepsilon \frac{\partial E}{\partial w_{12}} = \varepsilon \sum_{\mathbf{x}_i \in \mathbf{U}} [t_3(\mathbf{x}_i) - y_3(\mathbf{x}_i)] \frac{\mathrm{d}f_3}{\mathrm{d}s_3} w_{31} \frac{\mathrm{d}f_1}{\mathrm{d}s_1} x_{i2}$$

The adaptations of w_{20}, w_{21} and w_{22} are:

$$\Delta w_{20} = -\varepsilon \frac{\partial E}{\partial w_{20}} = \varepsilon \sum_{\mathbf{x}_i \in \mathbf{U}} [t_3(\mathbf{x}_i) - y_3(\mathbf{x}_i)] \frac{\mathrm{d}f_3}{\mathrm{d}s_3} w_{32} \frac{\mathrm{d}f_2}{\mathrm{d}s_2}$$

$$\Delta w_{21} = -\varepsilon \frac{\partial E}{\partial w_{21}} = \varepsilon \sum_{\mathbf{x}_i \in \mathbf{U}} [t_3(\mathbf{x}_i) - y_3(\mathbf{x}_i)] \frac{\mathrm{d}f_3}{\mathrm{d}s_3} w_{32} \frac{\mathrm{d}f_2}{\mathrm{d}s_2} x_{i1}$$

$$\Delta w_{22} = -\varepsilon \frac{\partial E}{\partial w_{22}} = \varepsilon \sum_{\mathbf{x}_i \in \mathbf{U}} [t_3(\mathbf{x}_i) - y_3(\mathbf{x}_i)] \frac{\mathrm{d}f_3}{\mathrm{d}s_3} w_{32} \frac{\mathrm{d}f_2}{\mathrm{d}s_2} x_{i2}$$

We can simplify the expressions above and at the same time gain some intuitive insight into the adaptation rule by introducing the concept of the *weighted output error* $\delta_i(\mathbf{x}_i)$ for some neuron i for a given input $\mathbf{x}_i$.

For the output neuron 3 we define the weighted output error as:

$$\delta_3(\mathbf{x}_i) = t_3(\mathbf{x}_i) - y_3(\mathbf{x}_i)$$

For neuron 1 we define the weighted output error as:

$$\delta_1(\mathbf{x}_i) = \delta_3(\mathbf{x}_i) \frac{\mathrm{d}f_3}{\mathrm{d}s_3} w_{31}$$

This weighted output error can be considered as the difference between the unknown (non-constant) target $t_1(\mathbf{x}_i)$ value the output of neuron 1 and its actual output $y_1(\mathbf{x}_i)$, i.e.

$$\delta_1(\mathbf{x}_i) = t_1(\mathbf{x}_i) - y_1(\mathbf{x}_i)$$

For neuron 2 we define the weighted output error as:

$$\delta_2(\mathbf{x}_i) = \delta_3(\mathbf{x}_i) \frac{\mathrm{d}f_3}{\mathrm{d}s_3} w_{32}$$

This weighted output error of neuron 2 can again be considered as the difference between the unknown (non-constant) target $t_2(\mathbf{x}_1)$ value of the output of neuron 2 and its actual output $y_2(\mathbf{x}_i)$, i.e.

$$\delta_2(\mathbf{x}_1) = t_2(\mathbf{x}_1) - y_2(\mathbf{x}_1)$$

We see that the weighted output error for the neurons in the first layer can be calculated from the error in the output layer (the error is *back-propagated*) and given the output error of some neuron we can adapt the input weights for any neuron j in the same way because we can rewrite the adaptation rule for the kth weight of neuron j now in a *general adaptation rule*:

$$\Delta w_{jk} = \varepsilon \sum_{\mathbf{x}_i \in \mathbf{U}} \delta_j(\mathbf{x}_i) \frac{\mathrm{d}f_j}{\mathrm{d}s_j} z_{ik}$$

with z_{ik} the kth input of neuron k.

The adaptation rule is in accordance with our intuitive ideas about adaptation. If, for instance, the target output $t_3(\mathbf{x}_i)$ of the output neuron 3 is greater than the actual output $y_3(\mathbf{x}_i)$, then one should increase the input weight w_{31} of the output neuron if the corresponding input $y_1(\mathbf{x}_i)$ is positive (which will be always the case for a sigmoid transfer function), and decrease the weight if $y_1(\mathbf{x}_i)$ is negative – and this is what is prescribed globally by the adaptation rule:

$$\Delta w_{31} = \varepsilon \sum_{\mathbf{x}_i \in \mathrm{U}} [t_3(\mathbf{x}_i) - y_3(\mathbf{x}_i)] \frac{\mathrm{d}f_3}{\mathrm{d}s_3} y_1(\mathbf{x}_i)$$

because $\mathrm{d}f/\mathrm{d}s$ is always positive (for a sigmoid transfer function). A similar explanation holds in case that the target is smaller than the actual output. The same reasoning holds for the adaptation of the weights in the first layer.

For the weighted error assigned to a neuron in the first layer one should intuitively reason as follows. If the target value $t_3(\mathbf{x}_i)$ is greater than the actual output $y_3(\mathbf{x}_i)$ and the weight w_{31} is positive, then one should increase the output $y_1(\mathbf{x}_i)$ to obtain a better result, because $y_3(\mathbf{x}_i)$ is a monotonically increasing function of $w_{31}y_1(\mathbf{x}_i)$. Thus the error assigned to the output of a neuron 1 must be proportional to the product $\{t_3(\mathbf{x}_i) - y_3(\mathbf{x}_i)\}w_{31}$, and because $\mathrm{d}f/\mathrm{d}s$ is always positive for a sigmoid transfer function. This is prescribed by the formula given above for the weighted output error of the neurons in the first layer.

Example 3.12

We consider a classification problem and use the single threshold labelling method. (We will see in the next section that the threshold labelling method gives a bad performance.)

The initial weights of the neurons are:

$$w_{10} = 0.0 \qquad w_{11} = 0.1 \qquad w_{12} = 0.1$$

$$w_{20} = 0.0 \qquad w_{21} = -0.05 \quad w_{22} = -0.1$$

$$w_{30} = -0.2 \quad w_{31} = -1 \qquad w_{32} = 1$$

All transfer functions are sigmoid functions: $f(s) = (1 + \varepsilon^{-s})^{-1}$. Assume the learning data sets are:

$$D_\mathrm{A} = \{A_1 = (0, 10), A_2 = (0, -10)\} \text{ with target } t(\mathbf{x}_i) > 0.5$$

$$D_\mathrm{B} = \{B_1 = (10, 0), B_2 = (-10, 0)\} \text{ with target } t(\mathbf{x}_i) < 0.5$$

In Figure 3.44 these data points are given together with the separating lines of the first and second input neurons. Looking in the direction of the arrows of these separating lines, the output of the neuron will be $y_1 > 0.5$ and $y_2 > 0.5$ to the right of these lines (the weight vectors are pointing in that direction) and < 0.5 on the other side. It will be clear that with these initial orientations of the separating lines, discrimination between points of data sets D_A and D_B is impossible.

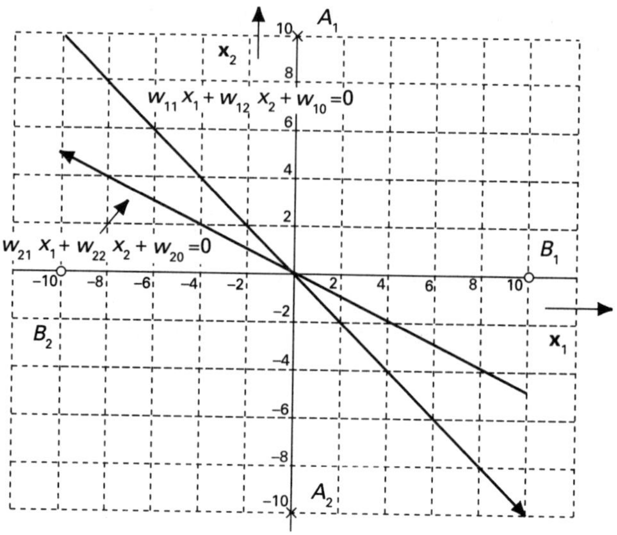

Figure 3.44 Data points and separating lines ($s=0$) of the first and second neuron in the first layer of Example 3.12

In Table 3.3 the values of $s_1(\mathbf{x}_i)$, $s_2(\mathbf{x}_i)$, $s_3(\mathbf{x}_i)$, $y_1(\mathbf{x}_i)$, $y_2(\mathbf{x}_i)$, $y_3(\mathbf{x}_i)$, $\mathrm{d}f_1/\mathrm{d}s$, $\mathrm{d}f_2/\mathrm{d}s$, $\mathrm{d}f_3/\mathrm{d}s_3$ and $t_3(\mathbf{x}_i)-y_3(\mathbf{x}_i)$ for the inputs of the data set are given.

For each input vector $\mathbf{x}_i$ there is a weighted input vector $\mathbf{s}_i$ obtained after the following transformation:

$$\begin{bmatrix} s_{i1} \\ s_{i2} \end{bmatrix} = \begin{bmatrix} w_{11} & w_{12} \\ w_{21} & w_{22} \end{bmatrix} \cdot \begin{bmatrix} x_{i1} \\ x_{i2} \end{bmatrix} + \begin{bmatrix} w_{10} \\ w_{20} \end{bmatrix}$$

In Figure 3.45 the data points after this transformation (in this case a linear transformation because w_{10} and w_{20} are zero) are represented in the *first-layer weighted input space* S_1. The weighted inputs $\mathbf{s}_i$ are subsequently transformed by a non-linear mapping:

$$\begin{bmatrix} y_{i1} \\ y_{i2} \end{bmatrix} = \begin{bmatrix} f_1(s_{i1}) \\ f_2(s_{i2}) \end{bmatrix}$$

In Figure 3.46 the data points, after this non-linear scalar mapping, are represented in the *first-layer output space* Y_1. In Figure 3.46 the separating line of the output neuron is also given.

From Figure 3.46 it will be clear that the transformed data points cannot be classified correctly by any separating line of the output neuron. The preceding transformation mentioned above must be changed first by changing the weights w_{10}, w_{11}, w_{12}, w_{20}, w_{21} and w_{22}.

Table 3.3

$\mathbf{x}_i$	s_1	y_1	s_2	y_2	s_3	y_3	df_1/ds	df_2/ds	df_3/ds	$t_3 - y_3$
(0, 10)	1	0.731	−1	0.269	−0.662	0.340	0.197	0.197	0.224	0.160
(0, −10)	−1	0.269	1	0.731	0.262	0.565	0.197	0.197	0.246	0.0
(10, 0)	1	0.731	−0.5	0.378	−0.554	0.365	0.197	0.235	0.232	0.0
(−10, 0)	−1	0.269	0.5	0.622	0.154	0.538	0.197	0.235	0.249	−0.038

First-layer weighted input space

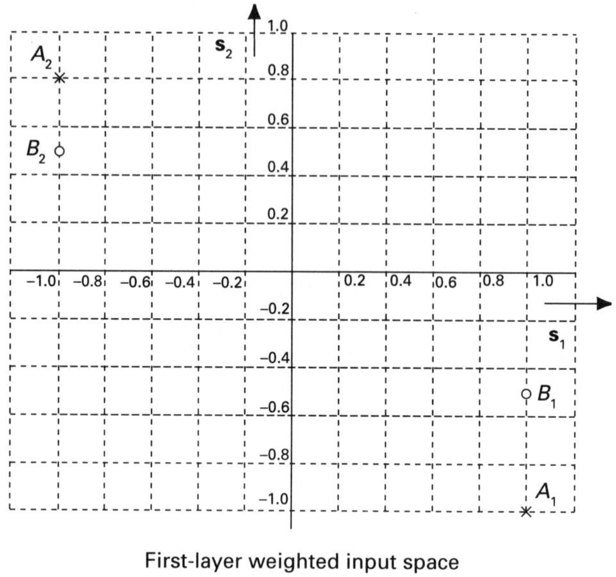

Figure 3.45 The first-layer weighted input space (s_1, s_2) of Example 3.12

First-layer output space

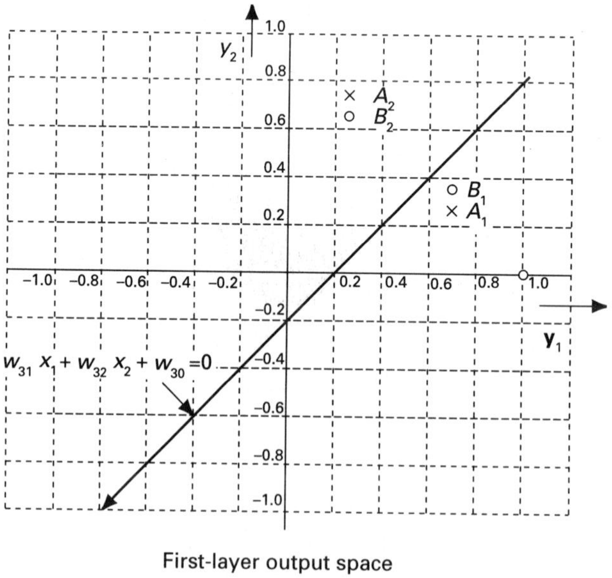

Figure 3.46 The first-layer output space (y_1, y_2) of Example 3.12

According to the formulas for adaptation of the weights of neuron 3, the weight vector of neuron 3 becomes (see also Table 3.3):

$$\begin{bmatrix}\Delta w_{30}\\ \Delta w_{31}\\ \Delta w_{32}\end{bmatrix}=\varepsilon(0.16)(0.224)\begin{bmatrix}1.000\\ 0.731\\ 0.269\end{bmatrix}+\varepsilon(-0.038)(0.249)\begin{bmatrix}1.000\\ 0.269\\ 0.622\end{bmatrix}=\varepsilon\begin{bmatrix}0.0264\\ 0.0237\\ 0.0156\end{bmatrix}$$

According to the formulas for adaptation of the weights of neuron 1, the weight vector of neuron 1 becomes:

$$\begin{bmatrix}\Delta w_{10}\\ \Delta w_{11}\\ \Delta w_{12}\end{bmatrix}=\varepsilon(0.16)(0.224)(-1)(0.197)\begin{bmatrix}1\\ 0\\ 10\end{bmatrix}+\varepsilon(-0.038)(0.249)(-1)(0.197)\begin{bmatrix}1\\ -10\\ 0\end{bmatrix}$$

Thus:

$$\begin{bmatrix}\Delta w_{10}\\ \Delta w_{11}\\ \Delta w_{12}\end{bmatrix}=\varepsilon\begin{bmatrix}-0.0052\\ -0.0186\\ -0.7060\end{bmatrix}$$

In the same way we will find for the adaptation of the weights of neuron 2:

$$\begin{bmatrix}\Delta w_{20}\\ \Delta w_{21}\\ \Delta w_{22}\end{bmatrix}=\varepsilon(0.16)(0.224)(1)(0.197)\begin{bmatrix}1\\ 0\\ 10\end{bmatrix}+\varepsilon(-0.038)(0.249)(1)(0.197)\begin{bmatrix}1\\ -10\\ 0\end{bmatrix}$$

Thus:

$$\begin{bmatrix}\Delta w_{20}\\ \Delta w_{21}\\ \Delta w_{22}\end{bmatrix}=\varepsilon\begin{bmatrix}0.0052\\ 0.0186\\ 0.0706\end{bmatrix}$$

After choosing some value of ε(e.g. $\varepsilon=0.5$) we can adapt the set of weights and repeat the whole procedure. After a great number of learning steps (e.g. 125) the MSE will become zero and one would expect the classification to be correct.

However, with the single threshold labelling method, a zero value for the MSE will not guarantee a correct classification. If at the final step of learning y_3 is a little bit smaller than 0.5 for the inputs A_1 and A_2, and a little bit greater than 0.5 for the inputs B_1 and B_2, then all inputs are wrongly classified, whereas the MSE is almost zero. For a correct classification one has to use the one–zero labelling method or the method of double threshold labelling by setting the target for the set D_A equal to, for example, 0.95 and for the set D_B equal to 0.05. ■

A general two-layer Perceptron may have n_1 neurons in the first layer and n_2 neurons in the second layer (see Figure 3.47). For the MSE for the finite set **U** of

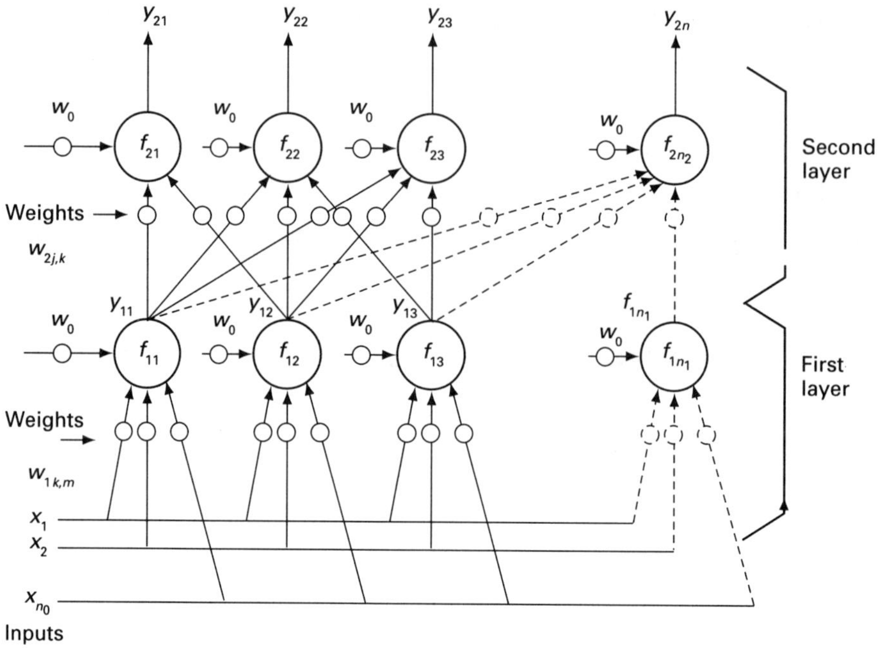

Figure 3.47 The general two-layer continuous Perceptron

input vectors of some given finite learning set **L** we have:

$$\mathbf{E}=\frac{1}{2}\sum_{\mathbf{x}_i\in \mathbf{U}}\sum_{j=1}^{n_2}\{t_{2j}(\mathbf{x}_i)-y_{2j}(\mathbf{x}_i)\}^2$$

According to Theorem 3.2 the learning implies for the adaptation of the weights connecting neuron k in the first layer to neuron j in the second layer:

$$\Delta w_{2j,k}=\varepsilon\sum_{\mathbf{x}_i\in \mathbf{U}}\{t_{2j}(\mathbf{x}_i)-y_{2j}(\mathbf{x}_i)\}\frac{\mathrm{d}f_{2j}}{\mathrm{d}s_{2j}}y_{1k}$$

If we represent all the outputs of the n_1 neurons in the first layer including the constant component $y_{10}=1$ with a vector $\hat{\mathbf{y}}_1$, then we can write, with the use of the extended weight vector $\hat{\mathbf{w}}_{2j}=[w_{2j,0}, w_{2j,1},\ldots, w_{2j,n_1}]^t$, for the adaptation of the weight vector $\hat{\mathbf{w}}_{2j}$ of output neuron j:

$$\Delta \hat{\mathbf{w}}_{2j}=\varepsilon\sum_{\mathbf{x}_i\in \mathbf{U}}\{t_{2j}(\mathbf{x}_i)-y_{2j}(\mathbf{x}_i)\}\frac{\mathrm{d}f_{2j}}{\mathrm{d}s_{2j}}\hat{\mathbf{y}}_1$$

With the notation $\delta_{2j}(\mathbf{x}_i)=t_{2j}(\mathbf{x}_i)-y_{2j}(\mathbf{x}_i)$ for the error of the output neuron j for an input $\mathbf{x}_i$, we can thus write the following theorem:

Theorem 3.7

The adaptation of the weight vector **w** of a neuron in the second layer of a two-layer Perceptron, in order to minimixe the MSE for a given set of n_2 target values $t_{2j}(\mathbf{x}_i)$ for a given finite set of input vectors **U**, is:

$$\Delta\hat{\mathbf{w}}_{2j} = \varepsilon \sum_{\mathbf{x}_i \in \mathbf{U}} \delta_{2j}(\mathbf{x}_i) \frac{\mathrm{d}f_{2j}}{\mathrm{d}s_{2j}} \hat{\mathbf{y}}_1$$

For the adaptation of the weight $w_{1k,m}$ connecting the net input x_m with neuron k in the first layer, we obtain according to Theorem 3.2:

$$\Delta w_{1k,m} = \varepsilon \sum_{\mathbf{x}_i \in \mathbf{U}} \sum_{j=1}^{n_2} \{t_{2j}(\mathbf{x}_i) - y_{2j}(\mathbf{x}_i)\} \frac{\mathrm{d}f_{2j}}{\mathrm{d}s_{2j}} w_{2j,k} \frac{\mathrm{d}f_{1k}}{\mathrm{d}s_{1k}} x_m$$

The sum of products over j can be considered as the error (from the output neurons back-propagated) of neuron k in the first layer. So with:

$$\delta_{1k}(\mathbf{x}_i) = \sum_{j=1}^{n_2} \{t_{2j}(\mathbf{x}_i) - y_{2j}(\mathbf{x}_i)\} \frac{\mathrm{d}f_{2j}}{\mathrm{d}s_{2j}} w_{2j,k}$$

we can write:

$$\Delta w_{1k,m} = \varepsilon \sum_{\mathbf{x}_i \in \mathbf{U}} \delta_{1k}(\mathbf{x}_i) \frac{\mathrm{d}f_{1k}}{\mathrm{d}s_{1k}} x_m$$

If we represent all the inputs of the network, including the constant component $x_0 = 1$, with a vector $\hat{\mathbf{x}}$, then we can write, with the use of the extended weight vector $\hat{\mathbf{w}}_{1k} = [w_{1k,0}, w_{1k,1}, \ldots, w_{1k,n_0}]^t$, the following theorem for the adaptation of the weight vector $\hat{\mathbf{w}}_{1k}$ of input neuron k:

Theorem 3.8

The adaptation of the weight vector $\hat{\mathbf{w}}$ of a neuron in the first layer of a two-layer Perceptron in order to minimize the MSE for a given set of n_2 target values $t_{2j}(\mathbf{x}_i)$ for a given finite set **U** of input vectors, is:

$$\Delta\hat{\mathbf{w}}_{1k} = \varepsilon \sum_{\mathbf{x}_i \in \mathbf{U}} \delta_{1k}(\mathbf{x}_i) \frac{\mathrm{d}f_{1k}}{\mathrm{d}s_{1k}} \hat{\mathbf{x}}_i$$

We define the *internal learning rate* for input vector $\mathbf{x}_i$ for the weights of neuron k in the first layer as:

$$\gamma_{1k}(\mathbf{x}_i) = \delta_{1k}(\mathbf{x}_i) \frac{\mathrm{d}f_{1k}}{\mathrm{d}s_{1k}}$$

With this definition we obtain for the adaptation of the weights of neuron k in the first layer due to input $\mathbf{x}_i$:

$$\Delta\hat{\mathbf{w}}_{1k} = \varepsilon \sum_{\mathbf{x}_i \in \mathrm{U}} \gamma_{1k}(\mathbf{x}_i) \cdot \hat{\mathbf{x}}_i$$

The sum of weighted – by the scalars $\gamma_{1k}(\mathbf{x}_i)$ – extended input vectors $\hat{\mathbf{x}}_i$ will be denoted by $\bar{\hat{\mathbf{x}}}$:

$$\bar{\hat{\mathbf{x}}} = \varepsilon \sum_{\mathbf{x}_i \in \mathrm{U}} \gamma_{1k}(\mathbf{x}_i) \cdot \hat{\mathbf{x}}_i$$

Thus we can write for the adaptation:

$$\Delta\hat{\mathbf{w}}_{1k} = \varepsilon \cdot \bar{\hat{\mathbf{x}}}$$

We see that after adaptation the weight vector of neuron k in the first layer will turn in the direction of the weighted sum $\hat{\mathbf{x}}$ of input vectors $\hat{\mathbf{x}}_i$. This implies that the separating hyperplane $\hat{\mathbf{w}}_{ik} \cdot \hat{\mathbf{x}} = 0$ in the extended input space, as well as the hyperplane $\mathbf{w} \cdot \mathbf{x} = 0$ in the non-extended input space, of neuron k will turn in a direction 'perpendicular' to the weighted sum of input vectors. For the two-dimensional case see Figure 3.48. [NB: the weighted input $s_{1k}(\mathbf{x}_i)$ of neuron k in the first layer is zero for an input $\mathbf{x}_i$ on that separating hyperplane.]

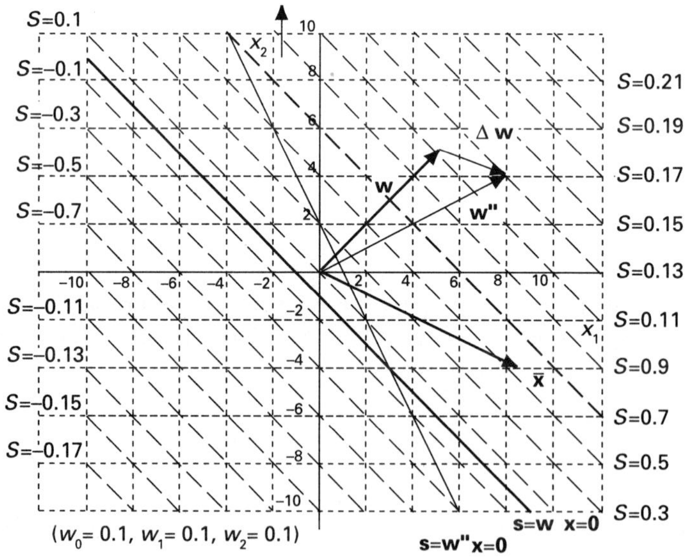

Figure 3.48 The input/weight space of a first-layer neuron with the adaptation $\Delta\mathbf{w}$ in the direction of the weighted sum $\bar{\mathbf{x}}$ of input vectors $\mathbf{x}_i$

The distance from the origin to the separating hyperplane is given by:

$$d = \frac{w_0}{|\mathbf{w}|}$$

Because of the adaptation of $\mathbf{w}$ and of the threshold $\Delta w_{1k,0} = \Sigma \gamma_{1k}(\mathbf{x}_i)$, there will be at the same time a translation of the separating hyperplane.

3.13 Under-fitting and over-fitting of a data set with a two-layer continuous Perceptron

A two-layer continuous Perceptron is frequently used to find a functional relationship behind a data set D of examples of pairs of arguments and function values: $D = \{(\mathbf{x}_1, t(\mathbf{x}_1)), ((\mathbf{x}_2, t(\mathbf{x}_2), \ldots, (\mathbf{x}_m, t(\mathbf{x}_m))\}$. However if we use a two-layer continuous Perceptron to approximate a given data set, then after learning, each output neuron j of the neural net will realize some function $g_{\mathbf{w}j}(\mathbf{x})$ that is *restricted* to the class of functions realizable by that net. The function $g_{\mathbf{w}j}(\mathbf{x})$ depends on the transfer functions of the neurons and the number of neurons in the first layer. So if one is going to accept the outcome of the neural net, one is *a priori* assuming that the functional relationship belongs to the restricted class of functions realizable by that particular configuration of the neural net.

If we use, for example, a neural net with two first-layer neurons and one output neuron and all neurons have a sigmoid transfer function (see Figure 3.49), then the class of functions with one argument value is restricted to types of the form given by y_{21} in Figures 3.50–3.52.

The weights for the neurons for the function y_{21} in Figure 3.50 are $w_{110} = -4$, $w_{111} = 1$, $w_{120} = 4$, $w_{121} = 1$, $w_{210} = 0$, $w_{211} = 1$ and $w_{212} = 1$. The weights for the neurons for the function y_{21} in Figure 3.51 are $w_{110} = 8$, $w_{111} = -2$, $w_{120} = -8$, $w_{121} = -2$, $w_{210} = 0$, $w_{211} = 1$ and $w_{212} = 1$. The weights for the neurons for the function y_{21} in Figure 3.52 are $w_{110} = 8$, $w_{111} = 2$, $w_{120} = -8$, $w_{121} = 2$, $w_{210} = 0$, $w_{211} = 1$ and $w_{212} = -1$.

Although we must conclude that the number of functions realizable by a two-layer net with a given number of first-layer neurons is restricted, we will, however, see in the next section that any continuous function can be approximated within a finite domain, up to any given accuracy, if we use a sufficient number of first-layer neurons.

Example 3.13

Assume we have a data set $D = \{-1, -0.5, 0.0, 0.5, 1\}$ with targets $t(-1) = 0$, $t(-0.5) = 1$, $t(0) = 0.5$, $t(0.5) = 0$ and $t(1) = 1$ generated by the unknown function $t(x) = 0.5 - 1.5x + 2x^3$. We want to find, with a two-layer Perceptron with one output neuron and two first-layer neurons, the unknown function behind the data set D. The examples of the data set are repeatedly presented to the learning neural net until

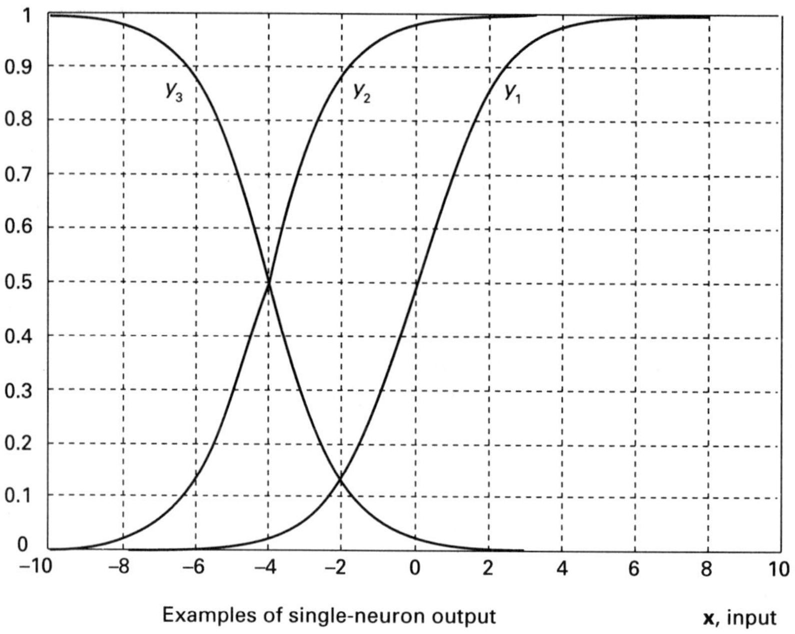

Figure 3.49 Reversed, translated and standard sigmoid functions

a minimum is reached for the MSE. Using the learning rules discussed before, we will find, depending on the initial random distribution of weights, three different functions $g_{\mathbf{w}}(\mathbf{x})$ realized by the neural net. The functions are presented in Figures 3.53–3.55 by the dashed lines. The solid line presents the function $t(x)=0.5-1.5x+2x^3$.

We see that we cannot find, even with more data points, the unknown function with the selected configuration of the neural net. ■

We found that if the number of first-layer neurons is too small, we cannot realize the approximate fitting of data points. On the other hand, we can choose too many neurons in the first layer. With a great number of neurons in the first layer the function realized by the neural net will go exactly through the data points (i.e. the MSE will be zero) but will fluctuate wildly in the intervals between the data points. We say the data points are *over-fitted.*

Example 3.14

We take the same data set as in Example 3.13: $D=\{-1, -0.5, 0.0, 0.5, 1\}$ with targets $t(-1)=0$, $t(-0.5)=1$, $t(0)=0.5$, $t(0.5)=0$ and $t(1)=1$ generated by the unknown function $t(x)=0.5-1.5x+2x^3$. Again we want to find with a two-layer Perceptron

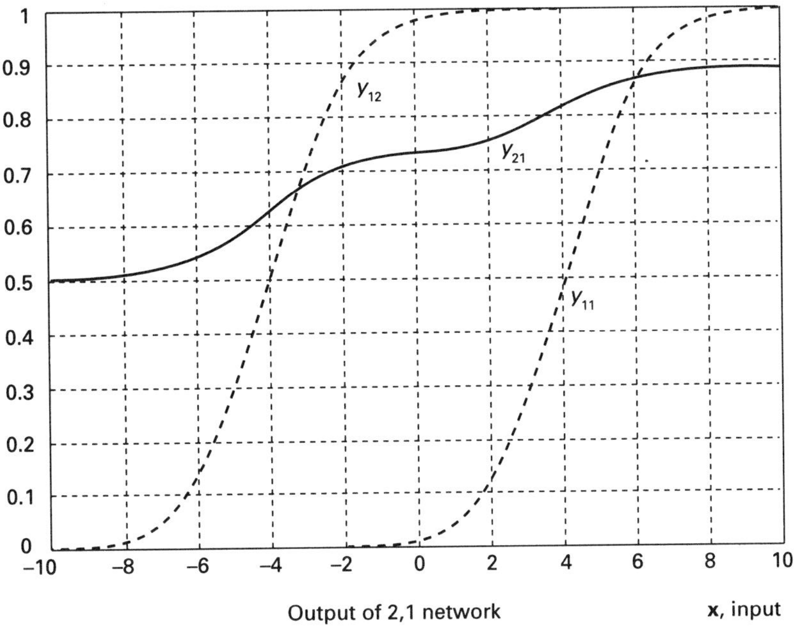

Figure 3.50 Example of the output y_{21} of a continuous Perceptron with two neurons in the first layer and one neuron in the second layer for a one-dimensional input

the unknown function behind the data set D. We take the number of first-layer neurons equal to the number of elements in the data set, i.e. five.

We will show that there exists a selection of weights such that the realized function fits almost exactly the data points. The MSE will then be zero.

For the fifth neuron in the first layer we select $w_{50} = -75$ and $w_{51} = 100$. This implies that the output y_5 of that neuron is almost equal to 1 for the fifth data point, 1, and will be zero for the other data points. For the fourth neuron we select $w_{40} = -25$ and $w_{41} = 100$. This implies that the output y_4 of that neuron is almost equal to 1 for the fourth and fifth data points, 0.5 and 1, and will be zero for the other data points. In the same way we select $w_{30} = 25$, $w_{31} = 100$, $w_{20} = 75$, $w_{21} = 100$, $w_{10} = 125$ and $w_{11} = 100$. In Table 3.4 the outputs for the first-layer neurons and the target value of the neuron in the second layer are given.

The outputs of the first layer are the inputs for the neuron in the second layer. If we select the weights for the output neuron in the second layer as $w_{20} = 0$, $w_{21} = -100$, $w_{22} = 200$, $w_{23} = -100$, $w_{24} = -100$ and $w_{25} = 200$, then the function (see Figure 3.56) produced by the neural network will go through the data points and the MSE will be zero but the function will fluctuate wildly in the interval between the data points. We can use the same method even if the number of data points is very large. ■

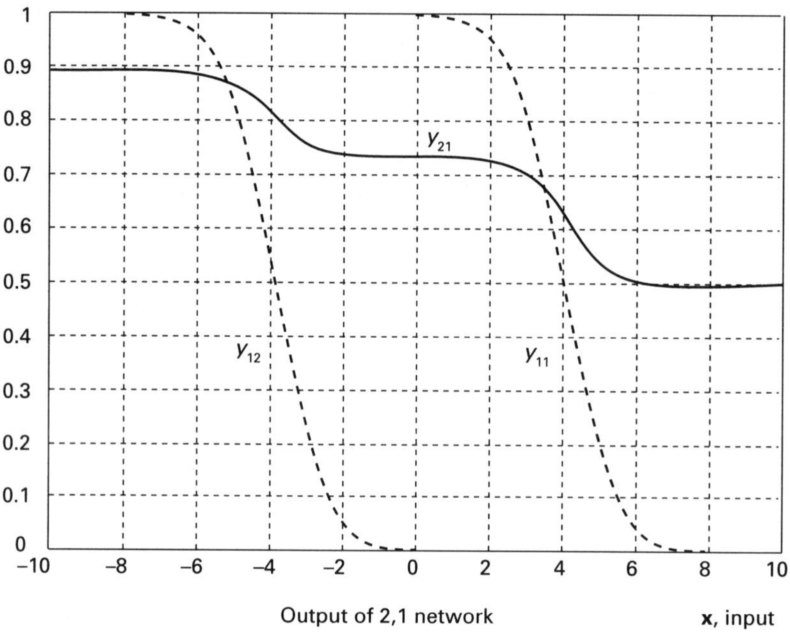

Figure 3.51 Example of the output y_{21} of a continuous Perceptron with two neurons in the first layer and one neuron in the second layer for a one-dimensional input

Table 3.4

x	y_1	y_2	y_3	y_4	y_5	t	$f^{-1}(t)$
−1	1	0	0	0	0	0	$-\infty$
−0.5	1	1	0	0	0	1	$+\infty$
0	1	1	1	0	0	0.5	0
0.5	1	1	1	1	0	0	$-\infty$
1	1	1	1	1	1	1	$+\infty$

From the discussion above we can conclude the following:

Practical statement 3.10

If we want to infer from a given data set the unknown functional relationship and select the number of first-layer neurons too low, then the unknown function will be under-fitted (will not go through all data points); if we select the number of neurons too high, then the unknown function will be over-fitted (the realized function will go through the data points but will fluctuate wildly in between).

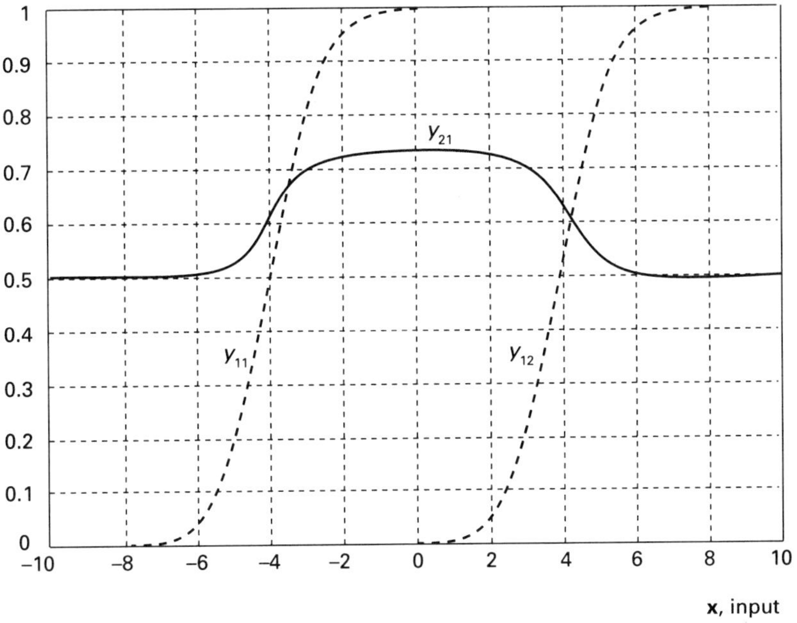

Figure 3.52 Example of the output y_{21} of a continuous Perceptron with two neurons in the first layer and one neuron in the second layer.

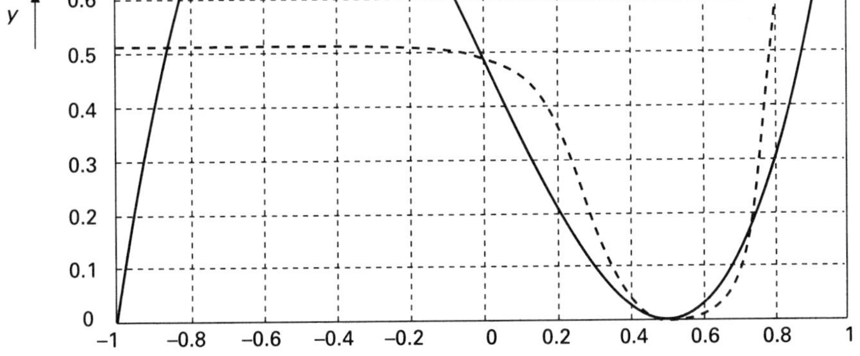

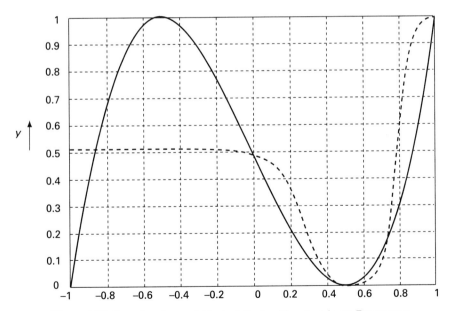

Figure 3.53 Function learned (dashed line) with a two-layer Perceptron.

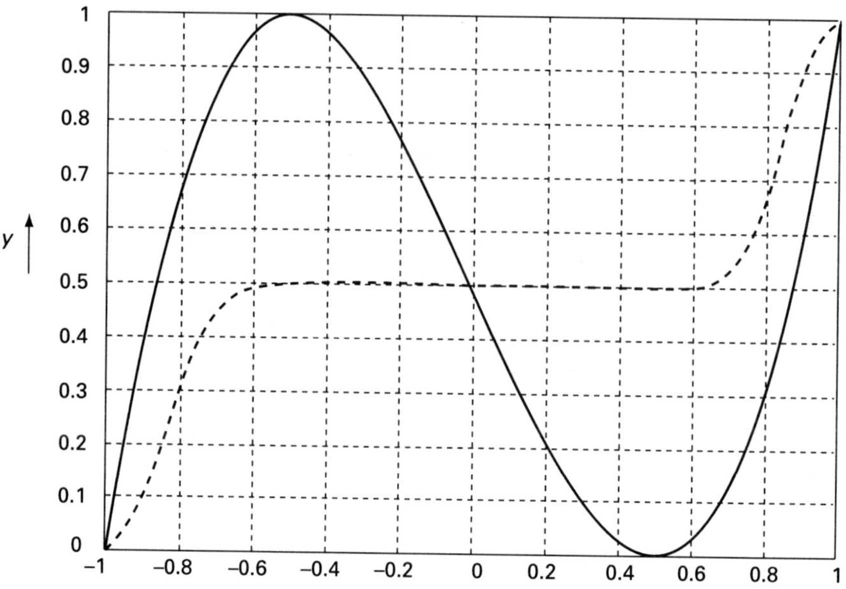

Figure 3.54 Function learned (dashed line) with a two-layer Perceptron.

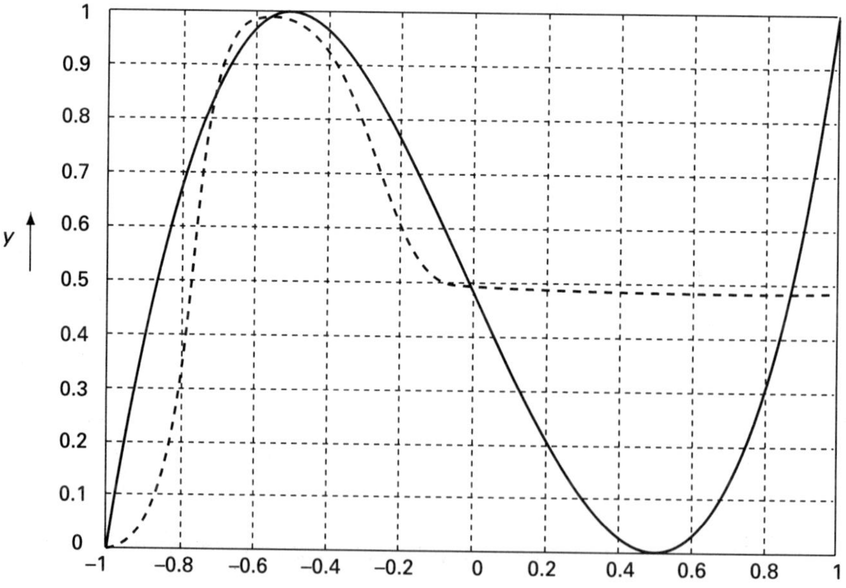

Figure 3.55 Function learned (dashed line) with a two-layer Perceptron.

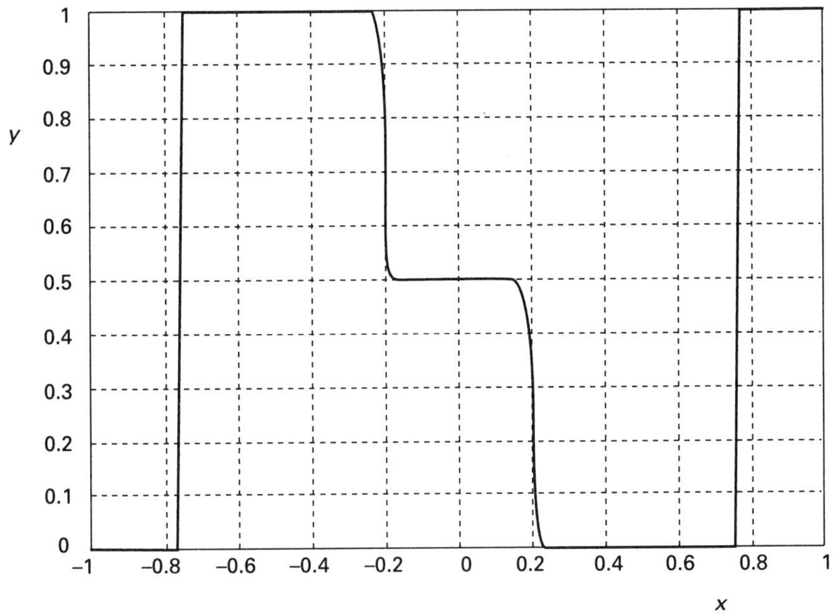

Figure 3.56 Function that can be realized by a continuous Perceptron with five neurons in the first layer and one output neuron that will be over-fitting the data of Example 3.13

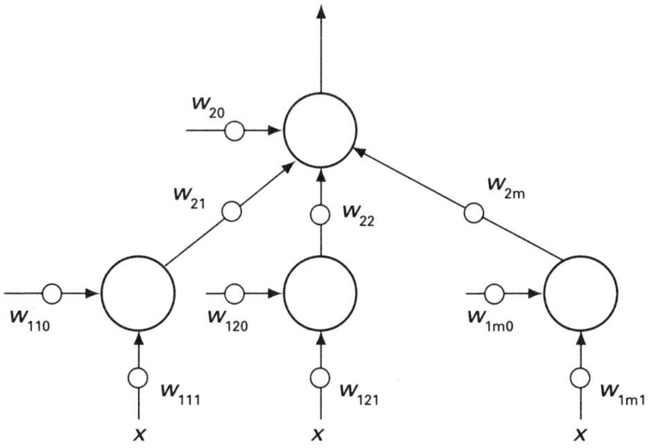

Figure 3.57 Two-layer continuous Perceptron with *m* first-layer neurons and one linear output neuron

3.14 The class of functions realizable with a two-layer Perceptron

In the previous section we found that the function realized with a two-layer continuous Perceptron will not go through the data point of the learning set if the number of first-layer neurons is too small; if, however, the number of first-layer neurons is too large, the function will go through the data point but will wildly fluctuate in between.

In this section we will show that a two-layer continuous Perceptron with a sufficient number of neurons in the first layer can approximate any continuous function arbitrarily well. This implies that there is no theoretical argument to use a Perceptron with more than two layers to identify some unknown function. We will, however, see in Section 3.15 that in some cases it might be more profitable to use a three-layer Perceptron because the total number of neurons may be less.

The basic idea behind the statement above is that we can approximate any continuous function with a Taylor series, and on the other hand we can also approximate the function realized by a two-layer Perceptron with a Taylor series. We will see that we can modify the coefficients of the Taylor series of the Perceptron by selecting appropriate weights in such a way that the coefficients of the Taylor series of the function realized by the Perceptron will be equal to the coefficients of the Taylor series of the given continuous function. We consider first the one-dimensional case.

Let $f(\mathbf{x})$ be a continuous function from $\mathbb{R}$ to $\mathbb{R}$, which is continuously differentiable round a certain point x_0. There then exists a unique series of reals a_i $(i=0, 1, 2, \ldots)$ for which:

$$f(\mathbf{x})=\sum_{i=0}^{n} a_i(x-x_0)^i+R_n$$

with $a_i=f(x_0)^i/i!$ with $f(x_0)^i$ the ith derivative of f in x_0, and R_n the remainder term with $\lim_{n\to\infty} R_n=0$.

According to Taylor's theorem, for every function f and for every domain $[x_0, x_1]$ and for every ε there exists an n such that:

$$\max_{x\in[x_0,\, x_1]}\left|f(x)-\sum_{i=0}^{n} a_i(x_0-x)^i\right|<\varepsilon$$

In other words we can approximate any function f in any domain $[x_0, x_1]$ arbitrarily well with the series expansion.

The same holds for the p-dimensional case with f a function from $\mathbb{R}^p\to\mathbb{R}$. The Taylor series in that case will be:

$$f(\mathbf{x})=\sum_{i=0}^{n}\frac{1}{i!}[(\mathbf{x}-\mathbf{x}_0)\cdot\nabla]^i f(\mathbf{x}_0)+R_n$$

with:

$$[(\mathbf{x}-\mathbf{x}_0)\cdot\nabla]^0 f(\mathbf{x}_0)=f(\mathbf{x}_0)$$

$$[(\mathbf{x}-\mathbf{x}_0)\cdot\nabla]^1 f(\mathbf{x}_0)=(\mathbf{x}-\mathbf{x}_0)\frac{\partial f}{\partial x_1}(x_0)+\cdots+(\mathbf{x}-\mathbf{x}_0)\frac{\partial f}{\partial x_p}(x_0)$$

$$[(\mathbf{x}-\mathbf{x}_0)\cdot\nabla]^k f(\mathbf{x}_0)=[(\mathbf{x}-\mathbf{x}_0)\cdot\nabla][(\mathbf{x}-\mathbf{x}_0)\cdot\nabla]^{k-1} f(\mathbf{x}_0)$$

and $\lim_{n\to\infty} R_n=0$.

We now return to the one-dimensional case and consider the function $g^m_{\mathbf{w}}(x)$ from $\mathbb{R}$ to $\mathbb{R}$, realized by a continuous Perceptron with m neurons in the first layer. Each neuron i in the first layer has a sigmoid transfer function:

$$f_{1i}[s(x)]=[1+\exp(w_{1i0}+w_{1i1}x)]^{-1}$$

with w_{1i0} the weight of the ith first-layer neuron connected to the constant input $x_0=1$, and w_{1i1} the weight of the ith first-layer neuron connected to input $x_1=x$ (see Figure 3.57).

We take one output neuron with a linear transfer function:

$$f_2(s)=s$$

with $s=w_{20}+\Sigma\, w_{2i}f_{1i}$ the weighted input of the output neuron. (We could take a non-linear transfer function as well, but it would only complicate our discussion.)

The function $g^m_{\mathbf{w}}(x)$ realized by the neural net will have the following form:

$$g^m_{\mathbf{w}}(x)=w_{20}+\sum_{i=1}^{m} w_{2i}f_{1i}(w_{1i0}+w_{1i1}x)$$

The Taylor expansion of this function round a point x_0 will be:

$$g^m_{\mathbf{w}}(x)=\sum_{i=1}^{n} a'_i(x-x_0)+R'_n$$

with

$$a'_0=g^m_{\mathbf{w}}(x_0)$$

$$a'_1=\frac{1}{1!}\frac{\mathrm{d}}{\mathrm{d}x}g^m_{\mathbf{w}}(x_0)$$

$$a'_2=\frac{1}{2!}\frac{\mathrm{d}^2}{\mathrm{d}x^2}g^m_{\mathbf{w}}(x_0)$$

etc.

with $g^m_{\mathbf{w}}(x_0)=w_{20}+\Sigma\, w_{2i}f_{1i}[s(x_0)]$; in short $g^m_{\mathbf{w}}=w_{20}+\Sigma\, w_{2i}f_{1i}$ and $\mathrm{d}f/\mathrm{d}s=f(1-f)$, and we obtain:

$$a'_0 = w_{20} + \sum_{i=1}^{m} w_{2i} f_{1i}$$

$$a'_1 = \frac{1}{1!} \sum_{i=1}^{m} w_{2i} \frac{\mathrm{d}f_{1i}}{\mathrm{d}s_{1i}} \frac{\mathrm{d}s_{1i}}{\mathrm{d}x} = \sum_{i=1}^{m} w_{2i} \{f_{1i} - f_{1i}^2\} w_{1i1}$$

$$a'_2 = \frac{1}{2!} \sum_{i=1}^{m} w_{2i} \frac{\mathrm{d}^2 f_{1i}}{\mathrm{d}s_{1i}^2} \frac{\mathrm{d}s_{1i}}{\mathrm{d}x} = \sum_{i=1}^{m} w_{2i} \{f_{1i} - 3f_{1i}^2 + 2f_{1i}^3\} w_{1i1}$$

etc.

Given for some n the Taylor approximation of a given function:

$$f(x) = \sum_{i=0}^{n} a_i (x - x_0)^i + R_n$$

we can select the weights of the Perceptron w_{2i}, w_{1i0} and w_{1i1} for $i = 1, 2, \ldots, m$ such that $a_j = a'_j$ for $j = 0, 1, 2, \ldots, n$.

If the number m of first-layer neurons is equal to the number of $n+1$ Taylor coefficients minus one, i.e. $m = n$, then we can take the weights w_{1i0} and w_{1i1} of the first-layer neurons almost arbitrarily, because we can still select the $m+1$ weights $w_{20}, w_{21}, \ldots, w_{2m}$ of the output neuron such that for the $n+1$ Taylor coefficients we have $a_j = a'_j$ for $j = 0, 1, 2, \ldots, n$.

We can put the requirement $a_j = a'_j$ and the expressions for a'_j in a matrix form:

$$\mathbf{a} = A\mathbf{w}_2$$

For $n = 3$ and $m = 2$ we obtain:

$$\begin{bmatrix} a_0 \\ a_1 \\ a_2 \end{bmatrix} = \begin{bmatrix} 1 & f_{11} & f_{12} \\ 0 & (f_{11} - f_{11}^2) w_{111} & (f_{12} - f_{12}^2) w_{121} \\ 0 & (f_{11} - 3f_{11}^2 + 2f_{11}^3) w_{111} & (f_{12} - 3f_{12}^2 + 2f_{12}^3) w_{121} \end{bmatrix} \cdot \begin{bmatrix} w_{20} \\ w_{21} \\ w_{22} \end{bmatrix}$$

The weights in the connections from the first layer to the output neuron can be found from:

$$\mathbf{w}_2 = A^{-1}\mathbf{a}$$

The matrix A can always be made non-singular by choosing suitable values of the weights w_{1i0} and w_{1i1} for $i = 1, 2, \ldots, m$. For a certain x_0 and w_{1i1}, the value of f_{1i}, occurring in matrix A, can be assigned any value between 0 and 1 by choosing w_{1i0}. Thus we can always make det $A \neq 0$.

The conclusion is that we can approximate any continuous function $f\colon \mathbb{R} \to \mathbb{R}$ in any domain $[x_0, x_1]$ arbitrarily well by a two-layer Perceptron with n input neurons and one linear output neuron if the function f is approximated by the first $n+1$ Taylor coefficients. The same holds for the p-dimensional case. The proof is similar and is based on the introduction of the $n(p-1)$ additional weights w_{1ip} which we can freely select. Note that for a close approximation the number of required input neurons can become quite large. For a full proof see Tromp (1993). Thus we have the following theorem:

Theorem 3.9

A two-layer continuous Perceptron with sigmoid transfer function for the neurons in the first layer and one linear neuron in the second layer can approximate any continuous function f: $\mathbb{R}^n \rightarrow \mathbb{R}$ in any domain with any given accuracy.

Although it is important to know that a two-layer Perceptron can approximate any continuous function and that we can calculate the required weights with the method described above, the weights obtained after learning with the descending gradient method from samples of the function will not be the same as the weights found with the method described above. With the Taylor series expansion the realized function will approximate the (known!) function very closely in the neighbourhood of the point x_0 used in the expansion, whereas the realized function obtained after learning will approximate the (unknown!) function over the entire domain interval of applied samples such that the MSE becomes as small as possible.

Example 3.15

Let the function we want to approximate be $f(x)=2x-x^2$. With the method described above we want to calculate the weights of a two-layer Perceptron. We approximate the function with the first three Taylor coefficients, so we need a Perceptron with two neurons in the first layer. We choose to approximate $f(x)$ around $x_0=1$. The first three Taylor coefficients are $a_0=1$, $a_1=0$ and $a_2=-1$.

We select the weights of the two first-layer neurons arbitrarily (we have to check that the matrix A is not singular):

$$w_{110}=0.3$$

$$w_{111}=0.1$$

$$w_{120}=0.4$$

$$w_{121}=0.2$$

For $x_0=1$ the values of the outputs of the first-layer neurons become:

$$f_{11}=0.690$$

$$f_{12}=0.731$$

The matrix:

$$A=\begin{bmatrix} 1 & f_{11} & f_{12} \\ 0 & (f_{11}-f_{11}^2)w_{111} & (f_{12}-f_{12}^2)w_{121} \\ 0 & (f_{11}-3f_{11}^2+2f_{11}^3)w_{111} & (f_{12}-3f_{12}^2+2f_{12}^3)w_{121} \end{bmatrix}$$

becomes

$$A = \begin{bmatrix} 1 & 0.690 & 0.731 \\ 0 & 0.064 & 0.188 \\ 0 & -0.004 & 0.016 \end{bmatrix}$$

For the weights connected with the output neuron we find with $\mathbf{w} = A^{-1}\mathbf{a}$:

$$w_{20} = \quad 56.788$$

$$w_{21} = -190.867$$

$$w_{22} = \quad 103.830$$

The function $g_{\mathbf{w}}(x)$ realized by the Perceptron with the weights given above is shown in Figure 3.58 together with the function $f(x)$.

The same configuration of the network was used to learn the function with the back-propagation gradient descent learning rule. The inputs were randomly chosen with a uniform distribution between -1 and 3. The initial weights were chosen at

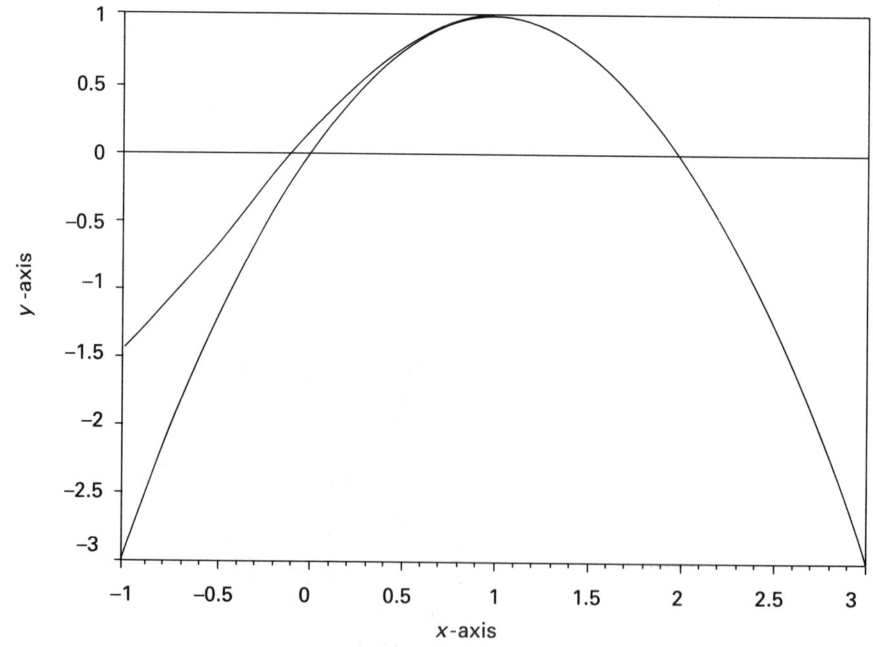

Figure 3.58 The function realized (upper curve) with a two-layer continuous Perceptron with two first-layer neurons and one linear output neuron and calculated weights with the Taylor approximation of $y = 2x - x^2$

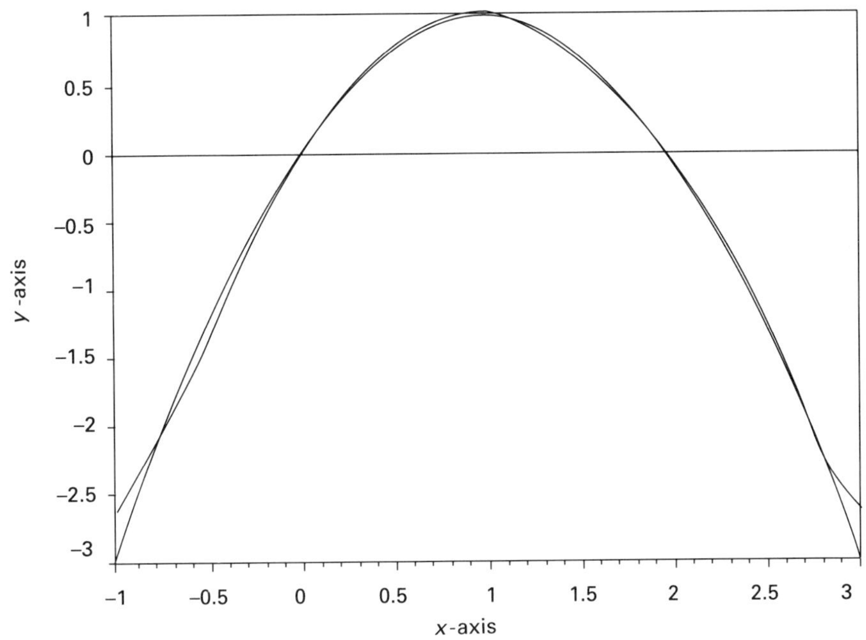

Figure 3.59 The function realized (upper curve) with a two-layer continuous Perceptron with two first-layer neurons and one linear output neuron learned with back-propagation with sampled data from the function $y = 2x - x^2$

random between -0.1 and 0.1. After 10 000 epochs (4.56 sec) the function had the form as shown in Figure 3.59. ■

3.15 The three-layer continuous Perceptron

Although a two-layer Perceptron can approximate any continuous function arbitrarily well, it may be profitable sometimes to use a three-layer Perceptron because the required number of neurons may be less, especially when the function to be approximated is expected to contain discontinuities.

We will show, for the one-dimensional case, that if a continuous or discontinuous function can be approximated by a piecewise linear function, it can be approximated by a three-layer Perceptron.

Assume we have a function $g(x)$ from $\mathbb{R}$ to $\mathbb{R}$ with function values between zero and one. We will see that this latter restriction will be eliminated. We divide the domain in intervals $[x_k, x_{k+1}]$ with $k = 0, 1, \ldots, N$, such that $[g(x_{k+1}) - g(x_k)]/(x_{k+1} - x_k)$ is finite, and in every interval we replace $g(x)$ by a linear function that goes through

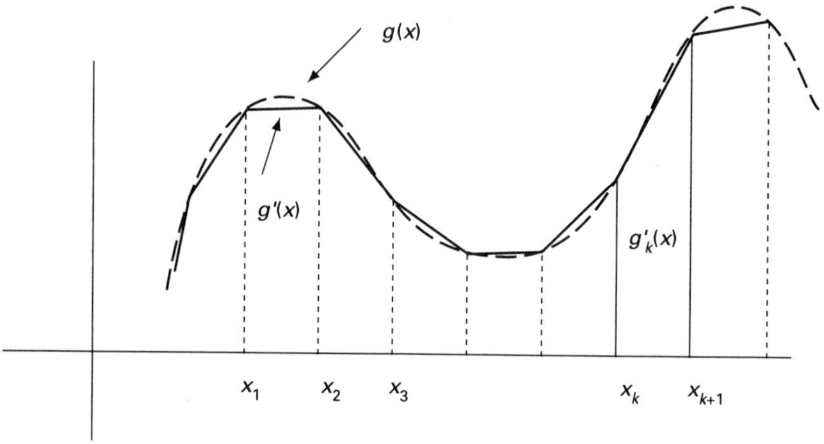

Figure 3.60 Piecewise linear approximation of some function $g(x)$

$g(x_k)$ and $g(x_{k+1})$. In this way we obtain a piecewise linear function $g'(x)$ that approximates $g(x)$ (see Figure 3.60). The length of the interval $x_{k+1}-x_k$ will be denoted by Δ_k. The approximation can be improved by making the interval Δ_k smaller. We will denote the approximating function in the domain $[x_k, x_{k+1}]$ by $g_k'(x)$, and so we can write:

$$g'(x)=\sum_{k=1}^{N} g_k'(x)$$

We first show that the functions $g_k'(x)$ with $k=1, 2, \ldots, N$ can be realized by a two-layer net with four neurons in the first layer and one neuron in the second layer. We will see that the sum Σ can be realized by one linear neuron in the third layer.

To realize a function $g_k'(x)$ (see Figure 3.61) we use the network configuration of Figure 3.62. All neurons have a sigmoid transfer function f. The first neuron realizes a decreasing step function at the point x_k (see Figure 3.63). We select w_{110} and w_{111} such that $-w_{110}/w_{111}=x_k$ and w_{111} is given a large negative value, e.g. $w_{111}=-100$. Thus the output y_{11} of the first neuron in the first layer will be equal to 1 for $x<x_k$ and will be 0 for $x>x_k$. The output is connected by a large negative weight w_{21} (e.g. -100) to the neuron in the second layer such that the output neuron in the second layer will always be 0 for $x<x_k$. The second neuron in the first layer realizes a rising step function (see Figure 3.63a). We select its weights such that $-w_{120}/w_{121}=x_{k+1}$ and we make w_{121} large and positive, e.g. $w_{121}=100$. Thus the output of neuron 2 in the first layer will 0 for $x<x_{k+1}$, and will be 1 for $x>x_{k+1}$. The output of this second neuron is connected by a large negative weight w_{22} to the neuron in the second layer such that the output neuron will always be zero for $x>x_{k+1}$.

The third neuron is used to realize the correct slope of $g_k'(x)$ (see Figure 3.63b). We select w_{130} and w_{131} such that $-w_{130}/w_{131}=(x_{k+1}+x_k)/2=\tilde{x}$.

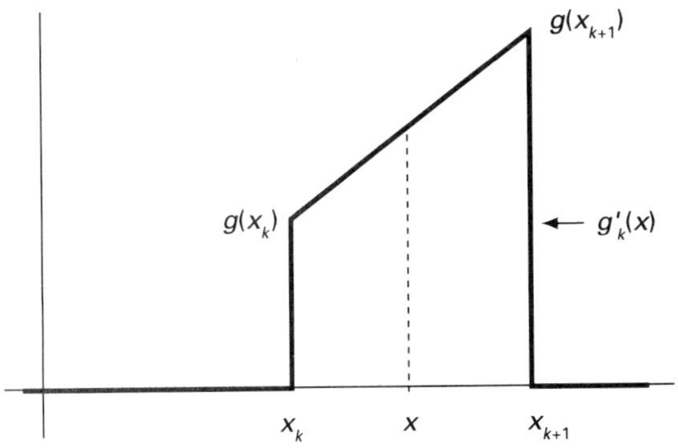

Figure 3.61 The function $g'_k(x)$

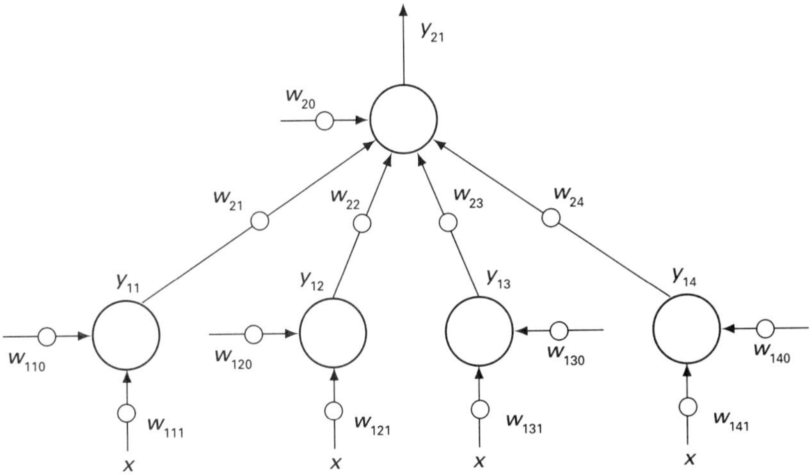

Figure 3.62 The network configuration to realize $g'_k(x)$

With the fourth neuron we can adjust the desired level of $g'_k(x)$ (see Figure 3.63c). We select w_{140} and w_{141} such that $-w_{140}/w_{141} < x_k$ and w_{141} has a large value. Thus neuron 4 realizes a rising step function at $x = -w_{140}/w_{141}$ and hence $y_{14} = 1$ in the domain $[x_k, x_{k+1}]$.

For the output y_{21} of the neuron in the second layer we obtain:

$$y_{21} = f_{21}(w_{20} + w_{21}y_{11} + w_{22}y_{12} + w_{23}y_{13} + w_{24}y_{14})$$

We select $w_{20} = 0$. Within the domain $[x_k, x_{k+1}]$, y_{11} and y_{12} are constant 0 and y_{14} is constant 1; thus with $y_{13} = f_{13}(w_{130} + w_{131}x)$ we obtain:

$$y_{21} = f_{21}[w_{23}f_{13}(w_{130} + w_{131}x) + w_{24}]$$

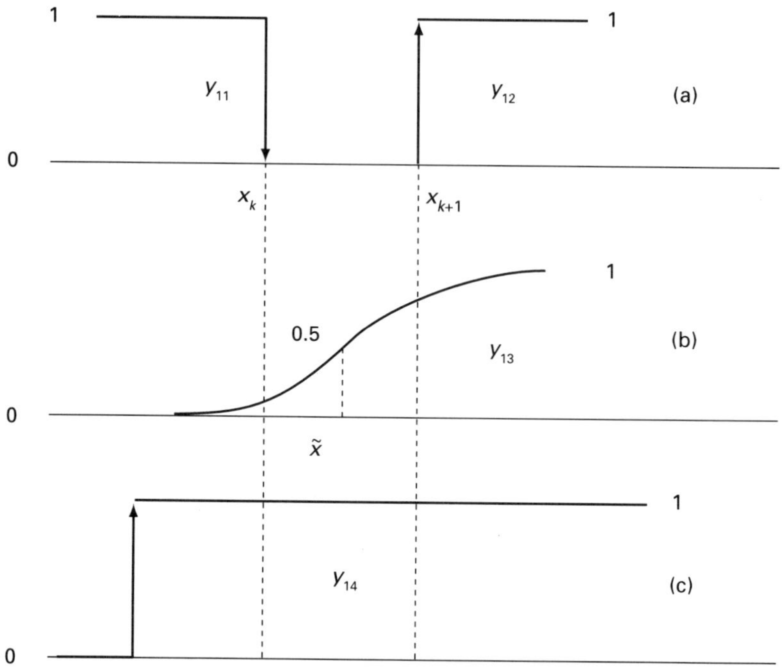

Figure 3.63 (a) The decreasing step function realized by the first neuron and increasing step function realized by the second neuron. (b) The output of the third neuron realizing the slope of $g'_k(x)$. (c) The output of the fourth neuron to adjust the level of $g'_k(x)$

If Δ is small enough we can approximate y_{21} around $\tilde{x}=(x_k+x_{k+1})/2$ by:

$$y_{21}(x)=y_{21}(\tilde{x})+\frac{dy_{21}}{dx}(x-\tilde{x})$$

At $\tilde{x}=-w_{130}/w_{131}$ we have $f_{13}(w_{130}+w_{131}x)=0.5$, thus:

$$y_{21}(\tilde{x})=f_{21}(w_{23}0.5+w_{24})$$

Because $y_{21}(\tilde{x})=g'_k(\tilde{x})=[g(x_k)+g(x_{k+1})]/2$ we obtain for the values of w_{23} and w_{24} the requirement:

$$f_{21}^{-1}(g'_k(\tilde{x}))=0.5w_{23}+w_{24}$$

For the derivative of the function realized by the neural net at $\tilde{x}$, we have:

$$\frac{dy_{21}}{dx}(\tilde{x})=\frac{df_{21}}{ds_{21}}\frac{ds_{21}}{dy_{13}}\frac{df_{13}}{ds_{13}}\frac{ds_{13}}{dx}$$

With $df_{21}/ds_{21} = f_{21}(1-f_{21})$, $s_{21} = w_{23}y_{13} + w_{24}$, $df_{13}/ds_{13} = 0.25$ and $s_{13} = w_{131}x$ we obtain:

$$\frac{dy_{21}}{dx}(\tilde{x}) = f_{21}(1-f_{21})w_{23}0.25w_{131}$$

Because we require that $y_{21} = f_{21}(\tilde{x}) = g'_k(\tilde{x})$ and

$$\frac{dy_{21}}{dx} = \frac{dg'_k}{dx}$$

we obtain for the values of w_{23} and w_{131} the requirement:

$$\frac{dg'_k}{dx} = 0.25g'_k(\tilde{x})[1-g'_k(\tilde{x})]w_{23}w_{131}$$

with $dg'_k/dx = [g(x_{k+1}) - g(x_k)]/\Delta$ and $g'_k(\tilde{x}) = [g(x_k) + g(x_{k+1})]/2$.

In conclusion we have for the values of the weights of neuron 3 and for w_{23} the requirements:

$$-w_{130}/w_{131} = (x_{k+1} + x_k)/2 = \tilde{x}$$

$$w_{23}w_{131} = \frac{4dg'_k/dx}{g'_k(\tilde{x})[1-g'_k(\tilde{x})]}$$

For the values of the weights of the neuron in the second layer we have the requirements:

$$w_{20} = 0$$

$$w_{21} < -10$$

$$w_{22} < -10$$

$$0.5w_{23} + w_{24} = f_{21}^{-1}(g'_k(\tilde{x}))$$

Example 3.16

Let $g'_k(x)$ be as shown in Figure 3.64 with $x_k = 5$, $x_{k+1} = 6$, $g(x_k) = 0.87$ and $g(x_{k+1}) = 0.73$.

We obtain $\tilde{x} = 5.5$ and $g'_k(\tilde{x}) = 0.8$. From Figure 3.24 we find $f^{-1}(g'_k(\tilde{x})) = 1.5$. For the derivative $dg'_k/dx = [g(x_{k+1}) - g(x_k)]/\Delta$ we obtain $dg_k/dx = -0.14$.

For the first neuron in the first layer we select $w_{111} = -200$. The requirement $-w_{110}/w_{111} = x_k$ is satisfied by $w_{110} = 1000$.

For the second neuron in the first layer we select $w_{121} = 200$. The requirement $-w_{120}/w_{121} = x_{k+1}$ is satisfied by $w_{120} = -1200$.

For the moment we skip the third neuron in the first layer.

For the fourth neuron in the first layer we have the requirements $-w_{140}/w_{141} < x_k$ and w_{141} large and positive. We select $w_{140} = -400$ and $w_{141} = 100$.

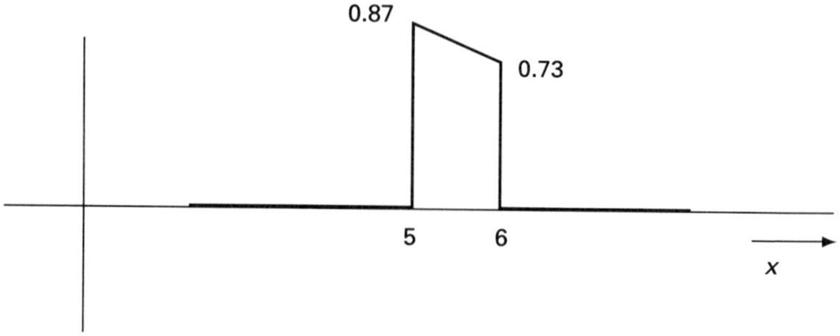

Figure 3.64 The function $g'_k(x)$ of Example 3.16 to be produced by the network configuration of Figure 3.62

For the neuron in the output layer we have the requirement:

$$w_{20}=0$$

$$w_{21}<-10$$

$$w_{22}<-10$$

$$0.5w_{23}+w_{24}=f_{21}^{-1}(g'_k(\tilde{x}))=1.5$$

We select $w_{20}=0$, $w_{21}=-100$, $w_{22}=-100$, $w_{23}=4$ and $w_{24}=-0.5$.

Finally we have for the requirements of the third neuron in the first layer:

$$-w_{130}/w_{131}=(x_{k+1}+x_k)/2=\tilde{x}=5.5$$

$$w_{23}w_{131}=\frac{4\mathrm{d}g'_k/\mathrm{d}x}{g'_k(\tilde{x})[1-g'_k(\tilde{x})]}=\frac{4(-0.14)}{0.8(1-0.8)}=-3.5$$

Because $w_{23}=4$ we obtain $w_{131}=-0.875$, and because of the first requirement we obtain $w_{130}=4.8125$.

A simulation of the neural net with the weights selected will give the function as shown in Figure 3.65. ■

The original function $g(x)$ was assumed to be approximated by a piecewise function $g'(x)$, which was constructed out of a sum of functions $g'_k(x)$:

$$g'(x)=\sum_{k=1}^{N} g'_k(x)$$

The summation can be realized by one linear neuron in a third layer:

$$g'(x)=w_{30}+\sum_{k=1}^{N} w_{3k}g'_k(x)$$

If the minimum value of original function $g(x)$ to be approximated is $g_{\min}$ and its maximum value is $g_{\max}$, we select $w_{30}=g_{\min}$ and for all k we select $w_{3k}=g_{\max}-g_{\min}$.

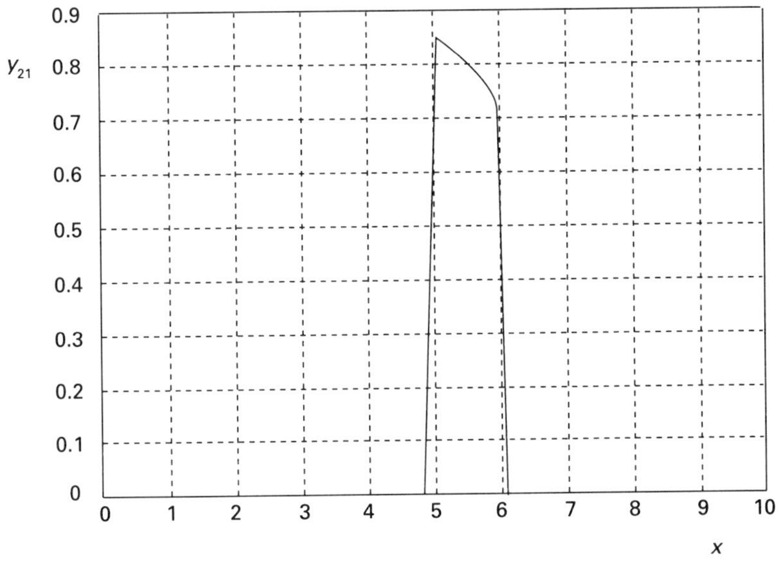

Figure 3.65 The function produced by the network of Figure 3.62 with the weights calculated in Example 3.16

We leave the proof for the case of a n-dimensional input to the reader. Our conclusion is the following theorem:

Theorem 3.10

Every function that can be approximated arbitrarily well by a piecewise linear function can be realized by three-layer continuous Perceptron with one linear neuron in the output layer.

3.16 Application of a two-layer continuous Perceptron to function identification

We found in Section 3.13 that if we want to identify, with a continuous two-layer Perceptron, some function from samples of that function, the number of first-layer neurons is critical. If we select the number to be too small, the function will be under-fitted, i.e. the function realized after learning will not go through the samples of the data set. If we select the number of first-layer neuron to be too large, the function will be over-fitted, i.e. the realized function will go through the samples of the data set but will fluctuate wildly in between. The last phenomenon will be observed when we determine the MSE for a test set much larger than the learning set.

We performed experiments with a learning set of 225 equally spaced samples from the domain $[-1, 1]*[-1, 1]$ of the non-linear two-dimensional function (see Figure 3.66):

$$g(x_1, x_2) = 0.125 + 0.125x_1 + 0.375x_1x_2 + 0.125x_2^2$$

In the first experiment we used a continuous Perceptron with two neurons in the first layer and one neuron in the second layer. All neurons had a sigmoid transfer function. After training we determined the MSE for 20 000 points uniformly chosen from the domain. We found a MSE of 0.014 11.

In a second experiment we used four neurons in the first layer. The MSE for the same test set turned out to be considerably smaller: MSE = 0.005 52. In the next experiment we used eight neurons in the first layer. After learning we found MSE = 0.006 11. Apparently the function was over-fitted. In a final experiment we used sixteen neurons in the first layer. The MSE turned out to be 0.007 56.

From these experiments we conclude that with respect to the MSE, under-fitting is much worse than over-fitting.

3.17 Application of a two-layer Perceptron to the mushroom classification problem

In Section 3.1 we promised to show that a continuous Perceptron can learn to classify two classes of mushrooms from samples.

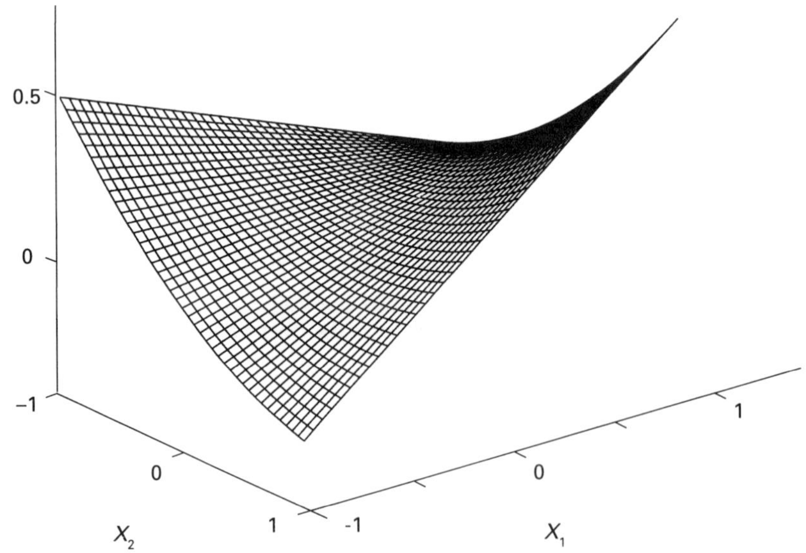

Figure 3.66 The function $g(x_1, x_2) = 0.125 + 0.125x_1 + 0.375x_1x_2 + 0.125x_2^2$

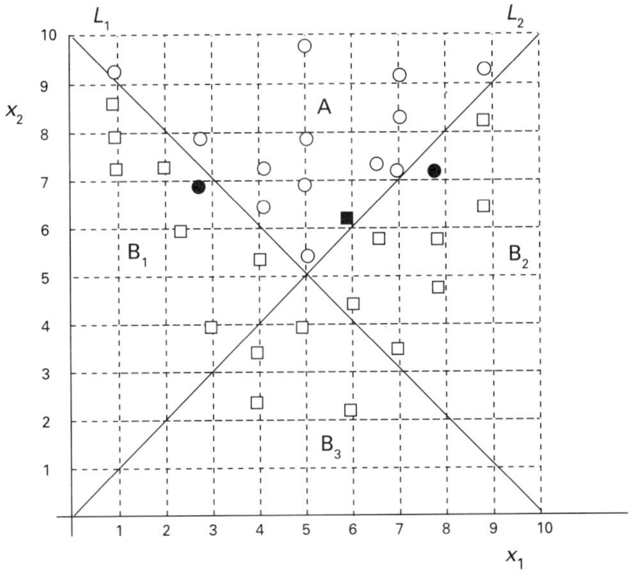

Figure 3.67 Data of the two-dimensional mushroom classification problem

We have a learning set consisting of a collection A of mushrooms that are very good medicine for some specific illness and a class B of poisonous mushrooms. The mushrooms of the two classes differ slightly in the length x_2 and in the thickness x_1 of the stem (see Figure 3.67).

The fifteen pairs (x_1, x_2) of class A are: (7.9, 7.2), (4.0, 6.3), (6.9, 8.3), (6.4, 7.3), (6.9, 7.2), (0.9, 9.3), (2.6, 6.9), (2.6, 7.9), (4.9, 9.8), (4.9, 6.9), (4.9, 7.8), (6.9, 9.2), (4.0, 7.3), (4.9, 5.5) and (8.8, 9.3). The target value for elements of class A during learning is $t=0$. The classification criterion for the output is $y<0.5$.

The nineteen points of class B are: (5.9, 2.3), (6.5, 5.8), (8.8, 8.3), (4.0, 5.4), (5.9, 6.3), (4.0, 2.4), (1.9, 7.3), (4.9, 3.9), (6.9, 3.5), (7.9, 5.8), (0.9, 8.6), (0.9, 7.3), (0.9, 7.9), (5.9, 4.4), (4.0, 3.4), (8.8, 6.4), (3.0, 3.9), (7.9, 4.8) and (2.4, 5.9). The target value for the elements of class B is $t=1$. The classification criterion for the output of the neural net is $y>0.5$.

In Section 3.1 we constructed a solution (not the best one) for the classification using the separating lines L_1 and L_2 of the two neurons in the first layer of a continuous Perceptron. All points to the right of L_1 and simultaneously to the left of L_2 belong to class A; the points in the remaining area are assumed to represent elements of class B. The wrongly classified elements for the constructed solution are represented by black spots in Figure 3.67.

We performed several experiments with a continuous Perceptron with two neurons in the first layer and one neuron in the second layer (see Figure 3.68). All neurons had a sigmoid transfer function. During training we used the hyperplane boundary classification method with one–zero labelling.

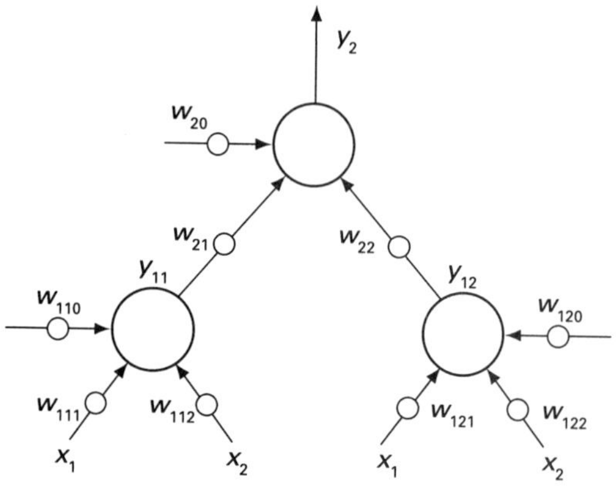

Figure 3.68 The neural network used for the mushroom classification problem

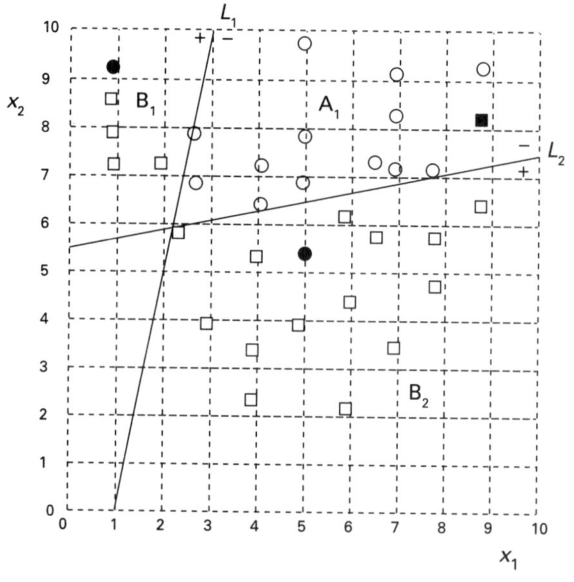

Figure 3.69 The first solution found by the neural network

After learning, we found several different solutions depending on the initial distribution of weights. In four solutions, thirty-one of the thirty-four samples of the learning set were correctly classified.

For the first solution the separating lines L_1 and L_2 realized by the two neurons in the first layer are shown in Figure 3.69. The wrongly classified elements of the learning set are represented by black spots. The weights of the neurons for this classification were:

$$w_{110}=5.196 \quad w_{111}=-4.791 \quad w_{112}=0.902$$

$$w_{120}=14.086 \quad w_{121}=0.444 \quad w_{122}=-2.459$$

$$w_{20}=-5.297 \quad w_{21}=6.499 \quad w_{22}=8.672$$

Thus the separating lines for the neurons in the first layer for this classification are:

$$L_1: \; x_2=5.312x_1-5.761$$

$$L_2: \; x_2=0.181x_1+5.728$$

This means that data points in the area B_1 in Figure 3.69 will give an output for the first neuron in the first layer: $y_{11}>0.5$, and for the second neuron in the first layer: $y_{12}<0.5$.

The weighted input of the output neuron is $s_2=-5.297+6.499y_{11}+8.672y_{12}$. For the elements of the learning set in the area B_1 the output y_2 of the output neuron will be >0.5 as required. For the area B_2 of Figure 3.69 $y_{11}<0.5$ and $y_{12}>0.5$. For the samples of the learning set in area B_2 this will give an output of the neuron in the second layer: $y_2>0.5$ as required. In area A_1 of Figure 3.69 we have $y_{11}<0.5$ and $y_{12}<0.5$. For the elements of the data set in area A_1 this will give: $y_2<0.5$ as required for the samples of class A.

The separating lines of the two first-layer neurons for the other three solutions, with thirty-one out of thirty-four points correctly classified, are given in Figures 3.70–3.72. Note that the solution of Figure 3.72 corresponds with our constructed solution of Section 3.1.

Another solution, with only twenty-five out of thirty-four correctly classified points, found by the learning process gives separating lines of the first-layer neurons defined by $x_2=0.806x_1-12.116$ and $x_2=0.460x_1+2.418$.

The optimal solution, with thirty-two out of thirty-four points correctly classified, represented by the separating lines in Figure 3.73 was not found by learning in our experiments.

We learn from this experiment that a classification problem can be solved in different ways and the neural net will give different solutions if we use a sufficient number of different distributions of the initial weights in the experiments. In general, the different solutions are difficult to predict from inspection of the data set. We also observe that we do not always obtain the optimal solution of the classification problem.

3.18 Application of a two-layer Perceptron to the detection of the frequency of a sine wave

Assume we are given a sine wave with a given constant amplitude A and arbitrary phase ϕ: $x(t)=A\cos(2\pi ft+\phi)$ and we want to detect with a multi-layer Perceptron

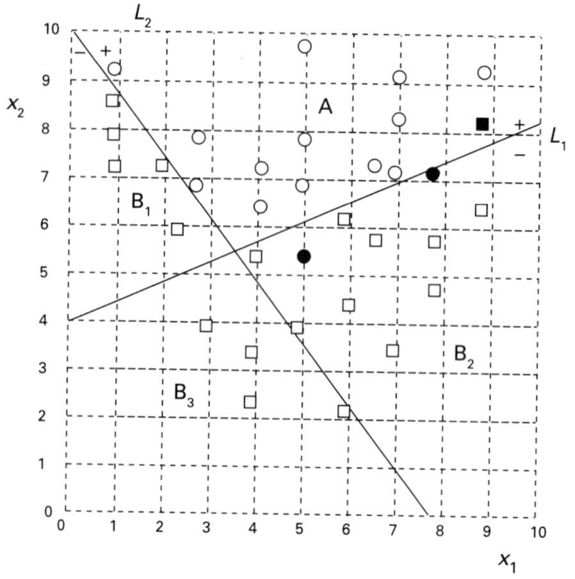

Figure 3.70 The second solution found by the neural network

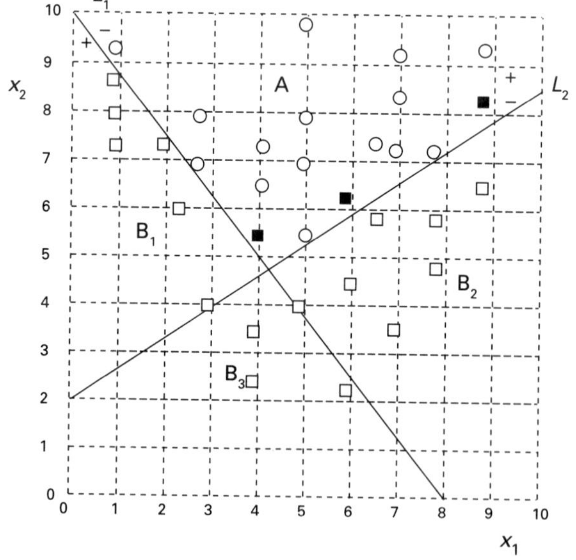

Figure 3.71 The third solution found by the neural network

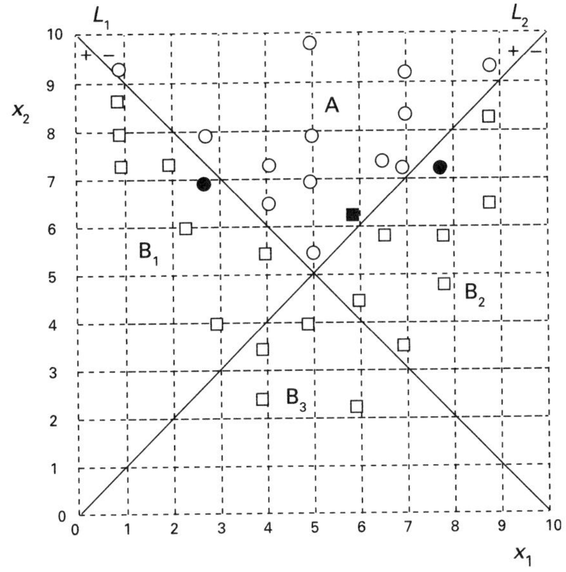

Figure 3.72 The fourth solution found by the neural network

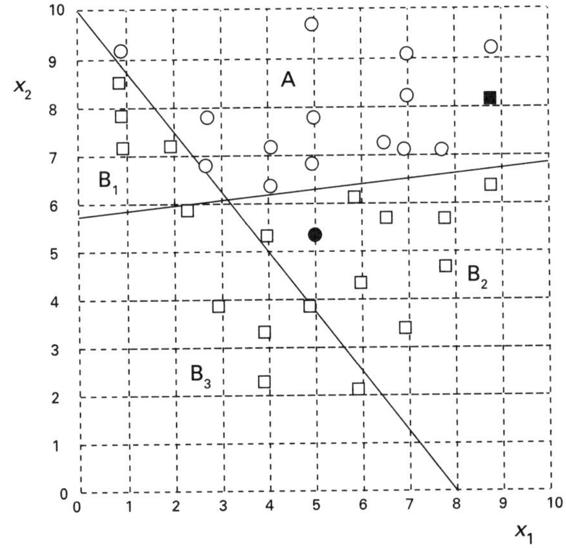

Figure 3.73 The optimal solution not found by the neural network

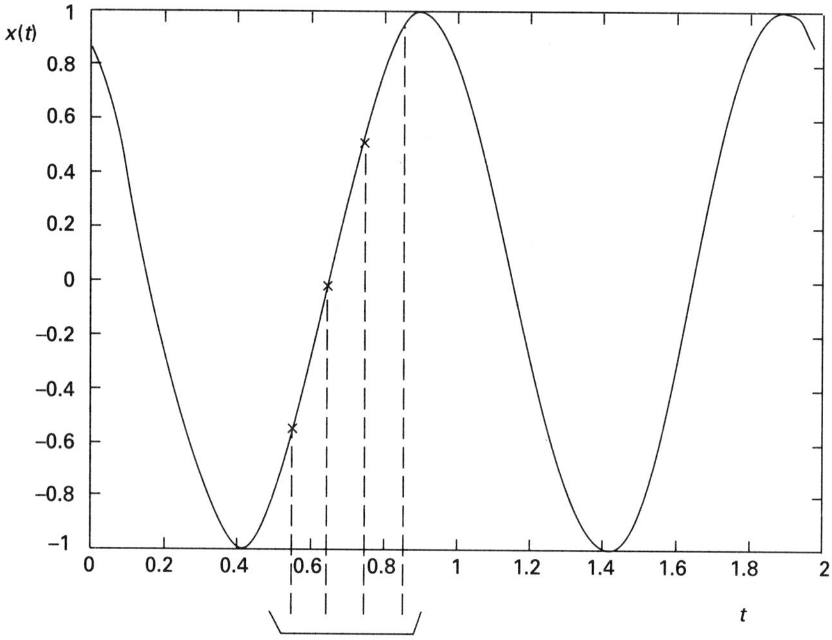

Figure 3.74 Sampling the sine wave with a window

whether the frequency f is equal to some frequency f_0 or not. Thus we want a neural net with one output neuron with, for instance, an output $y>0.5$ if $f \cong f_0$ and $y<0.5$ if not $f \cong f_0$. First we have to determine how we obtain observations from the given sine wave.

The neural net requires real-valued, finite-dimensional vectors as an input. For this purpose we can sample the given sine wave at equidistant intervals T_d during a finite interval w. We will call the finite observation interval the *window*. Let there be n samples in the window interval w; then the components of the *observation vector* will be:

$$x_i = A\cos(2\pi f T_d i + \phi) \text{ with } i = 0, 1, \dots, n$$

To obtain different observations of the sine wave, the window will be shifted over the sine wave to any position (see Figure 3.74).

We have to select the length of the window and the sample interval T_d. The length of the window must be such that if the frequency f of the sine wave differs from f_0, then the observation vectors obtained by sampling both sine waves must also be different.

Now consider the sine waves of Figure 3.75 one with frequency f_0 and one with a frequency slightly smaller than f_0. If the window is smaller than the period T_0 of the sine wave with frequency f_0, then there are positions of the window such that the sequence of samples in the window will be almost the same for both sine waves

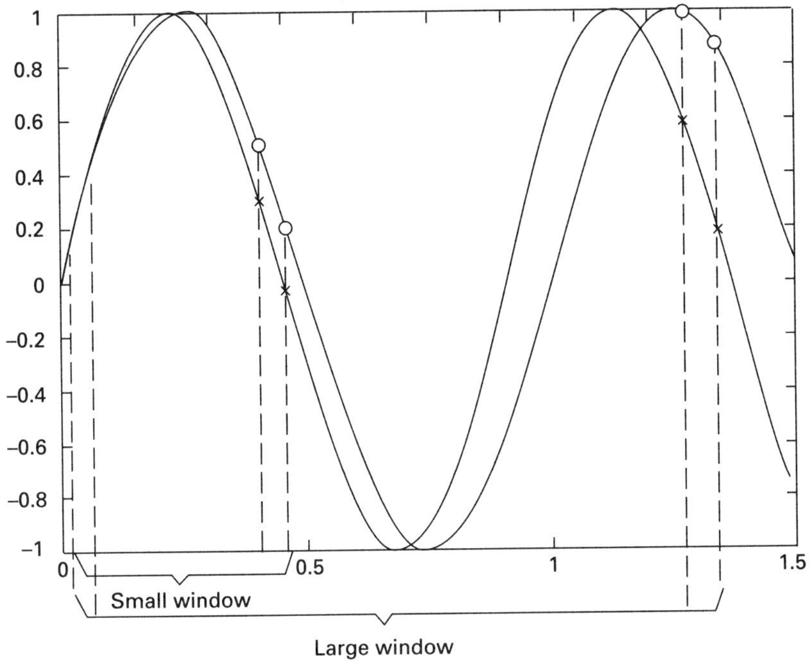

Figure 3.75 The influence of small and large windows on detecting small frequency differences

whatever the sample distance T_d might be. The observation vectors obtained by sampling both sine waves in the window will, however, differ more for all positions of the window, the longer the length of the window w is. This implies that if we want to detect the frequency f_0 with a high resolution we have to make the window length larger than the period T_0. In our experiment we selected $w = (5/4)T_0$.

If we sample a sine wave with frequency f_0 with sample distance T_d, then there are sine waves with the same amplitude but with different frequencies that will give the same set of observation vectors (see Figure 3.76).

The sample values of a sine wave with frequency f^* are:

$$x_i = A \sin(2\pi f^* T_d i)$$

The samples of a sine wave with frequency f_0 are:

$$x_i = A \sin(2\pi f_0 T_d i)$$

If $2\pi f^* R_d i = 2\pi f_0 T_d i \pm k2\pi$ for all $i = 0, 1, 2, \ldots, n$ and any integer k, the sample values of both sine waves will be the same.

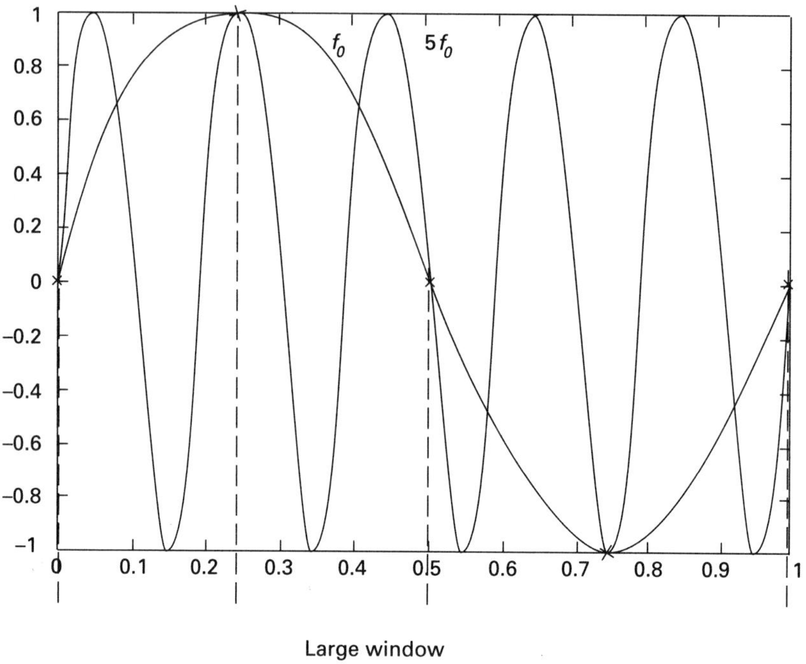

Figure 3.76 By sampling with sample frequency $f_d = 4f_0$ we can not distinguish between sine waves with frequency f_0 and $5f_0$

Thus the sample values will be the same if:

$$f^* = f_0 \pm \frac{k}{T_d} = f_0 \pm kf_d$$

with $f_d = 1/T_d$ the sample frequency.

In Figure 3.76 we have $f_d = 4f_0$ and thus with $k = 1$ we obtain the same samples for a sine wave with frequency $f^* = 5f_0$.

Thus after learning the response $y(f)$ of the neural network will be periodic: $y(f) = y(f \pm kf_d)$. Therefore if the sample frequency f_d is too low, we will not be able to discriminate between f_0 and a frequency slightly different from f_0: $f_0 + \Delta f_0 = f_0 + f_d$. In our experiment we selected $f_d = 12f_0$. With the selected window length $w = (5/4)T_0$ and $T_d = (1/12)T_0$ we obtain observations vectors of fifteen successive samples.

The learning set was obtained by sampling sine waves with ten different frequencies around f_0. Each sine wave was observed with our window at twelve equally spaced different sample positions, corresponding to twelve different phases between 0 and 2π. The result was a learning set of $10 \times 12 = 120$ different fifteen-dimensional observation vectors.

In Table 3.5 information is given about the frequencies of the sine waves in the learning set and about the targets for the observation vectors obtained by sampling

Table 3.5

ω/ω_0	0	0.3	0.6	0.9	1.1	1.4	1.7	2	2.5	3
Target	0.1	0.1	0.5	0.9	0.9	0.5	0.1	0.1	0.1	0.1

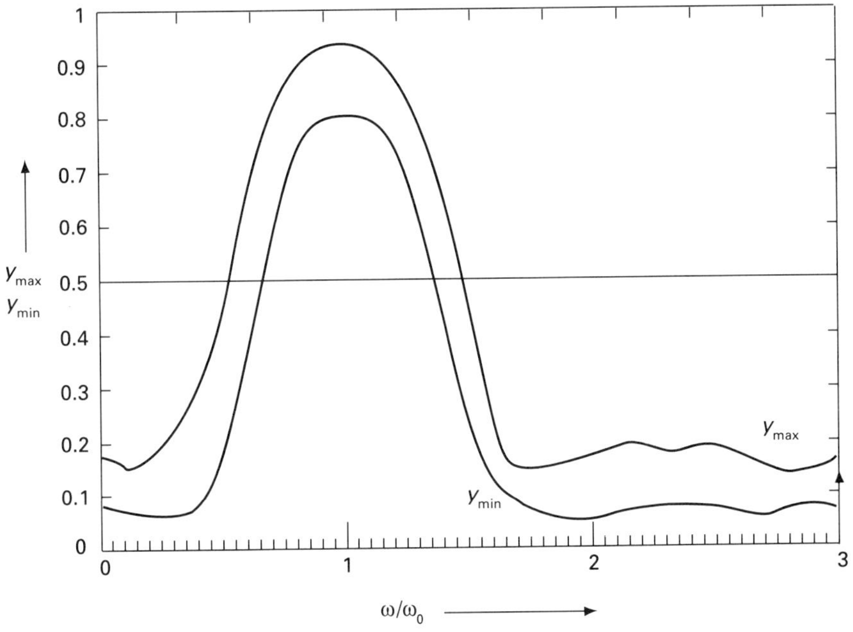

Figure 3.77 The maximum and minimum output of a two-layer network with one output neuron and four neurons in the first layer for sine waves with different frequencies and arbitrary phase shift

those sine waves. The test set of observation vectors was obtained with the same sampling window on arbitrary sine waves.

We performed experiments with a neural net with one output neuron and respectively two, three and four neurons in the first layer. Only the neural net with four neurons in the first layer gave a satisfactory result (see Figure 3.77). For a given frequency the output of the neural net varied between $y_{\min}$ and $y_{\max}$, depending on the phase of the particular sine wave.

For a particular distribution of the initial weights, and with the target 0.9 replaced by 1 and the target 0.1 replaced by 0, we obtained the very good result given in Figure 3.78.

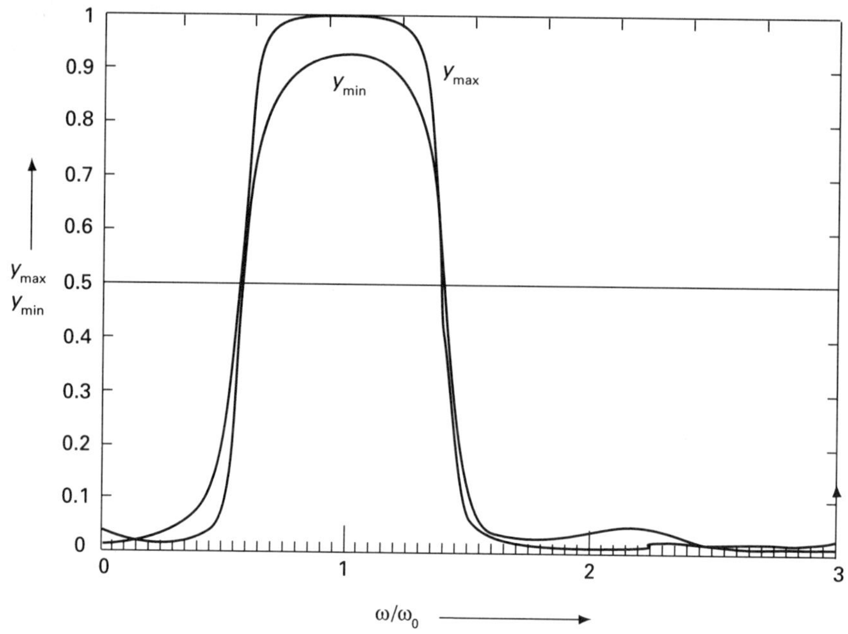

Figure 3.78 The maximum and minimum output of a two-layer network with one output neuron and four neurons in the first layer for sine waves with different frequencies and arbitrary phase shift. The targets during learning were respectively 1 and 0

3.19 Application of a multi-layer Perceptron to machine condition monitoring

It is important not to wait until a defective mechanical machine breaks down before repairing it. In an ideal maintenance strategy the machine would be taken out of service and repaired just moments before major damage occurs. To be able to predict the time when a machine needs maintenance we have to know its condition at each moment. One technique for monitoring machine condition could be analysis of the lubricating oils for the presence of particles that indicate wear. Another technique is vibration analysis: as the condition of a machine changes, so the vibration characteristics also change. We can take a set of consecutive samples of the vibration signal at some time during a certain interval and determine with fast Fourier transform (FFT) the frequency spectrum of the signal at that particular time interval. By analyzing the frequency spectrum of the vibration signal for a short time interval at different moments, we can monitor the condition of the machine or its parts. The frequency spectrum during a certain time interval can be represented by a vector $\mathbf{v}$ with components equal to the coefficients of the spectrum. We will call these vectors *spectral vectors.*

Figure 3.79 Illustration of the bearing system modelled in the text

We have to divide the set of spectral vectors into classes corresponding to different categories of machine condition. This classification problem can be learned by a multi-layer Perceptron.

In our application we had to learn the condition (damaged or not) of a ball-bearing in some machine (see Figure 3.79). We had at our disposal the vibration signals of four equal ball-bearings. One of the bearings was damaged due to an almost invisible small pit in the outer track. For every bearing we had a vibration signal at two different revolution speeds (1500 c/s and 3000 c/s) and with three different loads (no load, 2.5 kN and 5 kN). For every bearing the vibration signal was obtained with sensors in three different positions: horizontal, vertical and axial. Thus the number of vibration signals for each bearing was $2 \times 3 \times 3 = 18$.

The vibration signal with a length of about 1 sec was sampled with a frequency of 48 kHz. With a window of 128 samples we move with steps of 128 samples along the signal. In this way we can place the window in 256 different positions. With the FFT applied to every observation we obtain 256 different spectral vectors of sixty-four components for each observation signal.

We used a neural net with four neurons in the input layer each with sixty-four inputs and one output neuron with four inputs. Each neuron had a sigmoid transfer function. The target value of a spectral vector obtained from a vibration signal of the damaged bearing was 0; for the other spectral vectors the target value was 1.

In one experiment we trained the neural net with spectral vectors obtained from vibration signals of bearing no. 1 (not damaged) and bearing no. 2 (damaged) at a revolution speed of 1500 c/s, with the three different loads and with the sensor in both vertical and horizontal positions.

After learning, we tested the network with 256 different spectral vectors for each vibration signal. A spectral vector was classified as 'damaged' if the output of the

Table 3.6

Bearing number/train		Rev. speed (r.p.m.)		Load (kN)			Sensor			Classification	
		1500	3000	0	2.5	5	Horiz.	Vert.	Axial	Damaged	Not damaged
2	—		×	×				×		206	50
4	—		×	×				×		160	96
2	×	×		×				×		200	56
4	—	×		×				×		92	164
1	×	×		×				×		21	235
3	—	×		×				×		49	207
1	—		×	×				×		123	133
3	—		×	×				×		111	145
1	—		×		×			×		98	158
3	—		×		×			×		163	93
1	×	×			×			×		17	239
3	—	×			×			×		55	201
2	×	×			×			×		220	36
4	—	×			×			×		57	199
2	—		×		×			×		246	10
4	—		×		×			×		151	105

neural net was <0.5, and as 'not damaged' if the output was >0.5. A part of the results from the experiment is given in Table 3.6. Each row in the table corresponds to the observations of one vibration signal. The first column gives the number of the bearing and whether ($\times$) or not (–) the corresponding vibration signal was used during training. The next eight columns give the state of the bearing and the position of the sensor. The last two columns give the classification result of the 256 spectral vectors of the corresponding vibration signal.

We observe that correct classification (by majority voting) occurs, even on those vibration signals not used in the training phase, and certainly if the machine conditions are the same for training and testing.

In additional experiments, which included observations of the vibration signal of the damaged bearing at different revolution speeds in the training sets, we were able to improve the classification results.

3.20 The learning speed of a continuous multi-layer Perceptron

The learning time for a continuous multi-layer Perceptron can be very long because the adaptation of the weights can be very small due to the small value of the derivatives of the MSE:

$$\Delta\hat{\mathbf{w}} = -\varepsilon\nabla E$$

The error 'landscape' E may be very flat in certain areas and may have steep valleys in other regions. In the first areas the value of the internal rate ε may be chosen large, whereas at steep valleys and in the neighbourhood of a minimum the value of ε must be small.

How do we have to select ε in order to proceed rapidly through the error 'landscape'?

In Section 3.12 we found for the adaptation of the weights in the neural net: $\Delta\hat{\mathbf{w}} = -\varepsilon\nabla E$. For the adaptation of the extended weight vector of $\hat{\mathbf{w}}_{kj}$ of neuron j in layer we found:

$$\Delta\hat{\mathbf{w}}_{kj} = \varepsilon \sum_{\mathbf{x}_i \in \mathrm{U}} \delta_{kj}(\mathbf{x}_i)\frac{\mathrm{d}f_{kj}}{\mathrm{d}s_{kj}}\hat{\mathbf{z}}(\mathbf{x}_i)$$

with $\delta_{kj}(\mathbf{x}_i)$ the (back-propagated) error for input $\mathbf{x}_i$ assigned to the output of neuron j in layer k, and $\hat{\mathbf{z}}(\mathbf{x}_i)$ the input of that neuron if the input of the neural net is $\mathbf{x}_i$.

We can improve learning speed by adding to the calculated value of $\Delta\hat{\mathbf{w}}(t)$ (with $\hat{\mathbf{w}}(t)$ the vector containing all weights in the neural net) at learning step t a vector proportional to the calculated value $\Delta\hat{\mathbf{w}}(t-1)$ in the previous step:

$$\Delta\hat{\mathbf{w}}^*(t) = \Delta\mathbf{w}(t) + \alpha\Delta\mathbf{w}(t-1) \text{ with } \alpha \text{ between 0 and 1}$$

This method is called the *momentum method*, and α is called the momentum parameter.

A better method is the *line search method*. The adjustment of $\hat{\mathbf{w}}(t)$ at step t becomes:

$$\Delta\mathbf{w}(t) = -u\nabla E$$

with u selected such that MSE at the new value of the weight becomes minimal, i.e.

$$E[\hat{\mathbf{w}}(t+1)] = E[\hat{\mathbf{w}}(t) - u\nabla E] \text{ is minimal}$$

A simple way to find the value of u is to increase u in fixed steps δ until $E(\mathbf{w})$ no longer decreases. Although the line search method may take considerably fewer steps than the gradient descent method, we must bear in mind that each step may take many evaluations of the error function E. However, it turns out that the calculation of the error E can be simplified, though we will not deal with this subject.

Note that at the subsequent steps the direction vectors $-\nabla E[\mathbf{w}(t+1)]$ and $-\nabla E[\mathbf{w}(t)]$ are perpendicular, because for the optimal value of u we have:

$$0 = \frac{\mathrm{d}}{\mathrm{d}u} E[\mathbf{w}(t) - u\nabla E[\mathbf{w}(t)]] = -\nabla E[\mathbf{w}(t)] \cdot \nabla E[\mathbf{w}(t) - u\nabla E[\mathbf{w}(t)]]$$
$$= -\nabla E[\mathbf{w}(t)] \cdot \nabla E[\mathbf{w}(t+1)]$$

The approach to the minimum is therefore a zig-zag path. A still better strategy is to let the new search direction be a compromise between the gradient direction and the previous search direction $\mathbf{d}(t)$. If we write:

$$\mathbf{w}(t+1) = \mathbf{w}(t) + u\mathbf{d}(t)$$

then the search direction $\mathbf{d}(t)$ becomes:

$$\mathbf{d}(t) = -\nabla E[\mathbf{w}(t)] + \beta\mathbf{d}(t-1)$$

This method is called the *conjugate gradient method*.

3.21 Initialization of weights and scaling the input and output

If we use the sigmoid transfer function for all neurons in the neural net, then the output values will vary between zero and one. Thus if our training and test set contains target values beyond these boundaries, we have to rescale the target values $t(\mathbf{x}_i)$.

In the case of a linear scaling, the scaled values become:

$$t^*(\mathbf{x}_i) = \frac{t(\mathbf{x}_i) - t_{\min}}{t_{\max} - t_{\min}}$$

with $t_{\min} = \min_i\{t(\mathbf{x}_i)\}$ and $t_{\max} = \max_i\{t(\mathbf{x}_i)\}$. If we have a linear output neuron, no scaling of the targets is required.

From a theoretical point of view the scaling of the input vectors is not necessary, because the input of a neuron is not required to be bounded. However, large input

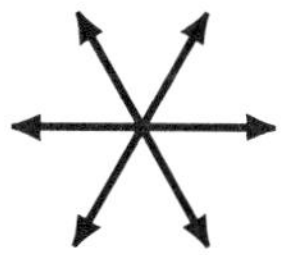

Figure 3.80 Example of the initial distribution of six initial non-extended weight vectors over the two-dimensional weight space

values together with large weights may result in large values of the weighted input $s(\mathbf{x}_i)=\Sigma w_j x_{ij}$ and the adaptation of weights may become almost zero because the derivative df/ds, occurring in the adaptation rule, is almost zero for large values of the weighted input $s(\mathbf{x}_i)$. Therefore the scaling of inputs depends on the value of the weights.

The initial weights may randomly be selected from any interval of real values. However, with a random selection of weights we may end up in a local minimum of the error function E, and we may then have to repeat the learning process many times with different initializations in order to determine whether the final solution is a local minimum or not. Even with random initializations it is possible that different initializations are almost the same, or the weights of different neurons in the same layer may be almost the same.

It is more profitable to distribute the initial weight vectors in some layer of neurons equally spaced over the weight space and to guarantee that subsequent initializations are different.

For instance, if we have a two-dimensional input with six neurons in the first layer, we can for a first initialization distribute the six (non-extended) weight vectors $\mathbf{w}_i$ over the two-dimensional weight space as illustrated in Figure 3.80. For a second initialization we can turn the set of vectors through an angle of $\pi/6$, etc. There remains the selection of the threshold weights w_{j0} for each neuron j. The extended weight vectors $\hat{\mathbf{w}}_j=[w_{j0}, w_{j1}, w_{j2}]$ determine the separating hyperplane $\hat{\mathbf{w}}_j\cdot\mathbf{x}=0$ realized by neuron j. To ensure that the inputs contribute to the output of a neuron, the weighted input must not be far from the separating hyperplane. This can be done by selecting the weight w_{j0} such that the separating hyperplane goes through the centre of gravity of the input data:

$$w_{j0}+w_{j1}x_{1c}+w_{j2}x_{2c}=0$$

with $x_{1c}=\Sigma x_{i1}/N$ and $x_{2c}=\Sigma x_{i2}/N$, where N is the number of examples in the training set.

This method for selecting the initial weights is not straightforward for the n-dimensional input. We invite the reader to develop a simple strategy based on the same method for the n-dimensional case.

Another, more brutal method for initialization is to give every weight (except w_{j0}) a value of 1, or -1. Thus the initial influence of an input x_{ij} is positive, zero or negative. In the case of an n-dimensional input we generate the set W_n of all diferent n-dimensional vectors with components 1, 0 or -1: $W_n=\{1, 0, -1\}^n$. If we have k

neurons in some layer, we select from W_n the collection C_n of all subsets containing k different vectors. For subsequent initializations we select the different subsets from C_n.

The value of the threshold weight w_{j0} of neuron j is selected as before by requiring that the separating hyperplane goes through the centre of gravity of the inputs:

$$w_{j0} + \Sigma\, w_{jk} x_{kc} = 0$$

with $x_{kc} = \Sigma_i x_{ik}/N$ and N the number of examples $\mathbf{x}_i$ in the training set.

Given those sets of initial weights we have to scale the input vectors such that the absolute weighted input $|s_j(\mathbf{x}_i)| < 5$ for every neuron j for every input $\mathbf{x}_i$ in order to guarantee that the initial derivatives $\mathrm{d}f/\mathrm{d}s$ will not become too small.

If $x_{\max} = \max\{|x_{ij}|\}$ over all i and j then:

$$\max_{ij}\{|s_j(\mathbf{x}_i)|\} \leqslant \max_j\{|w_{j0} \pm n x_{\max}|\}$$

Let w_0^* be the value of w_{j0} for which $|w_{j0} \pm n x_{\max}|$ is maximal; then every input vector will be rescaled to:

$$\mathbf{x}_i^* = \left\{\frac{5 - w_0^*}{n x_{\max}}\right\} \mathbf{x}_i \quad \text{if } \max_j\{w_{j0}\} + n \max_{ij}\{x_{ij}\} > 5$$

$$\mathbf{x}_i^* = \left\{\frac{5 + w_0^*}{n x_{\max}}\right\} \mathbf{x}_i \quad \text{if } \min_j\{w_{j0}\} + n \min_{ij}\{x_{ij}\} < 5$$

3.22 Exercises

1. Determine the weights for an optimal classification with one single neuron with sigmoid transfer function for the 'mushroom classification problem' as illustrated in Figure 3.4. We want the output for the healthy mushroom to be equal to 1 and for the other class equal to 0.
2. Explain why the adaptation of the extended weight vector $\hat{\mathbf{w}}$ containing all the weights of a neural net must be equal to $\Delta\hat{\mathbf{w}} = -\varepsilon \nabla E$ for a sufficiently small value of ε.
3. Check whether or not local learning is justified by our theory concerning the adaptation of weights.
4. Determine the adaptation of weights for a single-neuron Perceptron with sigmoid transfer function if the training set contains the elements [1, 1] with target 0.8, and [1, −1] with target 0.2. The initial distribution of weights is $w_0 = 1$, $w_1 = 0.5$. Use for your convenience Figures 3.25 and 3.26 for $f(s)$ and $\mathrm{d}f/\mathrm{d}s$.
5. Calculate a set of weights for the problem of Exercise 4 such that the MSE is zero.
6. Check whether or not a single neuron can realize a function $g_{\mathbf{w}}(\mathbf{x})$ with MSE $E = 0$ for the following elements: [1, 1] with target 0.8, [1, −1] with target 0.2, [−1, 1] with target 0.2, and [0, 0] with target 0.2.
7. Determine the weights of a single-neuron Perceptron with sigmoid transfer function such that the MSE is minimal for a training set with the elements [1, 1]

with target 0.8, [1, −1] with target 0.2, [−1, 1] with target 0.2, and [0, 0] with target 0.22.

8. With the hyperplane boundary classification with one–zero labelling it can happen that at a certain stage of training, the classification is correct for the training set but the weights must still be adapted. Explain why. What is the result of prolonged training where the classification is already correct?
9. Determine the first adaptation of the weight vector for Example 3.9 if the initial weights are $w_0 = -3$, $w_1 = -1$ and $w_2 = 1$. Take $\varepsilon = 10$.
10. Explain the two local minima with $E = 1$ in Figure 3.40 for $|\mathbf{w}| = 100$.
11. Given the data set $\{-1, -0.5, 0, 0.5, 1\}$ with targets $t(-1) = 0$, $t(-0.5) = 0.25$, $t(0) = 0.5$, $t(0.5) = 0.75$ and $t(1) = 1$. Determine the weights of all neurons with sigmoid transfer function such that the MSE is almost zero if we use a neural net with five neurons in the first layer and one neuron in the second layer.
12. Determine the weights of a neural network with four neurons in the first layer and one in the second layer, each with sigmoid transfer function such that the one-dimensional function $g_{\mathbf{w}}(x)$ is zero for $x < -1$ and $x > 1$ and will increase almost linearly from $g_{\mathbf{w}}(-1) = 0.73$ to $g_{\mathbf{w}}(1) = 0.87$ in the interval $[-1, 1]$.

4

THE SELF-ORGANIZING NEURAL NETWORK

4.1 Introduction

Among the different types of artificial neural networks, the self-organizing neural network discussed in this chapter resembles real biological neural networks more than the other types. The artificial neural network was first introduced by Kohonen (1982) as the 'self-organizing feature map'.

If it is true that the self-organizing neural network is a realistic, although very simplified, model of the human brain we can get some idea about how the brain might store pictures and how human beings could be able to recognize pictures.

In the next section we will show how pictures can be stored and recognized with a self-organizing neural network. The behaviour of the self-organizing neural network can be replaced by some equivalent algorithm which is easier to implement and is almost always used in applications of the self-organizing neural network. We will call the equivalent algorithm the *self-organizing neural net algorithm* (Kohonen, 1988). Due to the artificial and sophisticated mathematical operations in the algorithm, the resemblance with real biological neural networks is then lost.

As an introduction to the self-organizing neural network of Kohonen, we will explain in the next section the structure and behaviour of that neural net by describing the application of the neural net to the storage of visual images in a neural network.

Apart from the next section, we will deal with the equivalent self-organizing algorithm in the remaining sections of this chapter.

4.2 Anthropomorphic pattern recognition with a self-organizing neural network

If people were not able to perceive different sensory data as equivalent they would not be able to survive, e.g. children would not recognize their own mothers. The human ability to recognize, apparently without much effort, different pictures as equivalent, challenges scientists to copy the neurophysiological mechanisms involved in human pattern recognition. For this reason the processing of information by artificial neural networks has gained a lot of interest of many scientists in recent years.

In this section we will show how pattern recognition can be performed in a anthropomorphic way. The observation of a picture and the preprocessing of observed data incorporates to a certain extent neurophysiological phenomena. A self-organizing neural network is used to realize a retinotopic mapping from the 'retina' to the 'cortex'. The cortical representation of the observed pattern can then be used as a template for pattern recognition.

Besides using an artificial neural network for pattern recognition we will take into account some additional anthropomorphic mechanisms of visual perception to obtain a more human-like way of processing pattern information.

Important phenomena of human visual information processing are as follows:

1. Visual acuity.
2. Eye movement during visual perception.
3. Retinotopy.

By using these aspects of information processing by the human visual system we will demonstrate that we are able to recognize pictures in an elegant and straightforward artificial way if we adopt in addition the following neurobiological mechanisms:

4. Adaptation of neurosynaptic efficiency.
5. The self-organization of a neural network.

1. The visual ability of humans to distinguish the components of a pattern depends on the angular distane between the components and the eye axis. The resolving power, or acuity, is defined as the reciprocal of the visual angle subtended by the smallest details that the eye can distinguish. The resolving power declines sharply outside the central fovea (see Figure 4.1): at five arc minutes from the centre the acuity is reduced by 50 per cent. This implies that the information obtained by observing a pattern at one fixed point is far from complete and only at the centre of observation is the

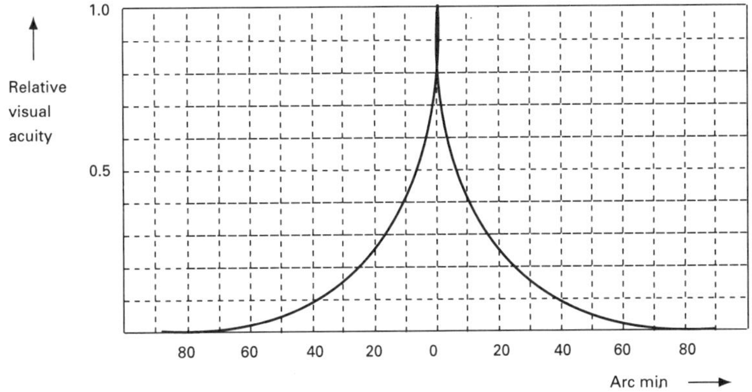

Figure 4.1 The relative visual acuity of man

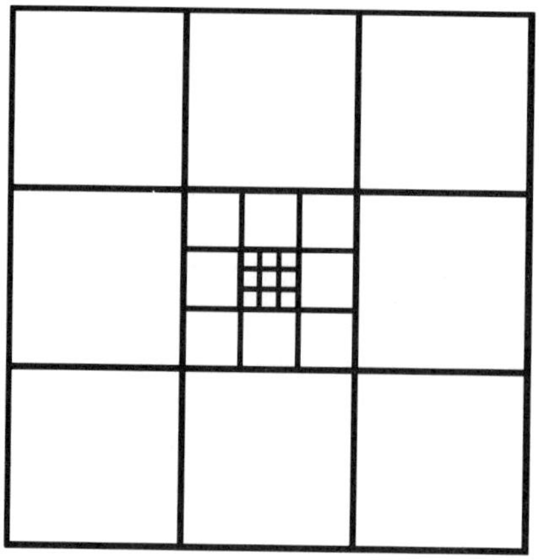

Figure 4.2 The rectangular observation window

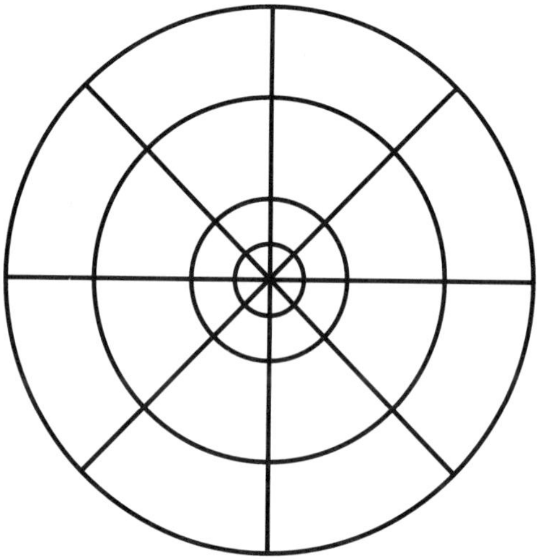

Figure 4.3 The circular observation window

information accurate; in peripheral areas only some diffuse and global structure of the pattern is perceived.

The human method of extracting information from a pattern can be simulated by observing that pattern through a window composed of observation fields of different sizes. Each field covers a different area of the pattern, and in each field only the average illumination of the pattern can be observed (Veelenturf, 1970). The greater the distance between the centre of the window and the centre of a field, the greater the area of the pattern covered by that field and hence the lower the resolving power. By using windows of the form of Figures 4.2 or 4.3 we can approximate this human way of extracting information from a pattern. We will give a more exact description of the observation window below.

In addition, our vision becomes sharper when the light of a pattern is brighter. Peripheral areas of the retina are more sensitive to light and less accurate, while central areas are less sensitive to light but more accurate. We can simulate this mechanism by weighting the observed illumination values of a window field by some appropriate factor.

2. When we look at a pattern, our eyes jump spontaneously about three times per second over about seven minutes of arc. These sudden jumps are called saccades. One of the functions of the saccades is to allow the visual system to use the most specialized area of the retina, the fovea, to process detailed visual information at successive fixation points. In the intersaccadic intervals only small amplitude movements, the drift, persists.

Further insight into the role of eye movement has been gained by investigation of stabilized images (e.g. Gerrits and Vendrik, 1972). The perception of well-stabilized retinal images disappears within a few seconds, showing the need for eye movement in normal vision to maintain visual perception.

Another psycho-physiological result shows that the size of saccadic movements increases with the size of the observed pattern (Stassen, 1980). Moreover it seems that the movements of the eye occur in a systematic fashion related to the individual stimulus and to elements of the pattern that contain the most information (Baker and Loeb, 1973) (see Figure 4.4). Semantic evocative memory will also play an important role in the directed scanning of a picture (Piaget, 1969).

The first phenomenon can be simulated artificially by successive random displacement of the observation window, mentioned above, across the pattern to be classified, followed by observation processing at each point.

Figure 4.4 Saccadic eye movement

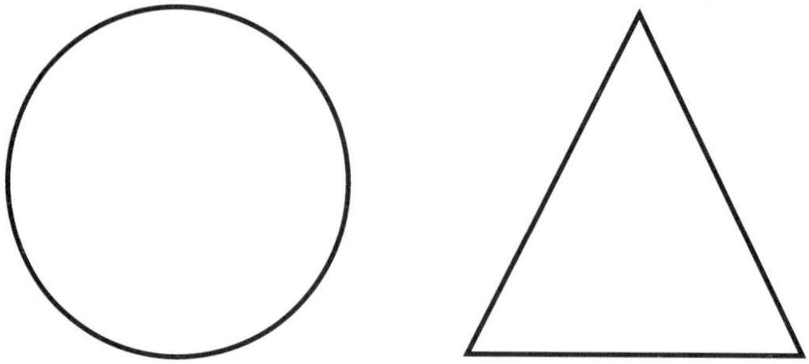

Figure 4.5 The circle and triangle used by the experiments of von Senden

The third phenomenon can be captured (without taking into account the semantics) to some extent, e.g. by moving the centre of the window at each step from one point of fixation to the centre of a highly illuminated field in that observation or to points with a great illumination contrast between neighbouring fields.

3. There exists some evidence for a more or less isomorphic mapping from the retina to some part of the cortex which is called *retinotopy*. This does not necessarily mean a photographic representation of the observed environment, or a metrically faithful copy, but must be understood as some feature preserving imaging, which results in a representation of topologically relevant features of the pattern observed.

The next experimental observation shows that the mapping from the retina to the cortex is not genetic but a result of acquired visual experiences; it also reveals that the mapping must preserve some metric properties.

When people with normal vision are confronted with the patterns of Figure 4.5, they can tell the difference spontaneously. When, however, the patterns are presented to adults who have been blind from birth and have then been given sight by an operation, these subjects are unable to detect immediately the difference between a triangle and a circle. After a few months some patients are able to recognize spontaneously the circle and the triangle separately without counting the corners of the triangle (von Senden, 1932).

One might suppose that retinotopy requires the establishment of well-ordered connections between the retina and the visual cortex. It is true that neural fibers grow according to some genetic plan, approximately to those places in which they are later needed, but the plasticity of the structure of connections is insufficient to explain the mechanism of learning by experience. It is more likely that retinotopy is due to the modifiability of the efficiency of information transmission by synapses. Repeated use of a synapse in a neural circuit increases the synaptic efficiency. Thus when many optic fibers originating from many different parts of the retina are connected to some cell, the synaptic efficiency of all incoming fibers may be altered such that the pertinent cell becomes optimally sensitive to the illumination of some specific parts of the retina. In this way the physical structure of connections is not changed, but rather the functional structure of connections, though with the same result.

4. As mentioned above, the behaviour of a neural net can be changed by modifying the synaptic efficiency of transmission of information from one cell to another. This phenomenon can be realized in an artificial neural net composed of artificial neurons. In our case of simulating visual information processing, each input of every neuron is connected to a different field of the observation window described above (replacing the retina). The output of each neuron is some monotone increasing function of all weighted input values. The impact of synaptic efficiency can be modelled by multiplying each value of observed illumination in a field by some real number, called the weight of that input. The value of a weight will be incremented if there exists a positive correlation between the value of the particular input value and the response of the neuron to whom the input fiber is connected. This way of modifying the weights is called the Hebb rule, after D. O. Hebb who was among the first to envisage the role of this learning mechanism in lasting synaptic changes (Hebb, 1949).

5. With the observation window we obtain isolated local information about the pattern at different (randomly selected) positions in the pattern. We have to realize, however, a cortical representation of the total pattern observed on the retina. This implies that neighbouring (similar) observations are mapped on neighbouring neurons, i.e. the neighbouring neurons will respond optimally to similar (neighbouring) observations. For pattern recognition this requirement is still insufficient because the relative position of pattern components with respect to other components in the pattern is crucial. Therefore the mapping must preserve the positional interrelationship between observations at different fixed points. Thus we have to mimic the topological feature while preserving mapping from the retina to the cortex. In addition the mapping must be learned by experience.

Using the self-organizing artificial neural net proposed by Kohonen, and scanning the pattern with the window mentioned above, we can simulate the learning of the 'topological feature preserving mapping'. At each observation all the synaptic weights (represented by a weight vector $\mathbf{w}$) of the 'winning neuron' (i.e. the neuron with the greatest response) and its neighbours will be adapted to the pertinent observation vector $\mathbf{v}$ such that the winning neuron and its neighbours become more sensitive to that observation $\mathbf{v}$. After adaptation the weighted input $\Sigma\, w_i v_i$ of the winning neuron and its neighbours will increase for the observation $\mathbf{v}$. The feature that preserves self-organizing mapping is mainly caused by this process of selective adaptation of synaptic weights and the lateral excitatory or inhibitory connections between the neurons. We will describe this structure in more detail below. Essential for the observation with the window is that, if $\mathbf{v}(x_i, y_i)$ is the vector of illumination values of the fields of the window, and the window is centered at position (x_i, y_i), then the following relation holds:

$$\text{if } d_V\{\mathbf{v}(x_i, y_i), \mathbf{v}(x_j, y_j)\} < d_V\{\mathbf{v}(x_i, y_i), \mathbf{v}(x_k, y_k)\}$$

$$\text{then } d_{\mathrm{E}}\{(x_i, y_i), (x_j, y_j)\} \leqslant d_{\mathrm{E}}\{(x_i, y_i), (x_k, y_k)\}$$

with d_V some distance measure in the n-dimensional input space V of observation vectors $\mathbf{v}$, and d_{E} being the Euclidean distance in the two-dimensional Euclidean picture space.

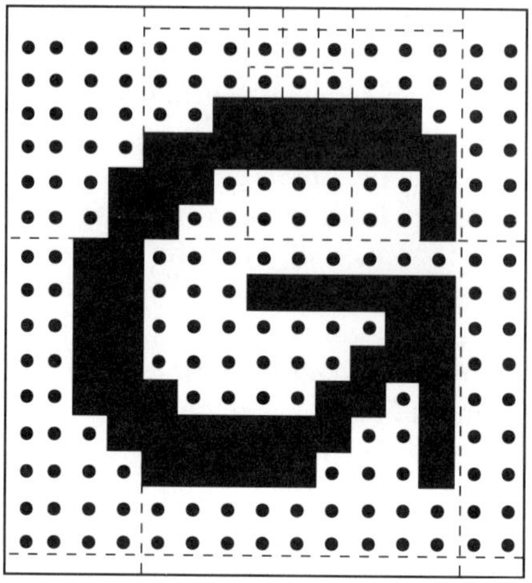

Figure 4.6 The capital 'G' used as a picture

In subsequent sections we will give a mathematical justification of our claims, which will turn out to be mainly a form of vector quantization together with an 'neighbourhood' ordering due to the lateral activation in the neural net.

We will now give a more detailed description of the anthropomorphical pattern recognizer.

Given some picture in a two-dimensional Euclidean object space O. We divide the plane in identical small squares: the *pixels*. The coordinates of a pixel are represented by a pair of integers (x, y). Each pixel in the plane has a *pixel value*: $p(x, y)$. When the picture covers a pixel we assign the pixel value 1, otherwise $p(x, y) = 0$.

The *picture P* is defined by a set of pairs: $P = \{[(x, y), p(x, y)] | (x, y) \in O\}$ (see Figure 4.6).

The picture is sampled at the position $\mathbf{e}_i = (x_i, y_i)$ with a square *window* $W_h(\mathbf{e}_i)$ (see Figure 4.7 with $h = 3$) which covers $3^h \times 3^h$ pixels, with its centre at position $\mathbf{e}_i$ in the plane. The window has *h resolution levels*. Each level k of the window consists of eight square *fields*, except the zero level which consists of one field. The total number of fields is equal to $8h + 1$. A field on level k covers a square of $3^{k-1} \times 3^{k-1}$ pixels of the plane. The field at level zero covers only the pixel at the central position $\mathbf{e}_i$ of the window.

Figure 4.7 represents what is observed through the window if the centre of the window is located in the second row and the ninth column of the picture of Figure 4.6.

With the jth field on the level k, there corresponds a *field value* $f_{k,j}$ defined by:

$$f_{k,j} = \frac{1}{w_k} \sum_{(x,y)} p(x, y)$$

with (x, y) in the jth field on level k, and w_k a constant for scaling the contribution of the illumination of a field at level k.

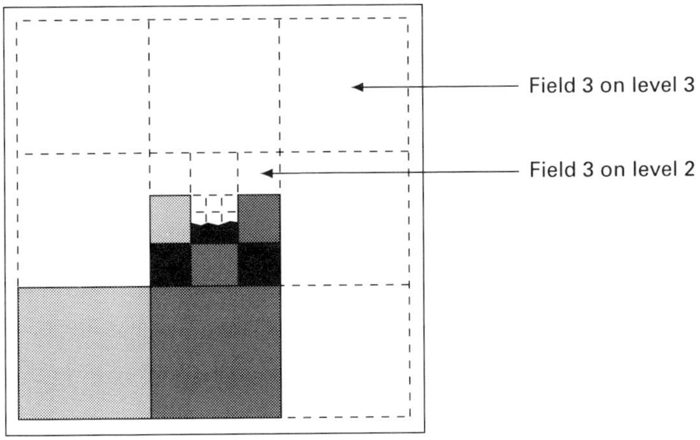

Figure 4.7 The result of the observation of the capital 'G' with the center of the window in the second row and at column 9

From the set of field values we construct an *observation vector* **v**. For each field value $f_{k,j}$ there exists a unique element v_i in the observation vector **v**. The observation vectors obtained by sampling the picture with the window at random positions are the input vectors of a self-organizing neural network.

Learning with the self-organizing neural network

The neural network consists of N artificial neurons in a two-dimensional lattice (see Figure 4.8). Each neuron u_r has one output line and the value of the output at time t will be denoted by $\eta_r(t)$. An observation vector **v**, obtained by sampling a picture with a window $W_h(\mathbf{e}_i)$, is the input vector for each neuron in the neural net.

Each neuron has the same set of $8h+1$ *external input lines* and the value of the jth external input at time t will be given by $v_j(t)$. (In Figure 4.8 the input lines are depicted for only one neuron.) The value $v_j(t)$ is the jth element of the observation vector $\mathbf{v}(t)$ mentioned above. Each external input value $v_j(t)$ of neuron u_r is multiplied by a *synaptic weight factor* $w_{r,j}(t)$.

If there are N neurons in the neural net, then there are in addition $N-1$ *internal input lines* for every neuron in the net that arrives from every neuron in the neural net. (In Figure 4.8 the internal input lines are depicted for only one neuron.)

The value $\eta_k(t)$ of neuron u_k, multiplied by a *lateral synaptic weight factor* γ_{rk}, constitutes the value of the kth internal input of neuron u_r. When $d(r, k)$ is the distance between neuron u_r and u_k, then the lateral weight factor γ_{rk} will depend on this distance. The lateral synaptic weight function may have the form given in Figure 4.9. When γ_{rk} is positive we say the lateral effect of neuron u_k on neuron u_r is *excitatory*;

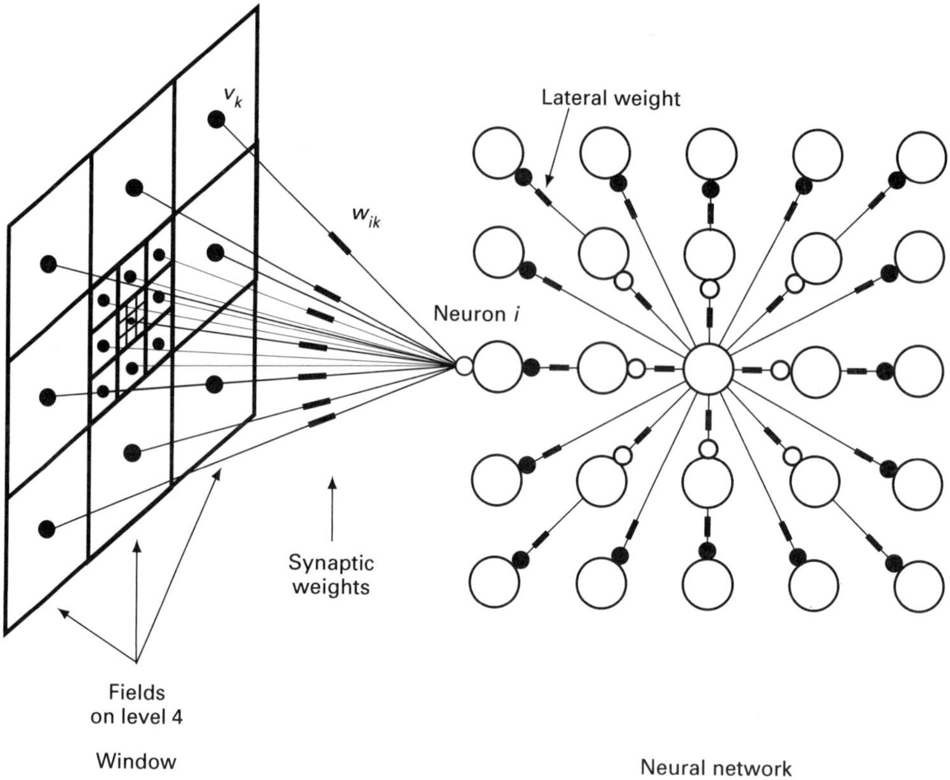

Figure 4.8 The observation window as an input for a two-dimensional neural network

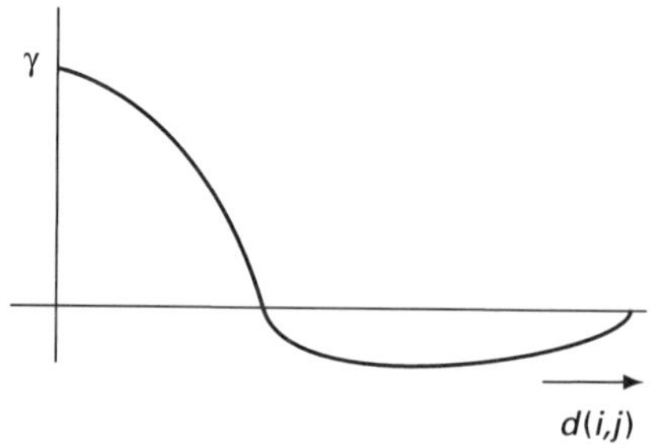

Figure 4.9 The lateral synaptic weight function

if γ_{rk} is negative we say the effect is *inhibitory*. This kind of lateral feedback has been known for a long time in neuroanatomy (e.g. Edelman and Finkel, 1985).

The value of the output $\eta_r(t+\Delta)$ of neuron u_r at time $t+\Delta$ is given by some non-linear monotone increasing function λ_r of the total weighted input $\phi_r(t)$ at time

t (Δ represents some delay between the input and output of a neuron):

$$\eta_r(t+\Delta)=\lambda_r\{\phi_r(t)\}$$

The weighted input at time y is defined by:

$$\phi_r(t)=\sum_j w_{rj}v_j(t)+\sum_r \gamma_{rj}\eta_j(t)$$

The mechanism of lateral excitation and inhibition yields for any observation vector $\mathbf{v}(t)$ a small region (depending on the shape of γ_{rj}) of neighbouring neurons with a large output value, whereas for all other neurons in the neural net the output values will be comparatively small.

By presenting an observation vector $\mathbf{v}(t)$ to the neural net, the outputs of the neural net will stabilize after some time. After stabilization the weights of the external inputs of all neurons will be changed according to the *Hebb adaptation rule*, which reflects the neurophysiological phenomenon that the synaptic efficiency is increased if there exists a positive correlation between presynaptic, i.e. $v_j(t)$, and postsynaptic, i.e. $\eta_r(t)$, activity of neuron u_r:

$$\mathbf{w}_r(\text{new})=\mathbf{w}_r(\text{old})+\varepsilon(t)\cdot\eta_r(t)\cdot\mathbf{v}(t)$$

with $0<\varepsilon(t)<1$ and decreasing with time.

This adaptation makes neuron u_r more sensitive to input vector $\mathbf{v}(t)$. There exists some evidence that the increment of synaptic efficiency of one synapse will be at the expense of the synaptic efficiency of the other synapses impinging on the same neuron. We can model this phenomenon by normalizing the weights such that $\Sigma_j w_{rj}=1$.

The main effect of the biologically plausible mechanism of lateral excitation and inhibition is the amplification of the output in a small region of the neural network and the suppression of all other outputs in the net. Subsequently the adaptation rule will then mainly adapt the weights of the 'winning' neuron and the weights of the neurons in the neighbourhood of the 'winning' neuron. As a consequence the neurons in the adapted neighbourhood will become more sensitive to observation vectors which are similar to the observation vector $\mathbf{v}(t)$ used in the adaptation step.

After the learning phase, described above, performed with a great number of samples of some picture P, for each observation of the pattern with the window there will be a neuron with the greatest response to that particular observation: the 'winning neuron'. Because there are more different observations than there are neurons, each neuron will be the winner for several similar observations. If neuron u_r is the winner for a set R_r of similar observations, then we call the set R_r the *receptive field* of neuron u_r. The weight vector $\mathbf{w}_r$ of neuron u_r will resemble the observation vectors $\mathbf{v}$ in its receptive field R_r.

In every weight vector $\mathbf{w}_r$ there is a component w_{r0} similar to the illumination values v_0 (corresponding to the central field of the observation window) of the observation vectors in the receptive field R_r of neuron u_r.

In Figure 4.10 we have given a visual representation of the 'cortical' image in a

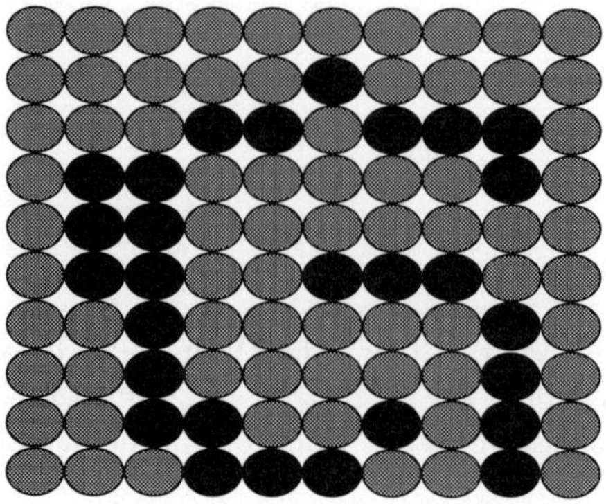

Figure 4.10 The representation of the capital 'G' by the value of w_{r0} of the weight vectors in the neural net after learning

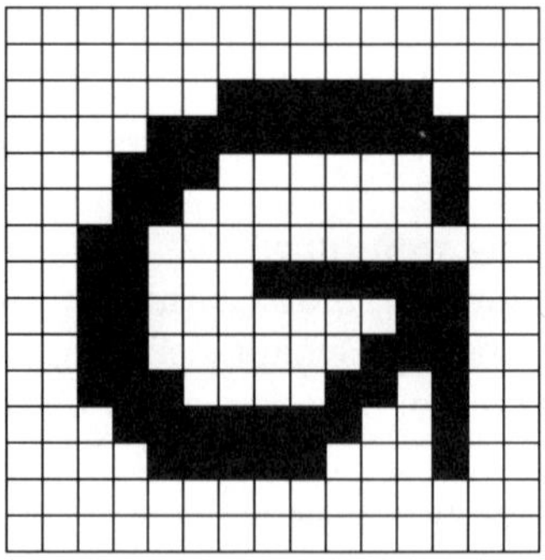

Figure 4.11 The capital 'G' to be learned

two-dimensional neural network of 10×10 neurons after learning with 1000 randomly located window observations $M_3(x, y)$ of the character 'G' given in Figure 4.11.

Each circle in Figure 4.10 represents a neuron in the neural network. A neuron u_r is black if its weight component $w_{r0} = 1$ (w_{r0} corresponds to the observed illumination value in the central field of the observation window) and is shaded if $w_{r0} = 0$. The initial distribution of all synaptic weights for all neurons was random in the interval [0, 1]. Similar results are obtained when the size m^2 of the picture is much larger than the size n^2 of the neural net. Clusters of similar observation vectors will then be represented by the weight vectors of neurons, achieving in this way a considerable reduction in the amount of data.

The shape of the collection of black circles resembles the pattern 'G'. The pattern of Figure 4.10 implies that if one scans, step by step, the pattern 'G' with the central position of the window on the pattern 'G', the neurons corresponding to the black circles in Figure 4.10 will give successively the largest response.

large in the beginning and both must be small at the end of the adaptation process. The *adaptation* function can be split into two parts: $g(r, s, t) = \varepsilon(t) \cdot h(r, s, t)$ with $\varepsilon(t)$ the learning rate:

$$\varepsilon(t) = (1+t)^{-1/2}$$

and with $h(r, s, t)$ the *region neighbourhood adaptation function*:

$$h(r, s, t) = \exp\left[\frac{-d_L(\mathbf{i}_r, \mathbf{i}_s)^2}{\sigma(t)^2}\right]$$

with $\sigma(t) = \sigma_0 \cdot (1+t)^{-1/2}$, and $d_L(\mathbf{i}_r, \mathbf{i}_s)$ the distance between neuron u_r and u_s.

The response $\eta_r(t)$ of a neuron u_r is some monotone decreasing function of $d_V[\mathbf{w}_r(t), \mathbf{v}(t)]$. The output $\eta_r(t)$ is, however, of no importance in applying the algorithmic adaptation rule because we will see that we are only interested in the values of the weights of the neurons. As in the case of the self-organizing neural net discussed in the previous section, at each observation $\mathbf{v}(t)$ the synaptic weights of the 'winning neuron' and its neighbours will be adapted to the pertinent observation vector. The feature that preserves self-organizing mapping is mainly caused by this process of selective adaptation of weights.

4.4 Vector quantization with the self-organizing algorithm

In this section we will derive the algorithmic adaptation rule presented in the previous section. We will see that with the adaptation rule we are minimizing some error function $E(W)$. By analyzing the error function $E(W)$, which we are minimizing during learning, we will obtain some understanding of what is happening during learning and what kind of properties can be contributed to the final result of learning.

The behaviour of the self-organizing neural net algorithm has a great resemblance to algorithms for vector quantization. In this section we will analyze the problem of vector quantization in terms of the preceding chapters.

According to the *adaptation rule* given in the previous section any weight vector $\mathbf{w}_r(t)$ in the net will be changed at time t into:

$$\mathbf{w}_r(t+1) = \mathbf{w}_r(t) + g(r, s, t)(\mathbf{v}(t) - \mathbf{w}_r(t))$$

The scalar-valued *adaptation function* $g(r, s, t)$ determines the region of adaptation, and the amount of adaptation (see Figure 4.13). The function $g(r, s, t)$ was written as $g(r, s, t) = \varepsilon(t) \cdot h(r, s, t)$ with $\varepsilon(t)$:

$$\varepsilon(t) = (1+t)^{-1/2}$$

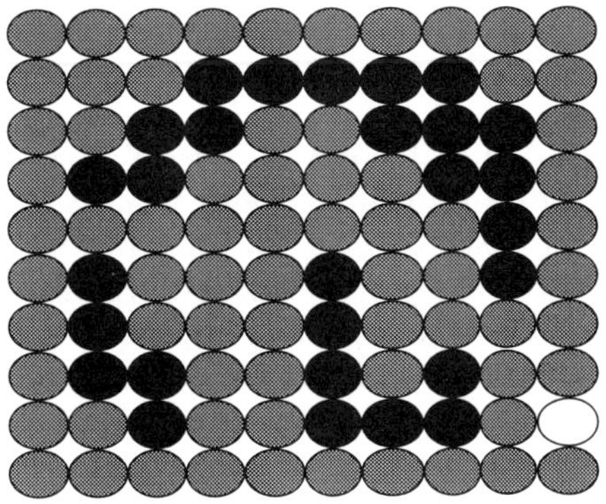

Figure 4.12 The rotated and flipped-over representation of a capital 'G' in the neural network

This result shows that we can realize a retinotopic or even to a certain extent a metric that preserves the mapping of a two-dimensional pattern into the neural network. It is particularly important to note that during learning no information was given to the neural network about the location of the observation window in the picture.

The sensitivity pattern of Figure 4.10 might be rotated or horizontally or vertically flipped over with respect to the original pattern. See Figure 4.12 for a simulation result obtained after 10 000 times of sampling the pattern 'G' of Figure 4.10.

This section has shown how with an artificial neural network we can realize a human-like way of pattern recognition.

By understanding the behaviour of this neural network, it will become clear in the next section that we can use instead a more artificial algorithm, which seems to have no relation to neural networks, but that will give almost the same result.

In the following sections of this chapter we will discuss the self-organizing algorithm, which is almost equivalent to the self-organizing neural network described above.

4.3 The equivalent self-organizing neural net algorithm

Typical for the neural network discussed in the previous section were the lateral inhibitory and excitatory connections between all neurons in the network. In this section we will present an almost equivalent algorithm implementing the behaviour of a self-organizing neural net without those lateral connections.

The main effect of the biologically plausible mechanism of lateral excitation and inhibition as used in the self-organizing neural net of the previous section is the amplification of the output, due to some input vector $\mathbf{v}(t)$, in a small region, and the suppression of all other outputs in the net. Subsequently the adaptation rule will then mainly adapt the weights of the 'winning' neuron and the weights of the neurons in the neighbourhood of the 'winning' neuron. As a consequence the neurons in the adapted neighbourhood will become more sensitive to observation vectors, which are similar to the observation vector $\mathbf{v}(t)$ used in the adaptation step.

The same result can, however, be obtained in a more artificial way (and one that consumes less simulation time if implemented on a sequential computer) by eliminating the lateral excitation and inhibition, and only detecting which neuron is maximally responding to the presented observation vector $\mathbf{v}(t)$. We can then subsequently change the weights of the winning neuron and of its neighbours according to the following *algorithmic adaptation rule* (Ritter and Schulten, 1986).

Given some observation vector $\mathbf{v}(t)$ at learning step t, and some distance measure d_V for the input space [e.g. $d_V(\mathbf{x}, \mathbf{y}) = |\mathbf{x} - \mathbf{y}|$], then:

1. Determine the winning neuron u_s, i.e.

$$d_V[\mathbf{w}_s(t), \mathbf{v}(t)] = \min_r d_V[\mathbf{w}_r(t), \mathbf{v}(t)]$$

2. Every weight vector $\mathbf{w}_r(t)$ in the net will be changed to:

$$\mathbf{w}_r(t+1) = \mathbf{w}_r(t) + g(r, s, t)[\mathbf{v}(t) - \mathbf{w}_r(t)]$$

with $g(r, s, t)$ a scalar-valued function with a value between 0 and 1 depending on the distance in the neural net between the winning neuron u_s and the neuron u_r

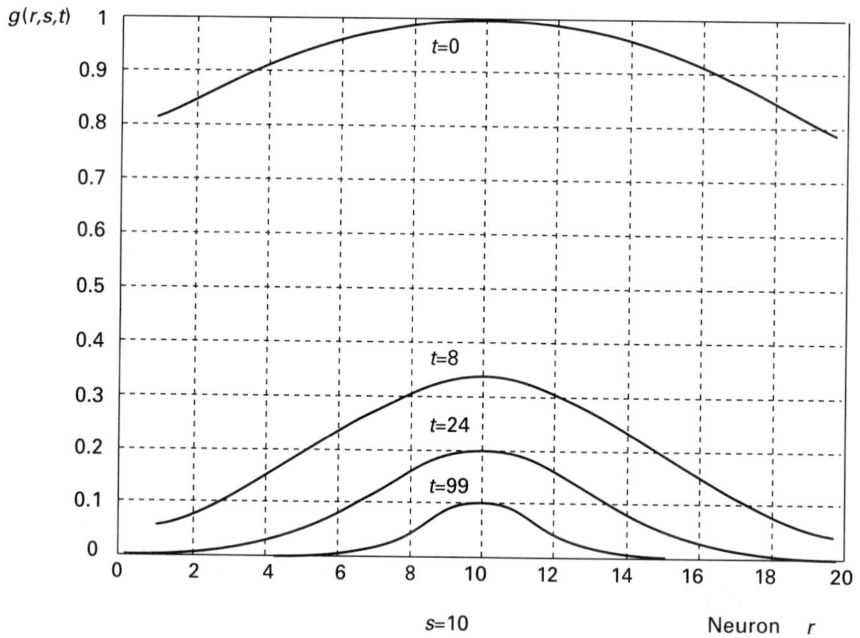

Figure 4.13 The function $g(r, s, t)$ for different moments of time and $s = 10$

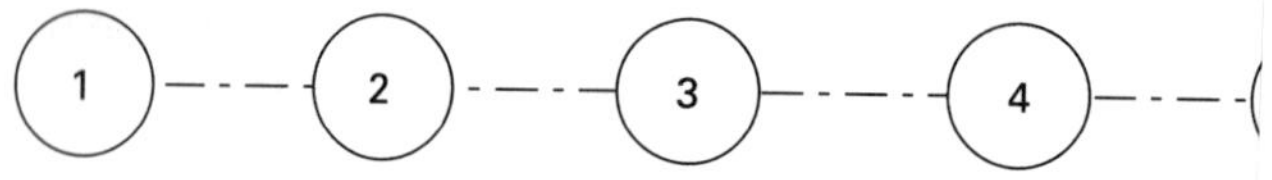

Figure 4.14 A one-dimensional neural network

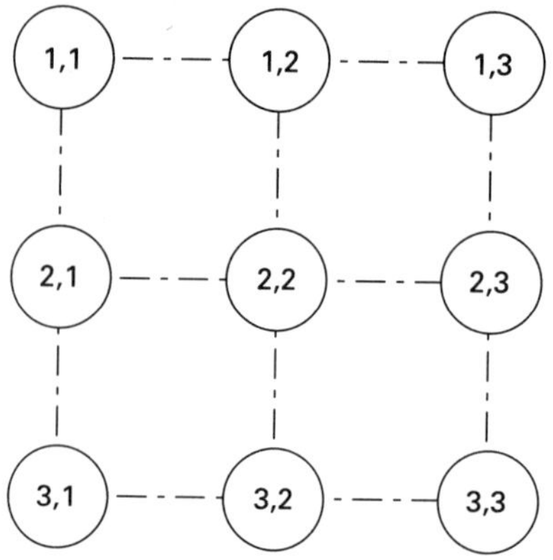

Figure 4.15 A two-dimensional neural network

to be adapted. See Figure 4.13 for the changing shape of $g(r, s$ in a one-dimensional neural network.

The neural network may be a one-dimensional (Figure 4.14 (Figure 4.15) or an n-dimensional lattice L. The neurons are loca point of such a lattice. With each neuron u_r in an n-dimensio associated an n-dimensional coordinate vector $\mathbf{i}_r$. The only thing distance measure d_L we use in the neural net. It may be any meas city block distance) as long as the distance measure satisfies the distance measure:

1. $d_L(\mathbf{i}_r, \mathbf{i}_s) \geq 0$ for all $\mathbf{i}_r$ and $\mathbf{i}_s$.
2. $d_L(\mathbf{i}_r, \mathbf{i}_s) = 0$ if and only if $\mathbf{i}_r = \mathbf{i}_s$.
3. $d_L(\mathbf{i}_r, \mathbf{i}_s) = d_L(\mathbf{i}_s, \mathbf{i}_r)$ for all $\mathbf{i}_r$ and $\mathbf{i}_s$.
4. $d_L(\mathbf{i}_r, \mathbf{i}_t) \leq d_L(\mathbf{i}_r, \mathbf{i}_s) + d_L(\mathbf{i}_s, \mathbf{i}_t)$

The adaptation function $g(r, s, t)$ determines the region of a amount of adaptation.

Different types of functions can be used for $g(r, s, t)$; how convergence the amount of adaptation as well as the region of

and $h(r, s, t)$:

$$h(r, s, t) = \exp\left[\frac{-d(r, s)^2}{\sigma(t)^2}\right]$$

with $\sigma(t) = \sigma_0 \cdot (1+t)^{-1/2}$ and $d(r, s) = d_L(\mathbf{i}_r, \mathbf{i}_s)$ the distance in the neural network between neuron u_r (with weight vector $\mathbf{w}_r$ and coordinate vector $\mathbf{i}_r$) and the winning neuron u_s (with weight vector $\mathbf{w}_s$ and coordinate vector $\mathbf{i}_s$).

In the final phase of the learning process we have $t \gg 1$ and thus $h(r, s, t) = 0$ for $r \neq s$ and $h(r, s, t) = 1$ for $r = s$. Thus in this final phase of the learning process the adaptation rule can be simplified to:

$$\mathbf{w}_s(t+1) = \mathbf{w}_s(t) + \varepsilon(t)[\mathbf{v}(t) - \mathbf{w}_s(t)]$$

This adaptation rule, which is used in the final phase of learning, can be explained by the process of vector quantization.

Suppose we have a finite set $D_V \subseteq V$ of data out of an d-dimensional input space V represented by d-dimensional real vectors $\mathbf{v}_i \in \mathbb{R}^d$.

A change $p(\mathbf{v}_i)$ of selecting vector $\mathbf{v}_i$ from the input space V for the data set D_V is associated with each input vector $\mathbf{v}_i$. The total number of data vectors in D_V is M. Assume we want to replace the data set D_V by another smaller set D_W with d-dimensional vectors $\mathbf{w}_j \in \mathbb{R}^d$. Let the number of elements in D_W be N, thus $N < M$. We want the set D_W to be such that the set D_V and D_W are in a certain way similar.

Given the sets D_V and D_W, we partition first the data set D_V into equivalent classes $R(\mathbf{w}_s)$. $R(\mathbf{w}_s)$ is called the *receptive field* of $\mathbf{w}_s$ in V and $R(\mathbf{w}_s)$ is defined by (see also Figure 4.16):

$$\mathbf{v}_i \in R(\mathbf{w}_s) \text{ if } d_V(\mathbf{v}_i, \mathbf{w}_s) = \min_k d_V(\mathbf{v}_i, \mathbf{w}_k)$$

with $d_V(\mathbf{v}_i, \mathbf{w}_k)$ the distance between $\mathbf{v}_i$ and $\mathbf{w}_k$.

We will say the set D_W of *reference vectors* $\mathbf{w}_s$ (weight vectors) and the given set D_V of data vectors are 'similar' if the set D_W is such that the sum over all reference vectors $\mathbf{w}_s$ of the expectation of the square distance $d_V^2(\mathbf{v}_i, \mathbf{w}_s)$ for every $\mathbf{v}_i \in D_V$ in the receptive field of $\mathbf{w}_s$ to the reference vector $\mathbf{w}_s$ is as small as possible. This implies that we want to minimize the following function:

$$E(W) = \sum_{\mathbf{w}_s} \sum_{\substack{\mathbf{v}_i \in R(\mathbf{w}_s) \\ \mathbf{v}_i \in D_V}} p(\mathbf{v}_i) \cdot d_V^2(\mathbf{v}_i, \mathbf{w}_s)$$

For simplicity of expression for the derivative to be used later, we add a factor 1/2 to $E(W)$:

$$E(W) = \frac{1}{2} \sum_{\mathbf{w}_s} \sum_{\substack{\mathbf{v}_i \in R(\mathbf{w}_s) \\ \mathbf{v}_i \in D_V}} p(\mathbf{v}_i) d_V^2(\mathbf{v}_i, \mathbf{w}_s)$$

Let us take for the distance measure $d_V(\mathbf{v}_i, \mathbf{w}_s)$ the Euclidean distance $|\mathbf{v}_i - \mathbf{w}_s|$, then

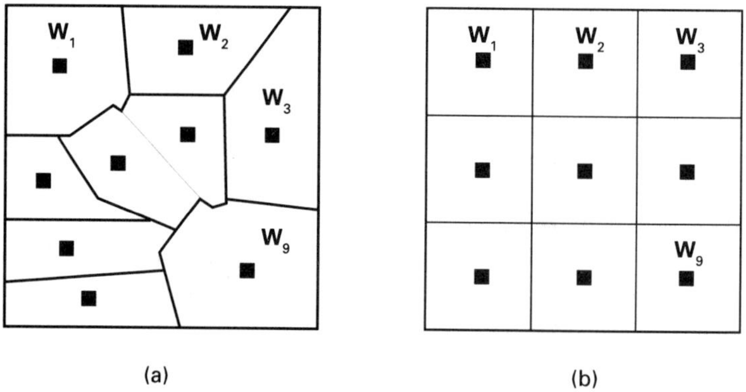

Figure 4.16 Distribution of reference vectors over the input space. (a) Random distribution. (b) Optimal distribution

$E(W)$ becomes:

$$E(W)=\frac{1}{2}\sum_{\mathbf{w}_s}\sum_{\substack{\mathbf{v}_i\in R(\mathbf{w}_s)\\ \mathbf{v}_i\in D_V}} p(\mathbf{v}_i)|\mathbf{v}_i-\mathbf{w}_s|^2$$

For the two-dimensional case and uniform distribution of elements of D_V over some bounded area A of V and nine reference vectors $\mathbf{w}_1, \mathbf{w}_2, \ldots, \mathbf{w}_9$ randomly uniformly distributed over the same area A, see Figure 4.16(a), and for the optimal distribution [$E(W)$ is minimal] of the reference vectors over A, see Figure 4.16(b).

Given some initial random distribution of N vectors $\mathbf{w}_s$, we want to know how we have to change the vectors $\mathbf{w}_s$ in order to reduce $E(W)$. The gradient descent method says that for a given data set D_V, change vector $\mathbf{w}_s$ by an amount:

$$\Delta\mathbf{w}_s=-\varepsilon\frac{\partial E(W)}{\partial \mathbf{w}_s} \text{ with } \varepsilon \text{ sufficiently small}$$

Hence the expectation $\Delta\bar{\mathbf{w}}_s$ of the increment of the reference $\mathbf{w}_s$ vector must be:

$$\Delta\bar{\mathbf{w}}_s=\sum_{\mathbf{v}_i\in R(\mathbf{w}_s)} \varepsilon p(\mathbf{v}_i)(\mathbf{v}_i-\mathbf{w}_s)$$

If we select at step t of the learning process a vector $\mathbf{v}(t)=\mathbf{v}_i$ with probability $p(\mathbf{v}_i)$ from the data set D_V, and change after the selection of $\mathbf{v}_i$ the reference vector $\mathbf{w}_s$ according to the *algorithmic adaptation rule*:

$$\Delta\mathbf{w}_s=\varepsilon(\mathbf{v}_i-\mathbf{w}_s) \text{ if } \mathbf{v}_i\in R(\mathbf{w}_s)$$

then the expectation $\bar{\mathbf{w}}_s(t+1)$ at learning step $t+1$ of any $\mathbf{w}_s(t)$ at learning step t becomes:

$$\bar{\mathbf{w}}_s(t+1)=\bar{\mathbf{w}}_s(t)+\sum_{\mathbf{v}_i\in R(\mathbf{w}_s(t))} \varepsilon p(\mathbf{v}_i)[\mathbf{v}_i-\mathbf{w}_s(t)]$$

as required by the gradient descent method for minimizing $E(W)$.

The gradient descent method requires that ε is infinitesimally small, especially in the neighbourhood of a minimum of $E(W)$. This can be realized by a decreasing function of time: $\varepsilon(t)$. Thus by applying at the final phase of learning [i.e. when $h(r, s, t)=0$ for $r \neq s$ and $h(r, s, t)=1$ for $r=s$] the algorithmic adaptation rule:

$$\mathbf{w}_r(t+1)=\mathbf{w}_r(t)+\varepsilon(t)g(r, s, t)[\mathbf{v}_i(t)-\mathbf{w}_r(t)]$$

we are minimizing:

$$E(W)=\frac{1}{2}\sum_{\mathbf{w}_s}\sum_{\substack{\mathbf{v}_i\in R(\mathbf{w}_s)\\ \mathbf{v}_i\in D_V}} p(\mathbf{v}_i)\cdot|\mathbf{v}_i-\mathbf{w}_s|^2$$

and thus the set D_W will become 'similar' to the set D_V. The amount of similarity depends on the distribution of weights at the beginning of the final phase of learning.

We can summarize our discussion by the following theorem:

Theorem 4.1

By using in the self-organizing algorithm the adaptation rule:

$$\mathbf{w}_r(t+1)=\mathbf{w}_r(t)+\varepsilon(t)h(r, s, t)[\mathbf{v}_i(t)-\mathbf{w}_r(t)]$$

we are minimizing at the final phase of learning:

$$E(W)=\frac{1}{2}\sum_{\mathbf{w}_s}\sum_{\substack{\mathbf{v}_i\in R(\mathbf{w}_s)\\ \mathbf{v}_i\in D_V}} p(\mathbf{v}_i)\cdot|\mathbf{v}_i-\mathbf{w}_s|^2$$

This means: we are performing, in the final phase of learning, vector quantization to the original data set D_V with respect to the set D_W of weight vectors $\mathbf{w}_s$ in the neural net at the beginning of the final phase.

We will call the final phase of learning when $h(r, s, t)=0$ for $r \neq s$ and $h(r, s, t)=1$ for $r=s$, the *quantization phase* of learning.

In order to evaluate the performance of vector quantization we can calculate, after learning, the *energy of quantization noise* defined by:

$$E_{DW}=\frac{1}{M}\sum_{\mathbf{w}_s}\sum_{\substack{\mathbf{v}_i\in R(\mathbf{w}_s)\\ \mathbf{v}_i\in D_V}} |\mathbf{v}_i-\mathbf{w}_s|^2$$

with M the number of elements in D_V.

The quantization noise is defined by:

$$Q_{DW}=\frac{1}{M}\sum_{\mathbf{w}_s}\sum_{\substack{\mathbf{v}_i\in R(\mathbf{w}_s)\\ \mathbf{v}_i\in D_V}} |\mathbf{v}_i-\mathbf{w}_s|$$

Although we are minimizing $E(w)$ during the quantization phase of learning with the algorithmic adaptation rule, we will have no guarantee that we will obtain the

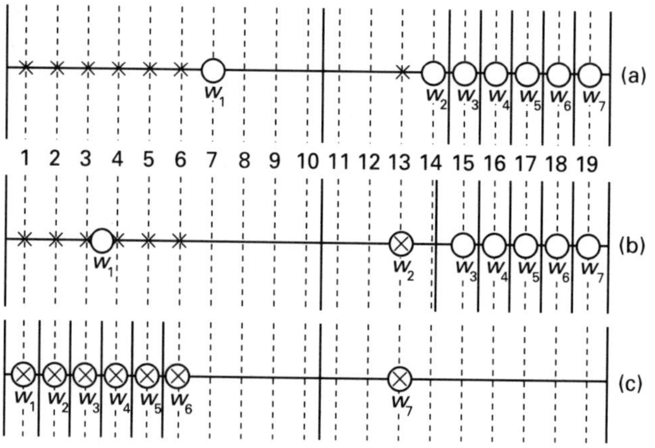

Figure 4.17 Distribution of one-dimensional data ($\times$) and weight vectors ($\bigcirc$) over the input/weight space. (a) Initial distribution. (b) Distribution after learning. (c) Optimal distribution

global minimum of $E(W)$ in the case of an arbitrary distribution of weights at the beginning of the quantization phase. (We will, however, see that the distribution at the beginning of the quantization phase is not arbitrary.) Depending on the distribution of weights at the beginning of the quantization phase we can get stuck in a local minimum.

Example 4.1

Assume we have a one-dimensional data set $D_V = \{1, 2, 3, 4, 5, 6, 13\}$. Let $p(\mathbf{v}_i) = 1/7$ for all $\mathbf{v}_i$ and let the set D_W of one-dimensional weight vectors be: $\mathbf{w}_1 = 7$, $\mathbf{w}_2 = 14$, $\mathbf{w}_3 = 15$, $\mathbf{w}_4 = 16$, $\mathbf{w}_5 = 17$, $\mathbf{w}_6 = 18$ and $\mathbf{w}_7 = 19$ (see Figure 4.17a). The receptive fields restricted to the input set D_V are $R(\mathbf{w}_1) \cap D_V = \{1, 2, 3, 4, 5, 6\}$, $R(\mathbf{w}_2) \cap D_V = \{13\}$ and the restricted receptive fields of the remaining weight vectors are empty. Applying repeatedly the algorithmic adaptation rule with $h(r, s, t) = 1$ for $r = s$ and $h(r, s, t) = 0$ for $r \neq s$ and $\varepsilon(t) \ll 1$:

$$\Delta \mathbf{w}_s = \varepsilon(t)[\mathbf{v}_i(t) - \mathbf{w}_s(t)] \quad \text{if } \mathbf{v}_i(t) \in R[\mathbf{w}_s(t)]$$

will finally give a set of weight vectors $\mathbf{w}'_1 = 3.5$, $\mathbf{w}'_2 = 13$, $\mathbf{w}'_3 = 15$, $\mathbf{w}'_4 = 16$, $\mathbf{w}'_5 = 17$, $\mathbf{w}'_6 = 18$ and $\mathbf{w}'_7 = 19$ because the restricted receptive fields of $\mathbf{w}_3$, $\mathbf{w}_4$, $\mathbf{w}_5$, $\mathbf{w}_6$ and $\mathbf{w}_7$ remain empty during learning (see Figure 4.17b). The final set of weights is, however, not optimal because $E(W) = 0$ for $\mathbf{w}'_1 = 1$, $\mathbf{w}'_2 = 2$, $\mathbf{w}'_3 = 3$, $\mathbf{w}'_4 = 4$, $\mathbf{w}'_5 = 5$, $\mathbf{w}'_6 = 6$ and $\mathbf{w}'_7 = 13$ (see Figure 4.17c).

We will see that local minima of $E(W)$ will be avoided by the self-organizing algorithm in the preceding ordering phase.

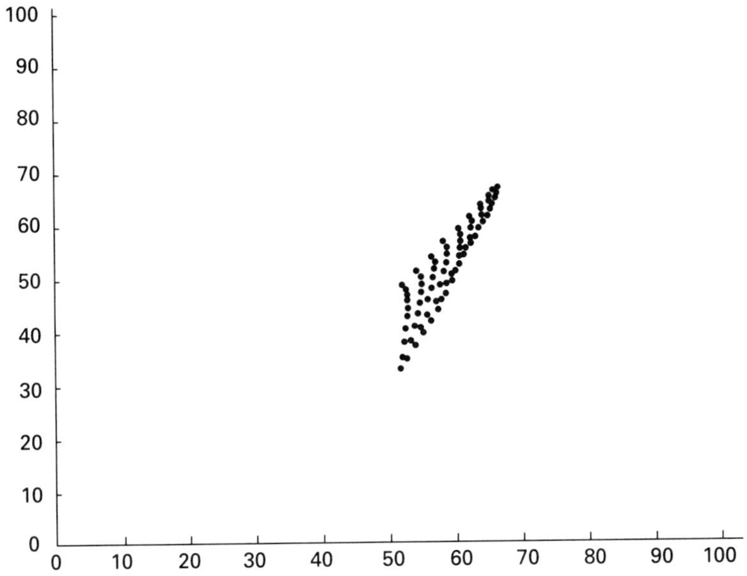

Figure 4.18 Result of a simulation experiment for the problem of Example 4.2 with uniform distributed data after ten learning steps

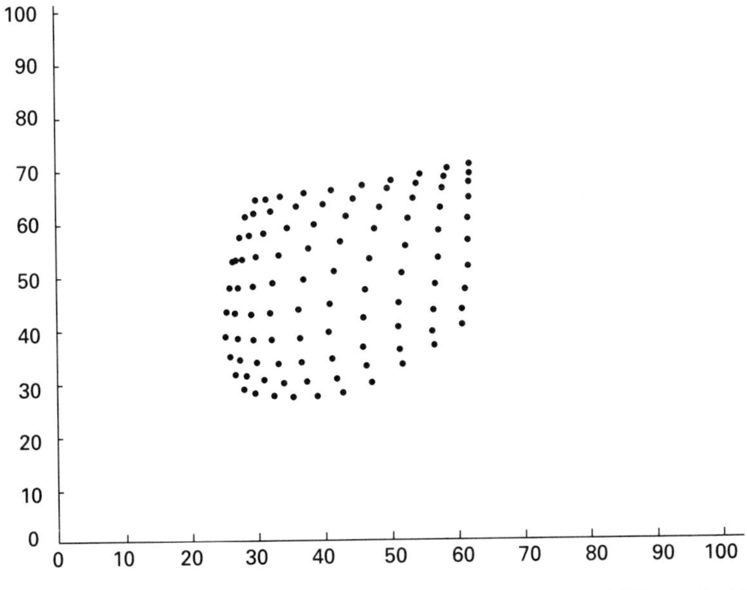

Figure 4.19 Result of a simulation experiment for the problem of Example 4.2 after fifty learning steps

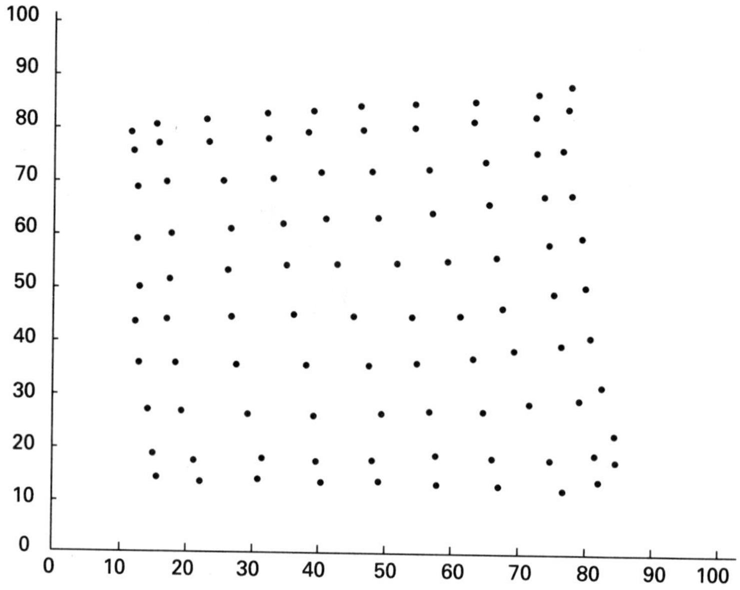

Figure 4.20 Result of a simulation experiment for the problem of Example 4.2 after 500 learning steps

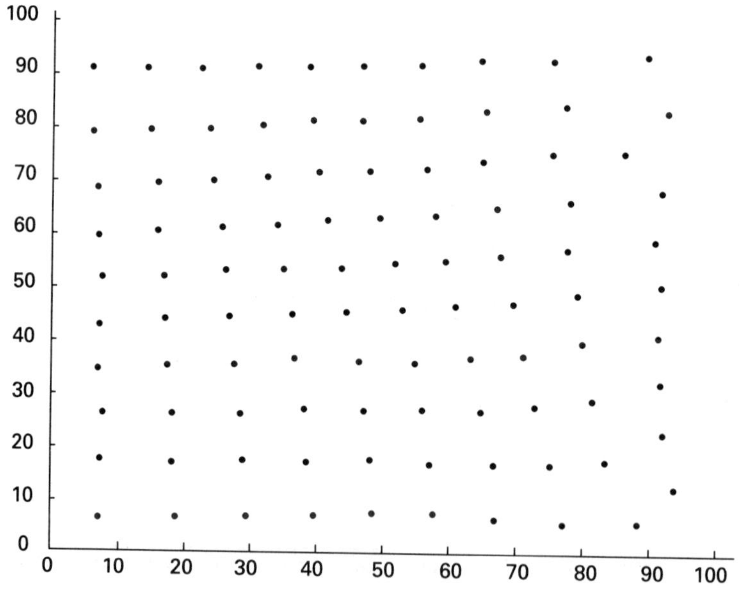

Figure 4.21 Result of a simulation experiment for the problem of Example 4.2 after 10 000 learning steps

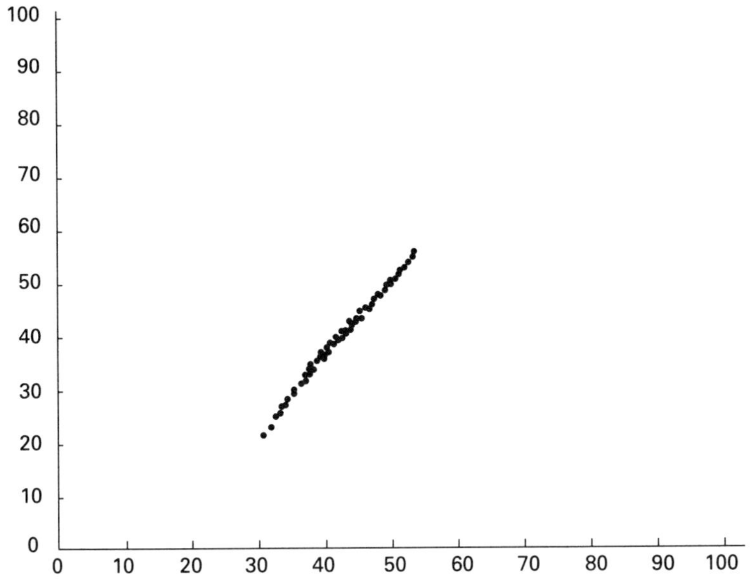

Figure 4.22 Result of a simulation experiment for the problem of Example 4.2 with non-uniform distributed data after ten learning steps

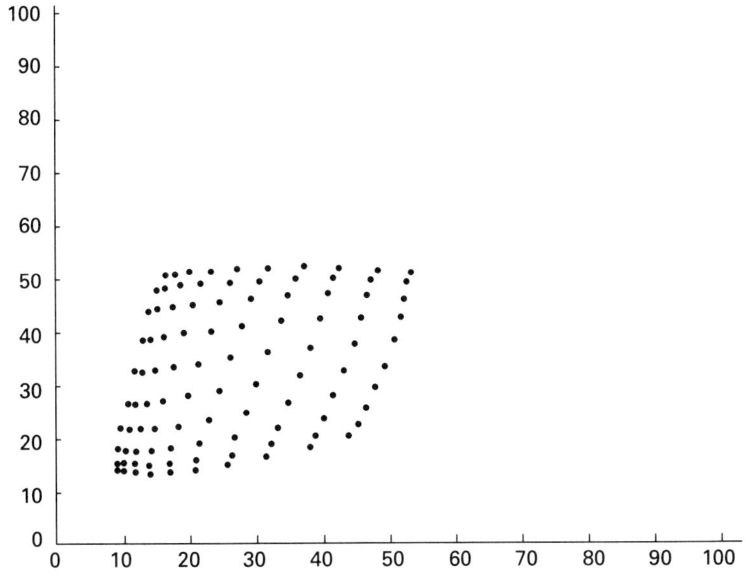

Figure 4.23 Result of a simulation experiment for the problem of Example 4.2 with non-uniform distributed data after fifty learning steps

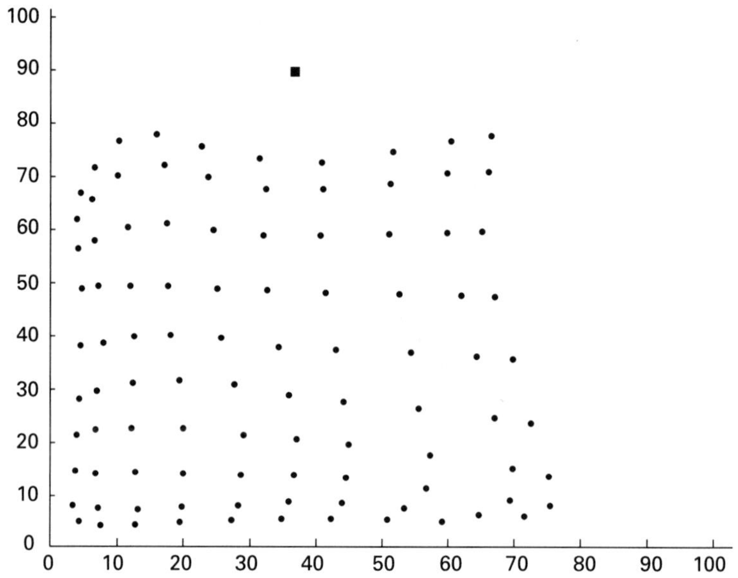

Figure 4.24 Result of a simulation experiment for the problem of Example 4.2 with non-uniform distributed data after 500 learning steps

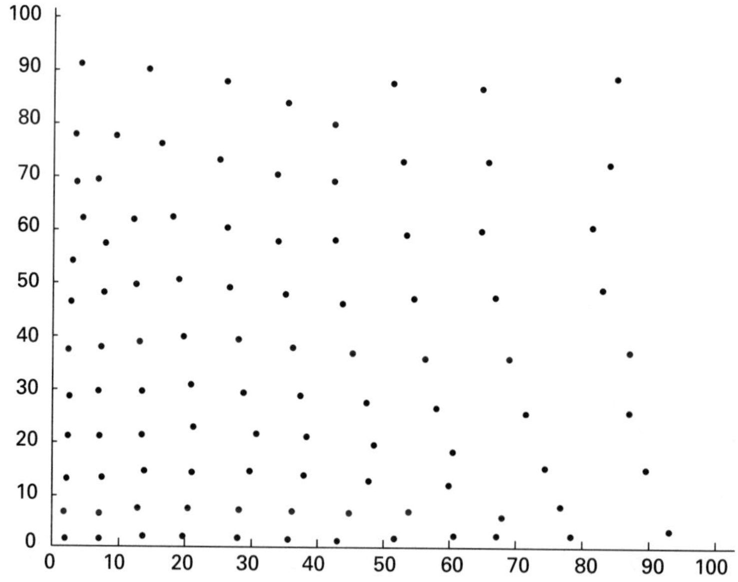

Figure 4.25 Result of a simulation experiment for the problem of Example 4.2 with non-uniform distributed data after 10 000 learning steps

Example 4.2

In a simulation experiment we used a data set D_V with 10 000 elements uniformly selected from a two-dimensional bounded integer input space: $V=[0, 99]\times[0, 99]$. We used a two-dimensional net with 10×10 neurons. We used the learning rate function $\varepsilon(t)$ and region adaptation function $h(r, s, t)$ with $\sigma_0=1000$ as mentioned in our theory. In Figures 4.18–4.21 the results are given after respectively 10, 50, 500 and 10 000 learning steps. Dots represent the values of the weights. (Note deviations from perfect vector quantization at the borders of the domain.) In a second experiment we used inhomogeneous distributed data. Again we used a data set with 10 000 two-dimensional vectors $\mathbf{v}_i$ with $[v_{i1}, v_{i2}]=[(x/10)^2, (y/10)^2]$ with $[x, y]\in[0, 99]\times[0, 99]$. The results after respectively 10, 50, 500 and 10 000 steps are given in Figures 4.22–4.25. The dots represent the values of weights. ■

4.5 Weight vector ordering with the self-organizing algorithm

We found that in the quantization phase of the learning process the vectors $\mathbf{v}_i$ in the receptive field of some weight vectors $\mathbf{w}_s$ will become 'similar' to $\mathbf{w}_s$. Assume we want in addition that the vectors in the receptive field of $\mathbf{w}_s$ are also to some extent similar to the weight vector $\mathbf{w}_r$ if neuron u_r is not a too distant 'neighbour' of neuron u_s in the lattice of neurons (see Figure 4.26). For this purpose we introduce a measure for neighbourhood of neurons in the neural network.

Let the degree of neighbourhood between neuron u_r and u_s be reflected by a

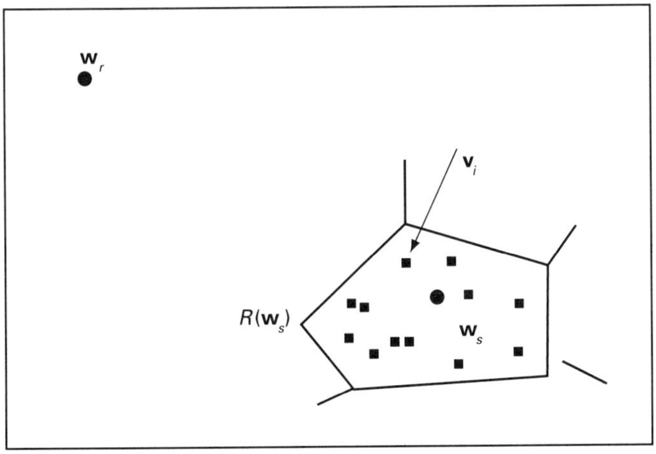

Figure 4.26 The receptive field $R(\mathbf{w}_s)$ of $\mathbf{w}_s$ in relation to weight vector $\mathbf{w}_r$

so-called *neighbourhood function* $h(r, s)$ defined by:

$$h(r, s) = \exp\left[\frac{-d^2(r, s)}{\sigma^2}\right]$$

with $d(r, s) = d_L(\mathbf{i}_r, \mathbf{i}_s)$ the distance (e.g. Euclidean distance) in the neural network between neuron u_r and neuron u_s, and with σ some constant.

We are now not only interested in a minimum value of:

$$E_s(W) = \frac{1}{2}\sum_{\mathbf{w}_s}\sum_{\mathbf{v}_i \in R(\mathbf{w}_s)} p(\mathbf{v}_i) d_V^2(\mathbf{v}_i, \mathbf{w}_s)$$

but we also want to have for any neuron u_r with weight vector $\mathbf{w}_r$ a small value of:

$$E_r(W) = \frac{1}{2}\sum_{\mathbf{w}_s}\sum_{\mathbf{v}_i \in R(\mathbf{w}_s)} p(\mathbf{v}_i) d_V^2(\mathbf{v}_i, \mathbf{w}_r)$$

depending on the neighbourhood value of neuron u_r with respect to neuron u_s. For that reason the error $E_r(W)$ will be weighted by the neighbourhood function $h(r, s)$: for close neighbours of neuron u_s the error is weighted by a factor $h(r, s)$ close to 1, and for distant neighbours of neuron u_s the error will be weighted by a factor $h(r, s)$ close to 0. Using the neighbourhood function $h(r, s)$ we consequently want a minimum for:

$$E(W) = E_s(W) + \sum_r h(r, s) E_r(W)$$

$$= \frac{1}{2}\sum_{\mathbf{w}_s}\sum_{\mathbf{v}_i \in R(\mathbf{w}_s)} p(\mathbf{v}_i) d_V^2(\mathbf{v}_i, \mathbf{w}_s) + \frac{1}{2}\sum_{\mathbf{w}_r \neq \mathbf{w}_s} h(r, s) \sum_{\mathbf{w}_s}\sum_{\mathbf{v}_i \in R(\mathbf{w}_s)} p(\mathbf{v}_i) d_V^2(\mathbf{v}_i, \mathbf{w}_r)$$

By choosing $h(r, s) = 1$ for $r = s$ we can incorporate the first summation in the second summation:

$$E(W) = \frac{1}{2}\sum_{\mathbf{w}_r}\sum_{\mathbf{w}_s}\sum_{\mathbf{v}_i \in R(\mathbf{w}_s)} p(\mathbf{v}_i) h(r, s) d_V^2(\mathbf{v}_i, \mathbf{w}_r)$$

By minimizing this error function the final result will be that for any weight vector $\mathbf{w}_s$, the vectors in $R(\mathbf{w}_s)$ will be 'similar' to weight vector $\mathbf{w}_s$ of neuron u_s, but also 'similar' to the weight vector $\mathbf{w}_r$ if neuron u_r is a close neighbour, measured by $h(r, s)$, of neuron u_s.

If we use the Euclidean distance for vectors in V, $d_V(\mathbf{v}_i, \mathbf{w}_r) = |\mathbf{v}_i - \mathbf{w}_r|$, then the expression for $E(W)$ becomes:

$$E(W) = \frac{1}{2}\sum_{\mathbf{w}_r}\sum_{\mathbf{w}_s}\sum_{\mathbf{v}_i \in R(\mathbf{w}_s)} p(\mathbf{v}_i) h(r, s) |\mathbf{v}_i - \mathbf{w}_r|^2$$

Given some initial random distribution of n vectors $\mathbf{w}_r$, we want to know how we have to change the weight vectors $\mathbf{w}_r$ in order to reduce $E(W)$. The gradient descent method prescribes: For a given data set D_V change vector $\mathbf{w}_r$ with an amount:

$$\Delta\mathbf{w}_r = -\varepsilon\frac{\partial E(W)}{\partial\mathbf{w}_r} \quad \text{with } \varepsilon \text{ sufficiently small}$$

Hence the expectation of the increment of the reference vector $\mathbf{w}_r$ must be:

$$\Delta\bar{\mathbf{w}}_r = \sum_{\mathbf{w}_s}\sum_{\mathbf{v}_i\in R(\mathbf{w}_s)} \varepsilon p(\mathbf{v}_i)h(r, s)(\mathbf{v}_i - \mathbf{w}_r)$$

Thus if we select with probability $p(\mathbf{v}_i)$ a vector $\mathbf{v}_i$ from the input space V and put it in the data set

$$D_V = \bigcup_{\mathbf{w}_s} R(\mathbf{w}_s)$$

and adjust according to the adaptation rule the value of $\mathbf{w}_r$ with an amount:

$$\Delta\mathbf{w}_r = \varepsilon h(r, s)(\mathbf{v}_i - \mathbf{w}_r) \quad \text{if } \mathbf{v}_i\in R(\mathbf{w}_s)$$

then the expectation $\bar{\mathbf{w}}_r(t+1)$ at learning step $t+1$, given $\mathbf{w}_r(t)$ at learning step t, becomes:

$$\bar{\mathbf{w}}_r(t+1) = \bar{\mathbf{w}}_r(t) + \sum_{\mathbf{v}_i\in R[\mathbf{w}_s(t)]} \varepsilon p(\mathbf{v}_i)h(r, s)[\mathbf{v}_i - \mathbf{w}_r(t)]$$

as required by the gradient descent method for minimizing:

$$E(W) = \frac{1}{2}\sum_{\mathbf{w}_r}\sum_{\mathbf{w}_s}\sum_{\mathbf{v}_i\in R(\mathbf{w}_s)} p(\mathbf{v}_i)h(r, s)|\mathbf{v}_i - \mathbf{w}_r|^2$$

The gradient descent method requires that ε be infinitesimally small, especially in the neighbourhood of a minimum of $E(W)$. This can be realized by a decreasing function of time: $\varepsilon(t)$. At the same time we can during learning contract the 'neighbourhood relationship' $h(r, s)$ to small regions such that close neighbours of a neuron in the lattice will gradually become the only neighbours. This can be realized by introducing a time dependence in the neighbourhood function: $h(r, s, t)$ (see Figure 4.13). These modifications give the *algorithmic adaptation rule* introduced in Section 4.3:

$$\mathbf{w}_r(t+1) = \mathbf{w}_r(t) + \varepsilon(t)h(r, s, t)[\mathbf{v}_i(t) - \mathbf{w}_r(t)]$$

We stated before that by using this adaptation rule we are minimizing the error function with the final result that for any weight vector $\mathbf{w}_s$ the vectors in $R(\mathbf{w}_s)$ will be similar to weight vector $\mathbf{w}_s$ of neuron u_s, but also similar to the weight vector $\mathbf{w}_r$ if neuron u_r is a close neighbour, measured by $h(r, s)$, of neuron u_s.

We can rephrase this statement as follows. If two neurons u_r and u_s in the neural net are neighbours:

$$d_L(\mathbf{i}_r, \mathbf{i}_s) = \min_k d_L(\mathbf{i}_r, \mathbf{i}_k)$$

then the final result of learning with the algorithmic adaptation rule will be (but not always) that the corresponding *receptive fields* $R(\mathbf{w}_r)$ and $R(\mathbf{w}_s)$ are *adjacent* (i.e. they have a common border):

$$R(\mathbf{w}_r) \cap R(\mathbf{w}_s) = \partial R_{rs} \neq \varnothing$$

∂R_{rs} is called the common border of $R(\mathbf{w}_r)$ and $R(\mathbf{w}_s)$.

If this property holds for every pair of neurons, we say that the neural net is *well ordered.* The phase of learning preceding the quantization phase will be called the *ordering phase* of learning.

Example 4.3

As an illustration of the ordering of weight vectors during learning we used in an experiment two-dimensional homogeneously distributed input data and a two-dimensional net with 10 × 10 neurons. In Figures 4.27–4.31 the weights are represented by dots. In addition lines are drawn between different weights if the weights belong to neighbouring neurons. At step $t=0$ there is no ordering at all because of the random initialization of weights (see Figure 4.27). After one learning step all weights become almost equal to the applied input vector (Figure 4.28). After 100 learning steps there already exists an almost perfect ordering but no correct vector quantization (Figure 4.29). In Figure 4.30 the intermediate result after 1000 steps is given. After 10 000 steps the net is well ordered and the vector quantization is almost correct (Figure 4.31). ■

Besides the vector quantization and ordering of weights, the learning process also implies that the weighted sum of all mutual distances between the final weights will

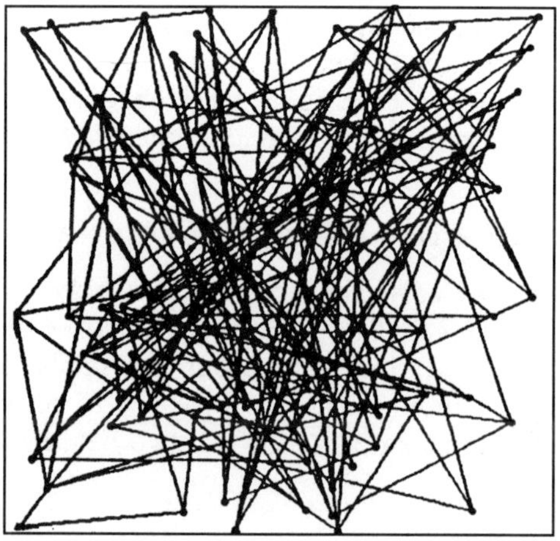

Figure 4.27 Initial weight ordering at $t=0$ for Example 4.3

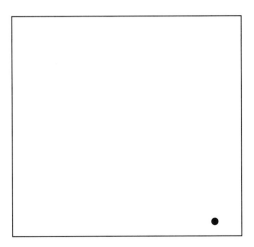

Figure 4.28 Weight ordering at $t=1$ of Example 4.3

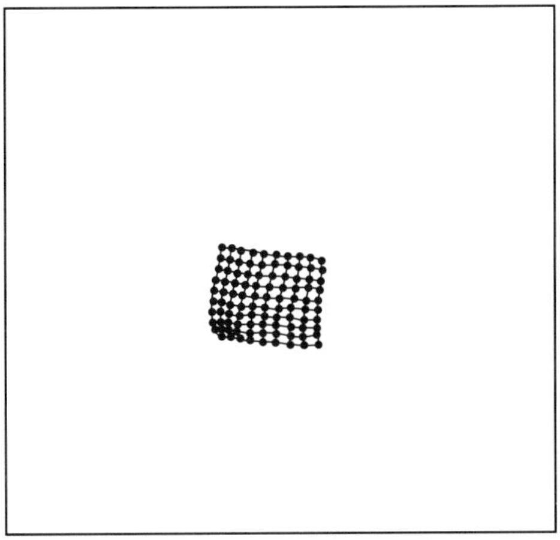

Figure 4.29 Weight ordering at $t=100$ of Example 4.3

be minimized. This phenomenon will become more evident if we simplify the expression for the error we are minimizing:

$$E(W, t)=\frac{1}{2}\sum_{\mathbf{w}_r}\sum_{\mathbf{w}_s}\sum_{\mathbf{v}_i\in R(\mathbf{w}_s)} p(\mathbf{v}_i)\varepsilon(t)h(r, s, t)|\mathbf{v}_i-\mathbf{w}_r|^2$$

We know that the vectors $\mathbf{v}_i$ that are members of a receptive field $R(\mathbf{w}_s)$ are almost equal to $\mathbf{w}_s$. Thus if $E(W, t)$ is reduced during learning we are at the same time reducing:

$$E_0(W, t)=\frac{1}{2}\sum_{\mathbf{w}_r}\sum_{\mathbf{w}_s}\sum_{\mathbf{v}_i\in R(\mathbf{w}_s)} p(\mathbf{v}_i)\varepsilon(t)h(r, s, t)|\mathbf{w}_s-w_r|^2$$

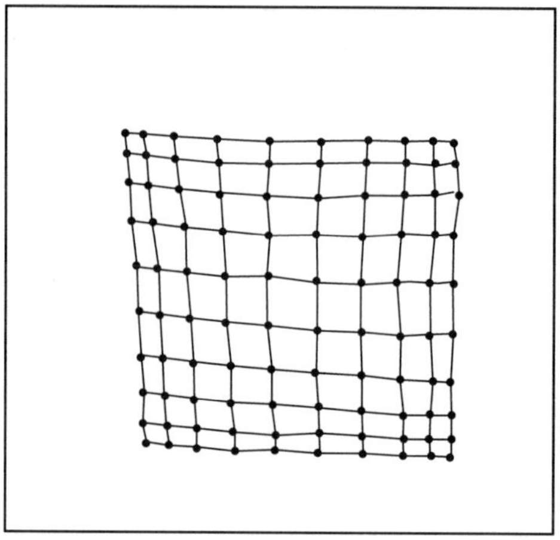

Figure 4.30 Weight ordering at $t = 1000$ of Example 4.3

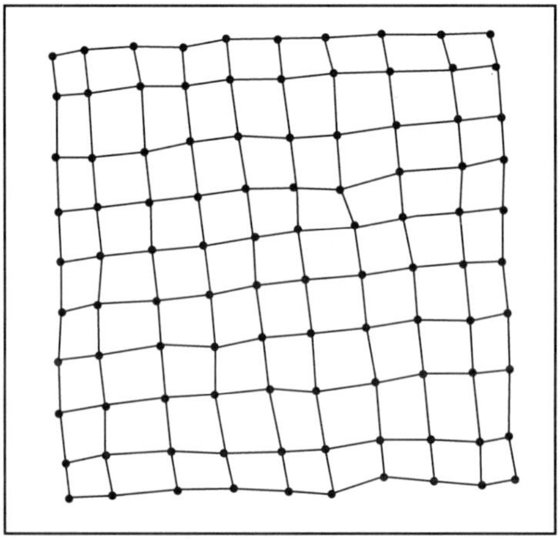

Figure 4.31 Weight ordering at $t = 10\,000$ of Example 4.3

The sum

$$\sum_{\mathbf{v}_i \in R(\mathbf{w}_s)} p(\mathbf{v}_i)$$

is equal to the probability that neuron u_s with weight vector $\mathbf{w}_s$ will be the winning neuron if we present some input vector $\mathbf{v}_i$ to the neural net. We will denote this probability by $p(\mathbf{w}_s)$.

We can now replace the expression for $E_0(W, t)$ by:

$$E_0(W, t) = \frac{1}{2} \sum_{\mathbf{w}_r} \sum_{\mathbf{w}_s} \varepsilon(t) p(\mathbf{w}_s) h(r, s, t) |\mathbf{w}_s - \mathbf{w}_r|^2$$

Or with $\varepsilon(t)h(r, s, t) = g(r, s, t)$:

$$E_0(W, t) = \frac{1}{2} \sum_{\mathbf{w}_r} \sum_{\mathbf{w}_s} g(r, s, t) p(\mathbf{w}_s) |\mathbf{w}_s - \mathbf{w}_r|^2$$

This expression reflects more directly that the final weight will be ordered. We can summarize our discussion by the following theorem:

Theorem 4.2

By using in the self-organizing algorithm the adaptation rule:

$$\mathbf{w}_r(t+1) = \mathbf{w}_r(t) + \varepsilon(t) h(r, s, t)[\mathbf{v}_i(t) - \mathbf{w}_r(t)]$$

the final result will be as follows:

1. *Vector quantization* of the data set D_V with respect to the weights of the neural net because we are minimizing in the final phase of learning:

$$E(W) = \sum_{\mathbf{w}_s} \sum_{\mathbf{v}_i \in R(\mathbf{w}_s)} p(\mathbf{v}_i) \cdot |\mathbf{v}_i - \mathbf{w}_s|^2$$

2. *The neural net will be ordered*: neighbouring neurons will have similar weight vectors, because we are minimizing at each step t:

$$E_0(W, t) = \frac{1}{2} \sum_{\mathbf{w}_r} \sum_{\mathbf{w}_s} g(r, s, t) p(\mathbf{w}_s) |\mathbf{w}_s - \mathbf{w}_r|^2$$

In the optimal situation of ordering, the net is called *well-ordered*: neighbouring neurons will have adjacent receptive fields.

$E(W, t)$ will be called the *quantization error*, and $E_D(W, t)$ will be called the *ordering error*.

Although the expression for the ordering error reflects also that the sum of weighted square distances between all weight vectors will be minimized, we can simplify the expression still further to obtain a more practical measure to evaluate the effect of reduction of the sum of squared distances between the weight vectors.

For the adaptation function we have $g(r, s, t) = \varepsilon(t)h(r, s, t)$. If $d_L(\mathbf{i}_s, \mathbf{i}_r)$ is small, then $h(r, s, t)$ is large (≈ 1) and if $d_L(\mathbf{i}_s, \mathbf{i}_r)$ is large, then $h(r, s, t)$ is small ($\rightarrow 0$). Moreover during learning the value of $\varepsilon(t)$ will decrease with time. At any given moment of time the value of $g(r, s, t)$ is the same for all pairs of neurons u_r and u_s with a given mutual distance δ_i. The set Δ of different distances in the neural network is finite, and thus the summation in the last expression for $E_0(W, t)$:

$$E_0(W, t) = \frac{1}{2} \sum_{\mathbf{w}_r} \sum_{\mathbf{w}_s} g(r, s, t) p(\mathbf{w}_s) |\mathbf{w}_s - \mathbf{w}_r|^2$$

can be replaced by the sum over the finite set of all distances:

$$E_0(W, t) = \frac{1}{2} \sum_{\delta_i \in \Delta} \sum_{d_L(\mathbf{i}_s, \mathbf{i}_r) = \delta_i} g(r, s, t) p(\mathbf{w}_s) |\mathbf{w}_s - \mathbf{w}_r|^2$$

For each distance δ_i the value of $g(r, s, t)$ will be large ($\leqslant 1$) at the beginning of the learning phase, and small ($\to 0$) at the final phase of learning. The greater the distance δ_i, the faster the value of $g(r, s, t)$ decreases. Now we replace for each value of the distance δ_i the time-dependent $g(r, s, t)$ by some constant 'effective mean value' g_i. It is hard to determine the value of g_i but we do not need the exact value; we only need to know that $g_i > g_j$ if $\delta_i < \delta_j$. With this crude replacement we obtain a measure $E_D(W)$, called the *approximate ordering error*, for the sum of weighted squared distances between weight vectors:

$$E_D(W) = \frac{1}{2} \sum_{\delta_i \in \Delta} \sum_{d_L(\mathbf{i}_s, \mathbf{i}_r) = \delta_i} g_i p(\mathbf{w}_s) |\mathbf{w}_s - \mathbf{w}_r|^2$$

We will use the property that the algorithm will reduce the value of the sum of weighted squared distances between weight vectors in the application of the algorithm to the 'travelling salesman problem' in Section 4.6.

We note that the distribution of weights will be such that the concentration of weights will be high (the receptive fields will be small) in areas of a high density of elements of the data set D_V, and the distribution of weights will be sparse (the receptive fields will be large) in areas of low density of elements of D_V. This phenomenon implies that the probability

$$p(\mathbf{w}_s) = \sum_{\substack{\mathbf{v}_i \in R(\mathbf{w}_s) \\ \mathbf{v}_i \in D_V}} p(\mathbf{v}_i)$$

is almost constant and can roughly be approximated by $1/N$ if N is the number of neurons in the net. In that case the expression for $E_D(W)$ becomes:

$$E_D(W) = \frac{1}{2N} \sum_{\delta_i \in \Delta} \sum_{d_L(\mathbf{i}_s, \mathbf{i}_r) = \delta_i} g_i |\mathbf{w}_s - \mathbf{w}_r|^2$$

The learning algorithm will minimize the value of $E_D(W)$, which means that for every value of the mutual distance δ_i between two neurons, the sum of the squared distances between weight vectors of those pairs of neurons will be minimized.

Thus we have on one hand the vector quantization – the algorithm will minimize (Theorem 4.1):

$$E(W) = \frac{1}{2} \sum_{\mathbf{w}_s} \sum_{\substack{\mathbf{v}_i \in R(\mathbf{w}_s) \\ \mathbf{v}_i \in D_V}} p(\mathbf{v}_i) |\mathbf{v}_i - \mathbf{w}_s|^2$$

and on the other hand we have the reduction of the sum of weighted squared distances between weight vectors because the algorithm will at the same time minimize:

$$E_D(W) = \sum_{\delta_i \in \Delta} \sum_{d_L(\mathbf{i}_s, \mathbf{i}_r) = \delta_i} c_i |\mathbf{w}_s - \mathbf{w}_r|^2 \text{ with } c_i = g_i/2N$$

The adaptation algorithm will try to minimize both error functions. It might, however, be that minimizing the quantization error $E(W)$ can be in conflict with minimizing the ordering error $E_0(W)$ – or the approximation thereof by $E_D(W)$. A better ordering can, for example, be made at the expense of vector quantization.

In case that the values of g_i are negligible for neurons that are not direct neighbours, we obtain for the ordering error the practical approximation:

$$E_D(W) = \sum_{d_L(\mathbf{i}_s, \mathbf{i}_r) = 1} c_1 |\mathbf{w}_s - \mathbf{w}_r|^2$$

(Note: for a Euclidean distance measure we have $\min\{d_L(\mathbf{i}_s, \mathbf{i}_r)\} = 1$.)

Example 4.4

We use a two-dimensional neural lattice with nine neurons and Euclidean distance measure d_L. For the input data set we take $D_V = \{[1, 1], [2, 2], [3, 3], [4, 4], [6, 6], [7, 7], [8, 8], [9, 9]\}$. For an optimal vector quantization one would expect the final values of the weights to be identical with the input data vectors ($E(W) = 0$) and distributed over the neurons in the neural lettice as, for example, depicted in Figure 4.32.

If we calculate the value of:

$$E_D(W) = \sum_{d_L(\mathbf{i}_s, \mathbf{i}_r) = 1} c_1 |\mathbf{w}_s - \mathbf{w}_r|^2$$

we obtain $E_D(W) = 132c_1$.

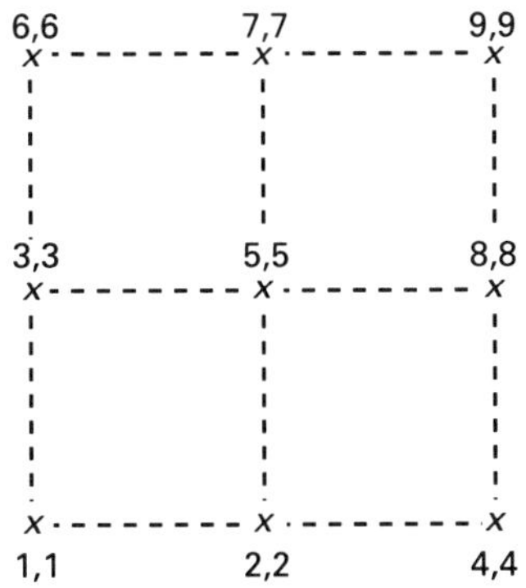

Figure 4.32 Expected values of weight vectors of Example 4.4

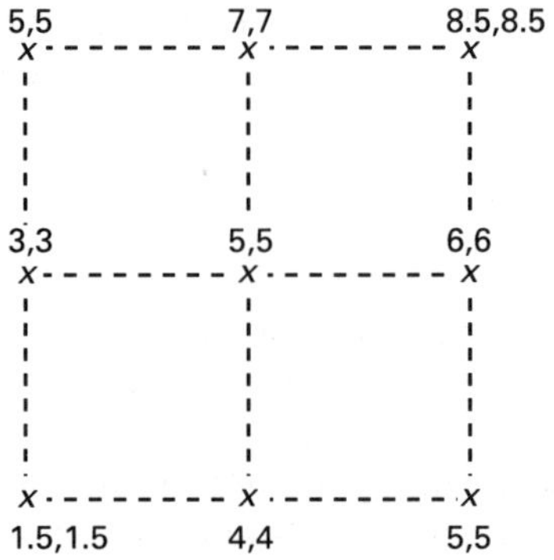

Figure 4.33 The obtained values of weight vectors of Example 4.4

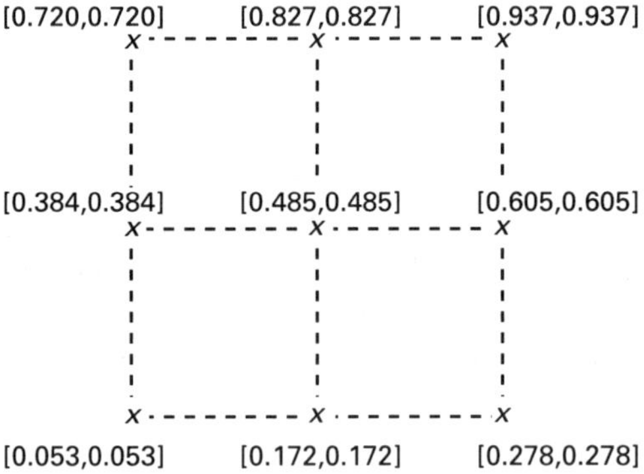

Figure 4.34 The neural lattice with its nine weight vectors

If, however, we apply the self-organizing algorithm we will not obtain the result given above. The nine input data vectors will not all be copied by the weights of the neural net and we will not get a perfect vector quantization. The weights will be as given in the neural lattice of Figure 4.33. The approximate ordering error $E_D(W)$ will, however, be lower than in case of perfect vector quantization: $E_D(W) = 51.5c_1$.

The phenomenon that vector quantization is overruled by the ordering of weight vectors occurs when the number of input vectors is of the same order as the number of neurons. In that case the measure of vector quantization $E(W, t)$ can have sharp local minima, and during the orderings phase of learning we can get stuck in such a minimum.

If, during learning, we apply for example 10 000 two-dimensional input vectors $\mathbf{v}_i$ with $v_{i1} = v_{i2}$ homogeneous distributed between [0, 0] and [1, 1], we will obtain perfect vector quantization of the data set D_V with nine equally spaced receptive fields with the weights in the centre. In a simulation experiment we obtained the weight vectors

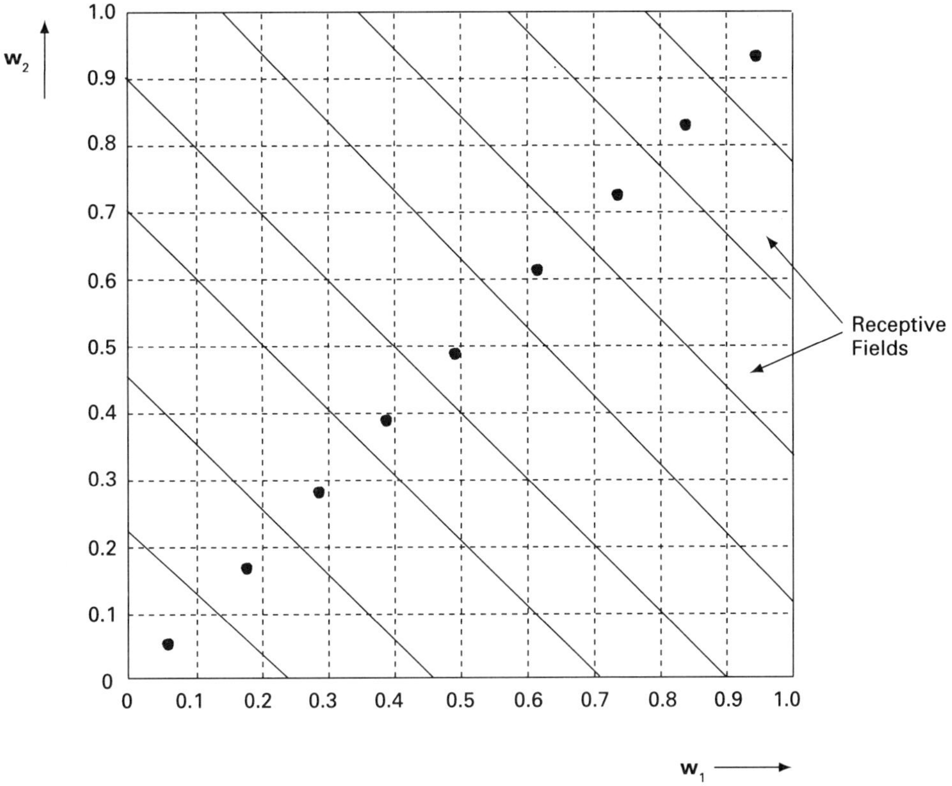

Figure 4.35 The input/weight space with the obtained weight vectors and their receptive fields in case of 10 000 homogeneous distributed input vectors with $v_{i1} = v_{i2}$

as given in the neural lattice of Figure 4.34. The weight vectors and their receptive fields are also given in the input space V restricted to $[0, 0] \times [1, 1]$ in Figure 4.35. ■

4.6 Application of the self-organizing neural-net algorithm to the travelling salesman problem

The travelling salesman problem can be solved reasonably well by the self-organizing algorithm. The solution gives at the same time some understanding of the behaviour of the algorithm.

A travelling salesman has to visit several towns located in a two-dimensional plane.

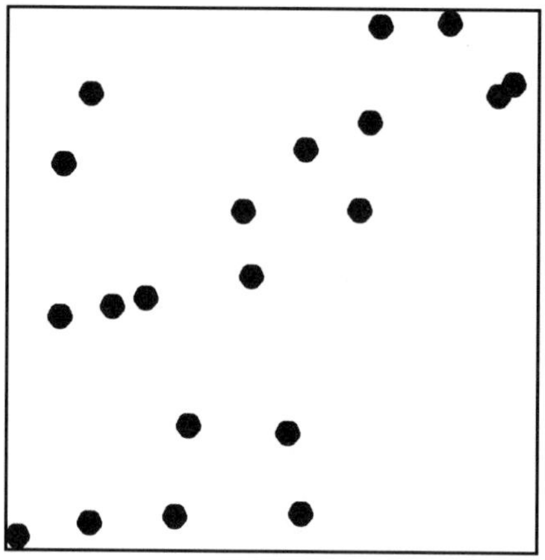

Figure 4.36 The towns to be visited by the travelling salesman

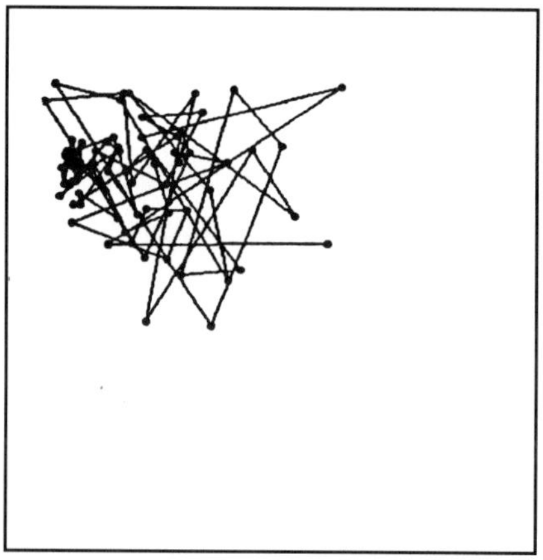

Figure 4.37 Initial distribution of weights (dots) in a one-dimensional net of 100 neurons. Weights of neighbouring neurons are connected with a straight line

The salesman wants to take the shortest route between all the towns given some town of departure (see Figure 4.36, where dots represent the towns).

We can represent the towns by two-dimensional vectors $\mathbf{v}_i$ containing the coordinates of the towns. Assume there are twenty towns to be visited, then our data set D_V contains twenty different vectors $\mathbf{v}_i$. We know that we can represent the vectors by twenty two-dimensional weights $\mathbf{w}_j$ in a neural net containing twenty neurons. Due to the vector quantization property of the self-organizing algorithm we know that the weight vectors in the net will become similar (or identical) to the input vectors $\mathbf{v}_i$ (Theorem 4.1). If we use a one-dimensional neural net we also know

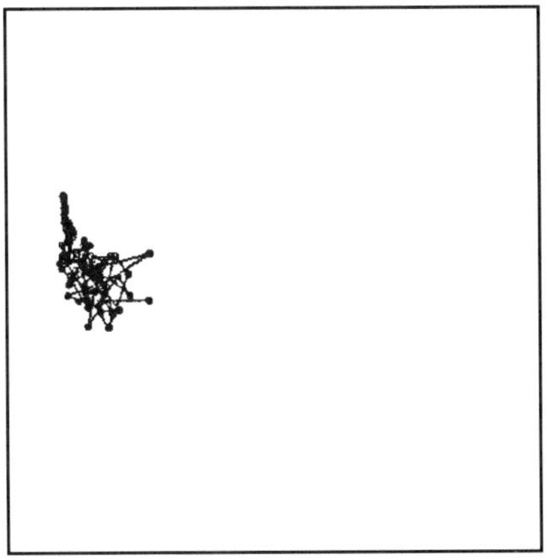

Figure 4.38 Distribution of weights (dots) after one learning step in a one-dimensional net of 100 neurons. Weights of neighbouring neurons are connected with a straight line

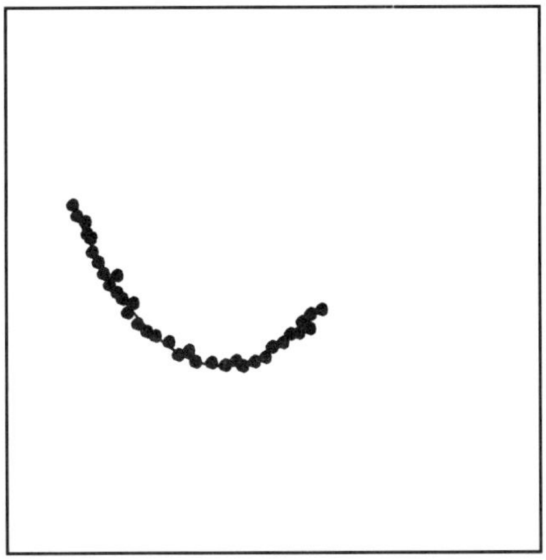

Figure 4.39 Distribution of weights (dots) after ten learning steps in a one-dimensional net of 100 neurons. Weights of neighbouring neurons are connected with a straight line

(Theorem 4.2) that the net will be well-ordered by one row of weights (representing towns) such that the distance between successive weights is as small as possible. We also found in the previous section that the sum of (weighted) squared mutual distances between all weights (representing towns) will be minimized.

Because the number of learning steps must be high (e.g. 5000) and the number of vectors in the data set is low (20) we have to present the data set several times (250 ×) to the self-organizing algorithm. Although twenty neurons are sufficient we used 100

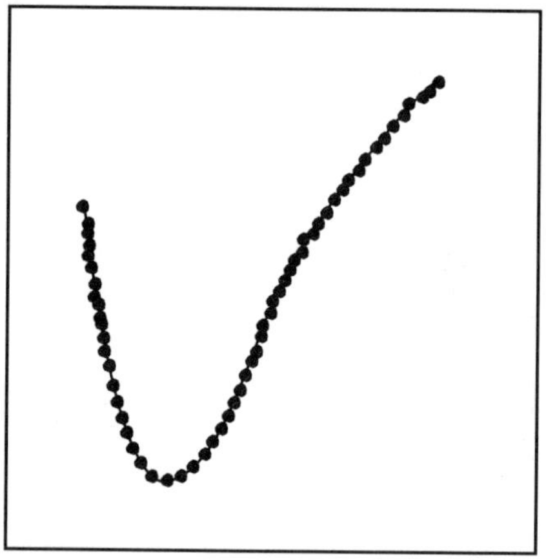

Figure 4.40 Distribution of weights (dots) after fifty learning steps in a one-dimensional net of 100 neurons. Weights of neighbouring neurons are connected with a straight line

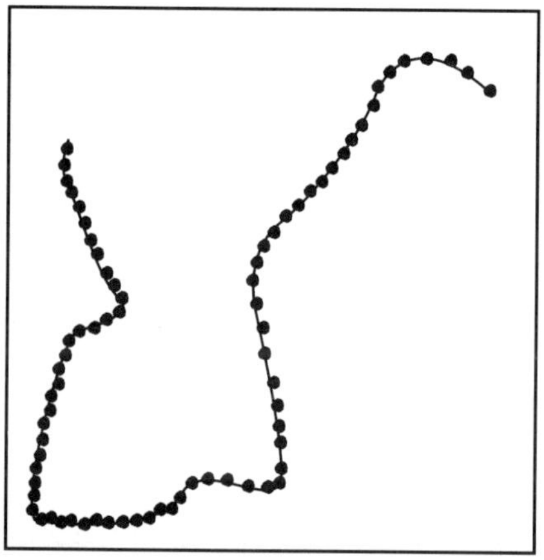

Figure 4.41 Distribution of weights (dots) after 500 learning steps in a one-dimensional net of 100 neurons. Weights of neighbouring neurons are connected with a straight line

neurons (and thus 100 weights) to ensure a smooth route in situations where towns are wildly distributed over the plane.

The results are given in Figures 4.37–4.41. Dots represent the weights in the neural net. If two weights belong to neighbouring neurons, then they are connected with a straight line. Figure 4.37 gives the initial distribution of weights. Figure 4.38 is the result after one learning step; for Figure 4.39 $t=10$; for Figure 4.40 $t=50$; and for

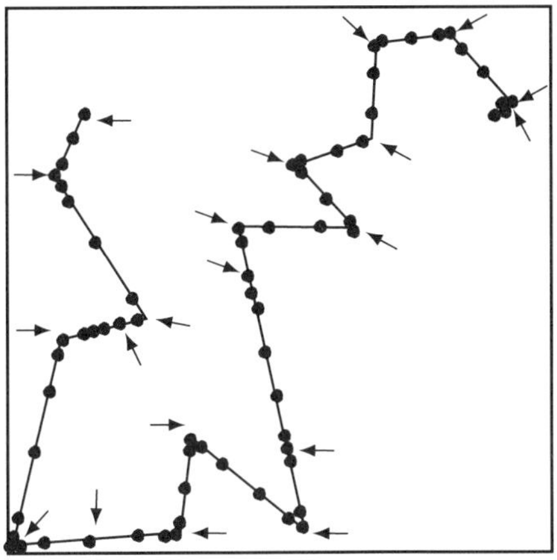

Figure 4.42 Final distribution of weights (dots) after 5000 learning steps. Arrows indicate the positions of the towns

Figure 4.41 $t = 500$. Figure 4.42 gives the final result after 5000 steps (arrows indicate the position of towns).

4.7 Application of the self-organizing neural net algorithm to picture colour quantization

Suppose we have a digitized picture of $m \times n$ pixels. This picture is said to be in true colour format if the three primary colours – red, green and blue (RGB) – for each pixel are coded separately with 8 bits. Thus the colour coding of one pixel requires 24 bits. This yields a total range of 16 million possible colours for each pixel. Many display devices, however, like PC monitors, are capable of displaying only 256 colors. Thus if we want to display a picture given in true colour format we have to reduce the number of colours. We also have to select the different elements of the 256 colours and to map each pixel colour on one of the selected 256 colours. The selected set of colours is called the palette of colours. The palette can be chosen such that the different colours are evenly distributed over the total range of RGB colours. We can, however, do better by selecting the palette colours corresponding to the probability of colours in the original picture. The colour of a pixel p in the picture can be represented by a three-dimensional vector $\mathbf{v}_p = [r_p, g_p, b_p]$ with $p = 1, 2, \ldots, m \times n$. The colours of the palette can be represented by 256 three-dimensional vectors $\mathbf{w}_s = [r_s, g_s, b_s]$ with $s = 1, 2, \ldots, 256$. We want to select the palette colours without losing too much colour information of the original picture. This means we have to select the vectors $\mathbf{w}_s$ such that:

$$E(W)=\sum_{\mathbf{w}_s}\sum_{\mathbf{v}_p\in R(\mathbf{w}_s)} p(\mathbf{v}_p)|\mathbf{v}_p-\mathbf{w}_s|^2$$

is minimal. This minimalization process can be realized by the self-organizing neural net algorithm (see Theorem 4.1).

In an experiment we used four different true colour format pictures:

1. 'Child', a scanned snapshot of a child.
2. 'PSLogo', an artificial picture with smoothly changing colours and a large black area.
3. 'Bridge', a scanned photograph of a bridge in a landscape.
4. 'Clay', a scanned photograph of several objects on a table.

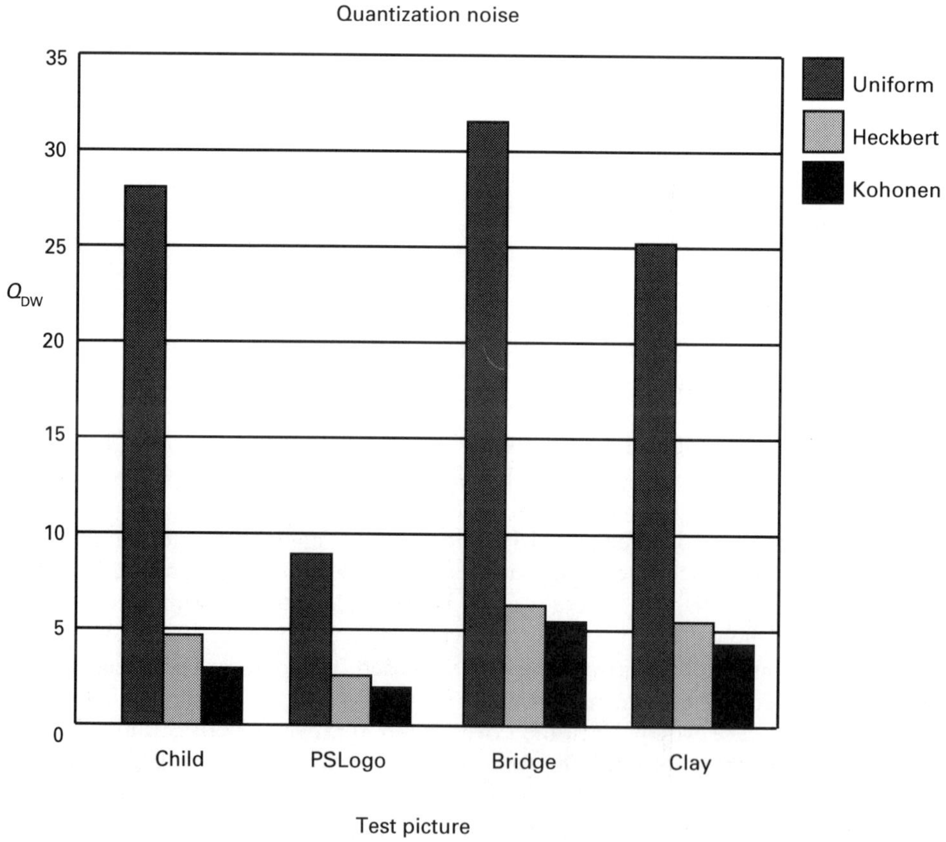

Figure 4.43 Quantization noise for different pictures and different quantization methods: uniform palette, Heckbert's algorithm and the self-organizing algorithm of Kohonen

In separate experiments we selected randomly 10 000 vectors from each picture. For each set of vectors we trained a different self-organizing neural network. After learning, the set of weight vectors in the four neural networks represents the palettes for the four different pictures. With the use of the neural network each colour vector $\mathbf{v}_p$ from a picture in true colour format was transformed to a palette colour vector $\mathbf{w}_s$: $\mathbf{v}_p \in R(\mathbf{w}_s)$. Once we have obtained the palette colour vectors for each pixel, the original picture can be reconstructed. The obtained pictures were of a very good quality.

The quality of picture colour vector quantization can also be measured by the *energy of quantization noise* defined by (see Section 4.4):

$$E_{DW} = \frac{1}{M} \sum_{\mathbf{w}_s} \sum_{\mathbf{v}_p \in R(\mathbf{w}_s)} |\mathbf{v}_p - \mathbf{w}_s|^2$$

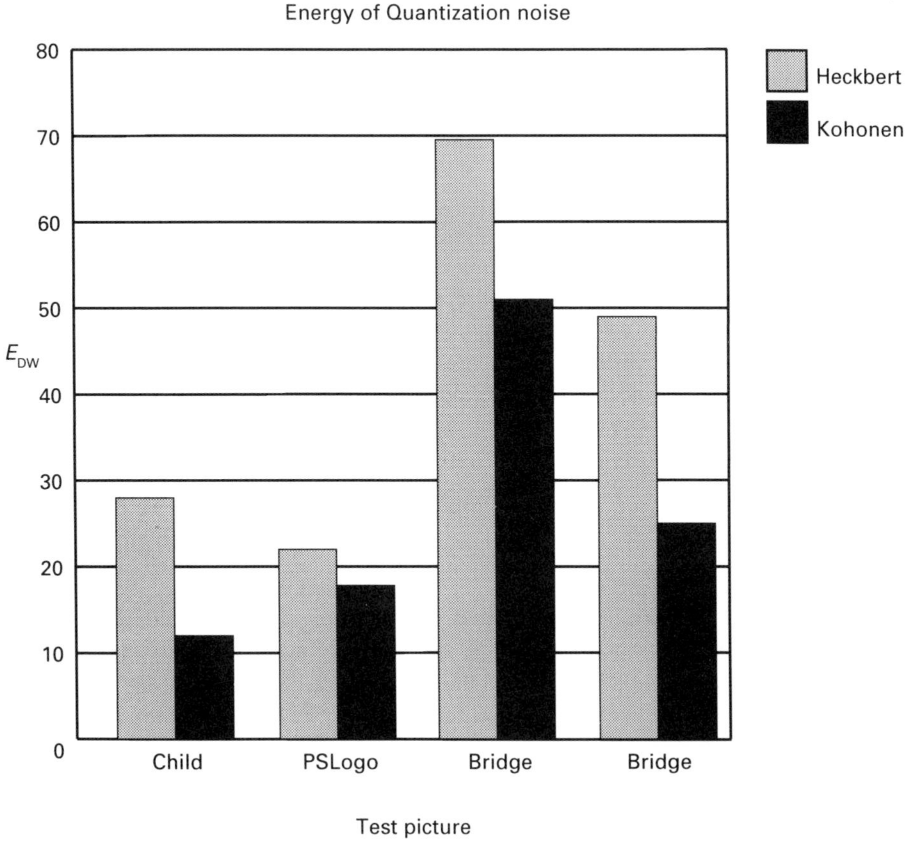

Figure 4.44 Energy of quantization noise for different pictures and different quantization methods: Heckbert's algorithm and the self-organizing algorithm of Kohonen

with M the number of elements in D_V; or by the *quantization noise* defined by:

$$Q_{DW} = \frac{1}{M} \sum_{\mathbf{w}_s} \sum_{\mathbf{v}_p \in R(\mathbf{w}_s)} |\mathbf{v}_p - \mathbf{w}_s|$$

We give the results in the histograms of Figure 4.43 (Q_{DW}) and Figure 4.44 (E_{DW}) in comparison with the method with uniform distributed colour palette and the well-known classical vector quantization method of Heckbert (1982).

4.8 Nearest-neighbour classification with a self-organizing neural net algorithm

If we have to classify objects represented by measurement vectors $\mathbf{v}_i$ into two different classes X_A and X_B, and we have only two finite sets of examples $D_A \subset X_A$ and $D_B \subset X_B$ and we have no other *a priori* information, then the best thing to do is to classify a new element $\mathbf{v}_i$ into class X_A if $\mathbf{v}_i$ is close to an element of D_A and into class X_B if $\mathbf{v}_i$ is close to an element of D_B, or more precisely: $\mathbf{v}_i$ is classified as an element of X_A if $|\mathbf{v}_i - \mathbf{w}_s| = \min_r |\mathbf{v}_i - \mathbf{w}_r|$ with $\mathbf{w}_r \in D_A \cup D_B$ and $\mathbf{w}_s \in D_A$ and $\mathbf{v}_i$ is classified as an element of X_B if $\mathbf{w}_s \in D_B$.

Example 4.5

In Figure 4.45 we have given the data sets of the mushroom classification problem mentioned in Section 3.1 and also the classification boundary between the classes X_A and X_B if we use the *nearest-neighbour method* presented above.

Suppose we do not want the elements of the data sets D_A and D_B as reference vectors for nearest-neighbour classification, but we prefer to have other (more reliable) data sets D'_A and D'_B with elements $\mathbf{w}'_s$ that represent clusters of elements of D_A and D_B respectively. Those sets can be obtained by using the self-organizing neural net algorithm applied to D_A and to D_B separately. ■

Example 4.6

We applied the self-organizing algorithm with the set D_A (represented by symbols 'a') of the previous example as a training set. We used a two-dimensional neural net A of 2×2 neurons. The weights obtained are given in Figure 4.46 by small circles. We did the same with the data set D_B with a neural net B. The weights are represented by small squares in Figure 4.47.

We can now use both neural nets to classify new vectors. We can determine the weight vector $\mathbf{w}_s$ such that $|\mathbf{v}_i - \mathbf{w}_s| = \min_r |\mathbf{v}_i - \mathbf{w}_r|$ with $\mathbf{w}_r$ the weight vectors of neural nets A and B. If the 'winning neuron' belongs to neural net A, then we classify $\mathbf{v}_i$ as an element of class A, and if $\mathbf{w}_s$ is an element of neural net B, then we classify $\mathbf{v}_i$ as belonging to class X_B. The boundary between X_A and X_B is given in Figure 4.48. ■

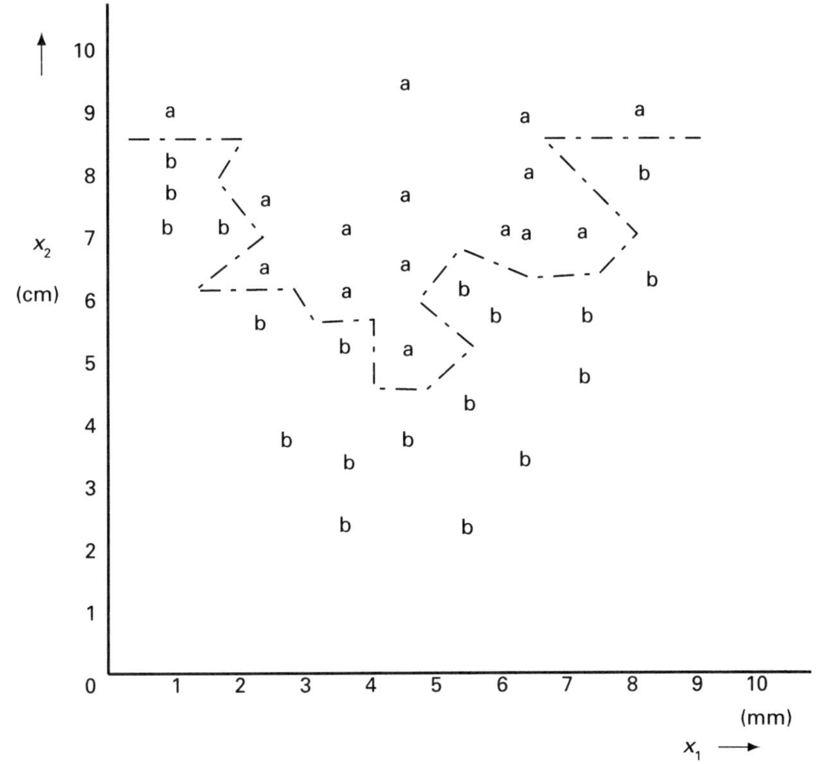

Figure 4.45 Classification boundary for the nearest-neighbour method

In Figure 4.49 one can see what is happening to the classification boundary if we use two neural networks with 4×8 neurons. The number of neurons in each net is larger than the number of samples in the data sets D_A and D_B. Note that almost every example of the data sets is represented by some weight vector in the neural nets. The redundant weights are automatically placed by interpolation between examples of the data sets.

If besides the examples of the data sets $D_A \subseteq X_A$ and $D_B \subset X_B$, one has some additional information about the classes X_A and X_B, then the nearest-neighbour classification might not be optimal. Suppose, for instance, one knows (or assumes) that there are some class conditional probability density functions $f(\mathbf{v}|A)$ and $f(\mathbf{v}|B)$, then, in the case of equal cost of misclassification and equal class probabilities $p(A)$ and $p(B)$, one must assign an input $\mathbf{v}$ to class X_A if $f(\mathbf{v}|A) > f(\mathbf{v}|B)$ and to class X_B if $f(\mathbf{v}|B) > f(\mathbf{v}|A)$. This will not, however, be the case if we use the nearest-neighbour method. If one assumes the existence of overlapping probability distributions by which the examples are generated, it is better to use the Bayes method that we will discuss in the next section.

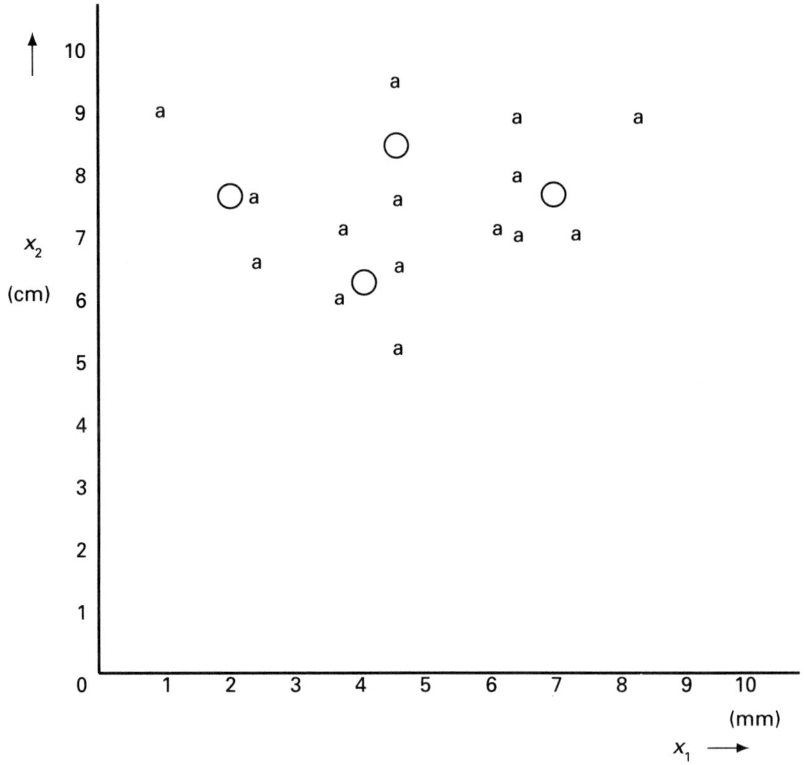

Figure 4.46 Result of vector quantization for class X_A with a 2×2 neural Kohonen network A

Example 4.7

We consider a one-dimensional case. We have two finite data sets D_A and D_B. The elements of D_A are generated according to some Gaussian distribution density function with a mean of $\mu_A = 0.0$ and a deviation of $\sigma_A = 1.00$. The other set is generated by distribution function with $\mu_B = 2.00$ and deviation $\sigma_B = 1.50$ (see Figure 4.50). The optimal boundary for classification is $t = 1.09$. The other boundary is $t = -4.29$ (not given in Figure 4.50).

We train with the data set D_A a one-dimensional self-organizing neural network algorithms with five neurons. The values of the final weights of this network A are given in the first row at the bottom of Figure 4.50. We do the same for the data set D_B. The final weights of network B are given in the second row. If we now apply the nearest-neighbour method, all inputs in the receptive field of the leftmost weight of neural net B are wrongly classified as elements of class X_B (see the last row of Figure 4.50). In the same way all inputs in the receptive field of the rightmost weight of the neural net A are wrongly classified as members of class X_A. ■

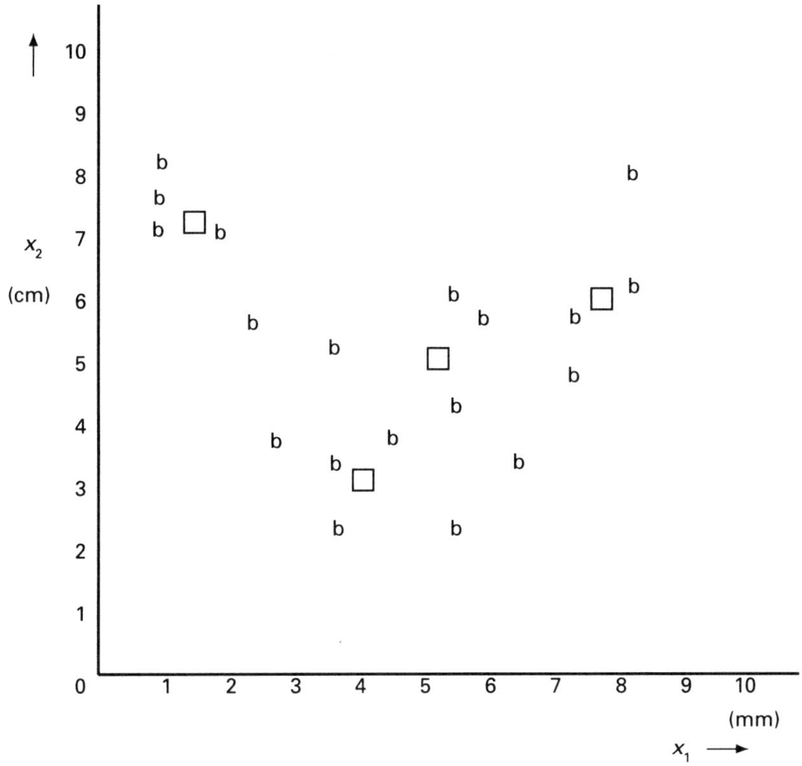

Figure 4.47 Result of vector quantization for class X_B with a 2×2 neural Kohonen network B

4.9 The Bayes classification with a self-organizing neural net algorithm

If one knows or assumes that the examples of data sets of different classes are generated according to some underlying probability distribution functions, then the best thing to do is to estimate the parameters of those distributions from the given data sets, and use this information to determine a threshold for classifying new data.

In case of a two-class classification problem (with equal class probabilities $p(A)$ and $p(B)$ and equal costs for misclassification) one assigns an input $\mathbf{v}_i$ to class X_A if the class conditional density function $f(\mathbf{v}_i|A)$ is larger than the class conditional density function $f(\mathbf{v}_i|B)$. This method of classifying inputs can be realized by a self-organizing neural net without separately estimating the parameters of the underlying probability distributions, because the neural net will do the job for us.

First we have to make a slight modification of the algorithmic adaptation rule. We append both the input vectors $\mathbf{v}_i$ (hereafter called the *master input vector*) and the

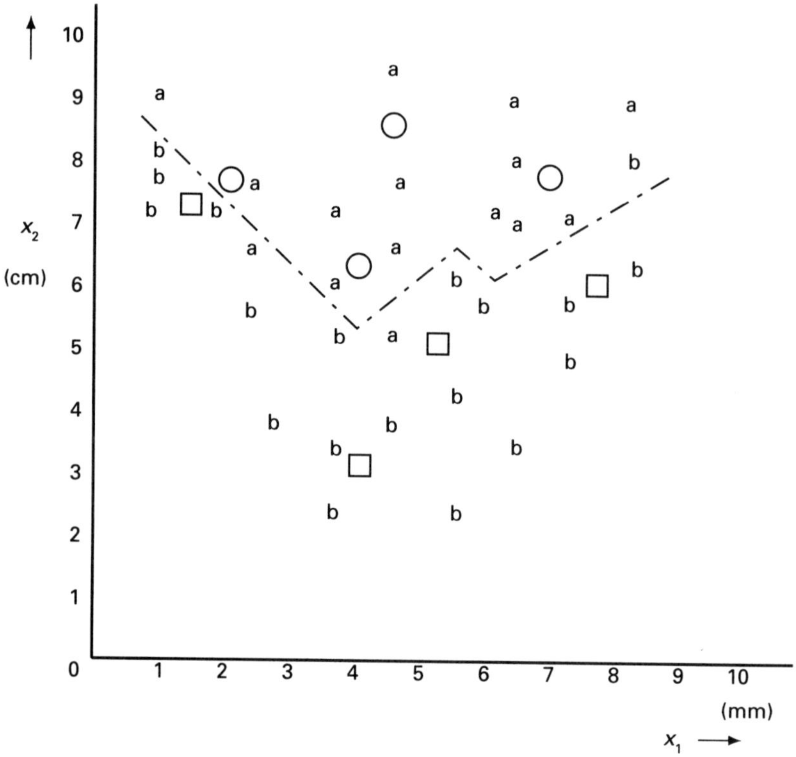

Figure 4.48 Classification boundary with the nearest-neighbour method for the representations in the neural networks A and B

weight vectors $\mathbf{w}_r$ (hereafter called the *master weight vectors*) with a so-called *slave vector* $\tilde{\mathbf{v}}_i$ (respectively $\tilde{\mathbf{w}}_r$) with a number of components equal to the number of classes. The newly appended input vector will be denoted by $\hat{\mathbf{v}}_i = [\mathbf{v}_i, \tilde{\mathbf{v}}_i]$. The appended weight vector will be denoted by $\hat{\mathbf{w}}_r = [\mathbf{w}_r, \tilde{\mathbf{w}}_r]$.

If an input vector is an element of the data set of the kth class, then we make the kth element of the slave input vector equal to 1; if it does not belong to the kth class, then we make that component of the slave input vector equal to 0. The components of the initial master weight vector and of the initial slave weight vector are randomly chosen between 0 and 1. The algorithmic adaptation rule is then as follows.

Given some master input vector $\mathbf{v}(t) = \mathbf{v}_i$ with $\hat{\mathbf{v}}(t) = [\mathbf{v}_i, \tilde{\mathbf{v}}_i]$.

1. Determine the winning neuron u_s for the master vector $\mathbf{v}(t)$, i.e.

$$d_V[\mathbf{w}_s(t), \mathbf{v}(t)] = \min_r d_V[\mathbf{w}_r(t), \mathbf{v}(t)]$$

2. Every weight vector $\hat{\mathbf{w}}_r(t) = [\mathbf{w}_r, \tilde{\mathbf{w}}]$ in the net will be changed to:

$$\hat{\mathbf{w}}_r(t+1) = \hat{\mathbf{w}}_r(t) + g(r, s, t)[\hat{\mathbf{v}}(t) - \hat{\mathbf{w}}_r(t)]$$

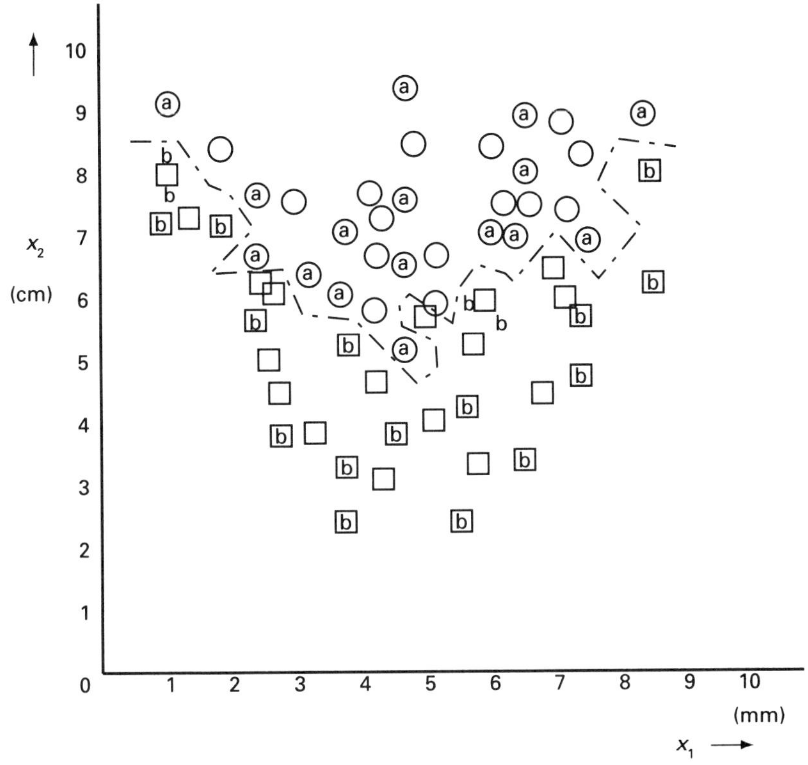

Figure 4.49 Classification boundary with the nearest-neighbour method for the representations in two 4×8 neural networks A and B

with $g(r, s, t)$ a scalar-valued function with a value between 0 and 1 depending on time and on the distance in the neural net between the winning neuron u_s and the neuron u_r to be adapted.

To explain the final result and what is happening during the learning phase we confine ourselves to the two-category classification problem. In this case we can use a one-dimensional slave vector. The one-dimensional slave vectors for master vectors of D_A will be given a value 0 and the one-dimensional slave vectors for master vectors of D_B will be given the value 1.

Due to the vector quantization property of the algorithm the final weight vectors on the one hand will become similar to the elements $\hat{\mathbf{v}}_i$ corresponding to the master vectors of the data set $D_A \subseteq X_A$, and on the other hand similar to vectors $\hat{\mathbf{v}}_i$ corresponding to the examples $\mathbf{v}_i$ of the data set $D_B \subseteq X_B$. In a region of the input space (= weight space) where there are only elements of D_A, the slave element of the weight vectors will be permanently adapted to a value 0, and in a region dominated by elements of D_B the slave elements of the weight vectors will be mainly adapted

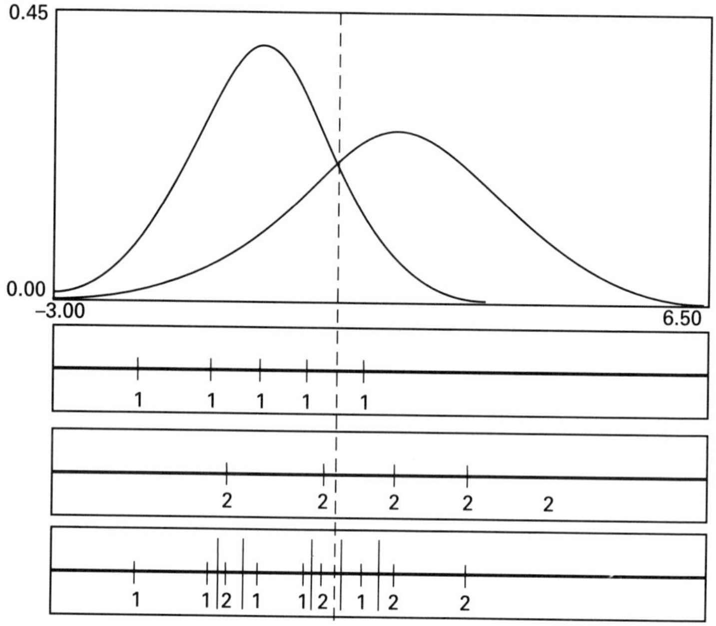

Figure 4.50 Top diagram: two Gaussian-distributed, one-dimensional classes. First row: weights obtained after vector quantization with a one-dimensional net of five neurons for data with the left distribution. Second row: weights obtained after vector quantization with a one-dimensional net of five neurons for data with the right distribution. Third row: classification result with nearest-neighbour method

to a value 1. In a region of the input space where the unknown class conditional probability density function are the same, $f(\mathbf{v}|\mathrm{A})=f(\mathbf{v}|\mathrm{B})$, the number of elements of D_{A} and D_{B} will be almost the same. In that region the slave elements of the weight vectors will be adapted as many times to value 1 as to value 0, thus in those regions the slave elements of weight vectors will become equal to 0.5.

The final result will be that a master input vector $\mathbf{v}_i \in X_{\mathrm{A}}$ will belong to the receptive field $R(\mathbf{w}_r)$ of a master weight vector $\mathbf{w}_r$ with a slave element with a value smaller than 0.5, and if $\mathbf{v}_i \in X_{\mathrm{B}}$, then $\mathbf{v}_i$ belongs to the receptive field of a weight vector with a slave element value larger than 0.5.

After the learning phase the neural net can be used as a classifier: given some input vector $\mathbf{v}_i$, one can determine the 'winning neuron', then the slave value of the corresponding weight vector indicates whether (<0.5) or not (>0.5) the vector belongs to X_{A}.

The weight vectors with a slave value of ≈ 0.5 will be located near the optimal discrimination curve. We applied the method to the two-dimensional classification

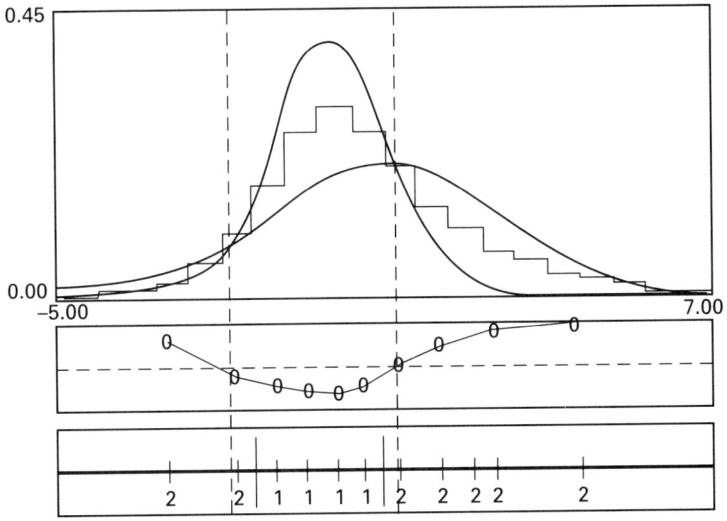

Figure 4.51 Top diagram: two Gaussian-distributed, one-dimensional classes. First row: the value of the slave element of the weight vectors obtained after vector quantization with a one-dimensional net of ten neurons for data of both distributions. Third row: classification result with the Bayes method

problem discussed in Example 3.8 with a two-dimensional neural net with 10×10 neurons. We found a classification error of 5.45 per cent.

Example 4.8

We consider a two-category classification problem with a one-dimensional input. The data set D_A is generated by a Gaussian distribution function with a mean of $\mu_A = 0.00$ and deviation $\sigma_A = 1.00$. The data set D_B is generated by a Gaussian distribution with $\mu_B = 1.00$ and $\sigma_B = 2.00$. A histogram for the frequency of the elements of $D_A \cup D_B$ in the training set, together with the class conditional density function, is given in Figure 4.51.

We used a single one-dimensional neural network with ten neurons. In the first row below the histogram in Figure 4.51 we have plotted the value of the slave elements of the ten weight vectors. We observe that for input elements $\mathbf{v}$ with $f(\mathbf{v}|B) > f(\mathbf{v}|A)$, the slave value is larger than 0.5 and thus will be classified correctly. The final row in Figure 4.51 gives the values of the master weights of the ten neurons after learning. ■

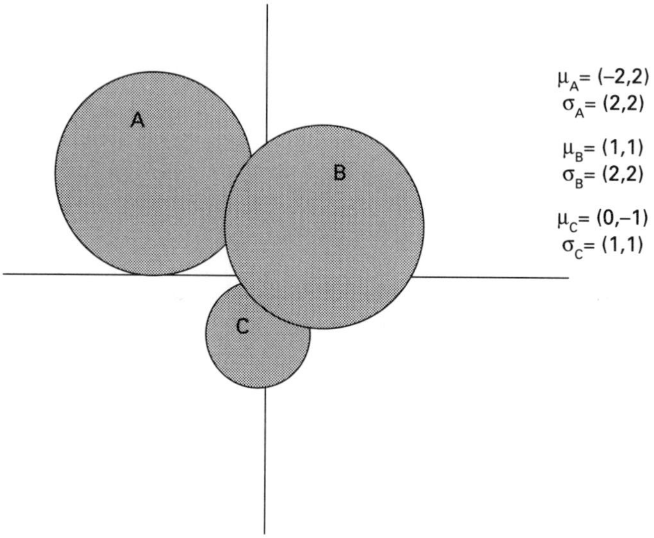

Figure 4.52 A three-class, two-dimensional data classification problem

Example 4.9

We consider a three-category two-dimensional classification problem. The three classes were generated with the following Gaussian probability distributions (see Figure 4.52):

$$\text{class A with } \mu_A=(-2, 2) \text{ and } \sigma_A=(2, 2)$$

$$\text{class B with } \mu_B=(1, 1) \text{ and } \sigma_B=(2, 2)$$

$$\text{class C with } \mu_C=(0, -1) \text{ and } \sigma_c=(1, 1)$$

The neural network was two-dimensional with 10×10 neurons. The input vectors of the data set were extended with a slave vector with three components. The first component is equal to 1 if the input vector is an element of class A; if not, then the value will be zero. The second component is only equal to 1 for input vectors from class B and the third component is only 1 for elements of class C. The weight vectors of the neurons were five-dimensional with the first two representing the master vector and the remaining three the slave vector. All weight values were randomly initialized.

The results after training with 1000 examples (1000 learning steps) are represented in Figure 4.53. The x, y coordinates of a symbol (A, B or C) give the values of the first and second weight components. These master weight vectors represent the quantized data set. If a symbol is equal to 'A', then the first component of the pertinent slave weight vector has the largest value. The same holds for the symbols 'B' and 'C' with respect to the second and third slave components. The four bold face symbols in Figure 4.53 are incorrect. The three rightmost vectors with symbol **A** have to be

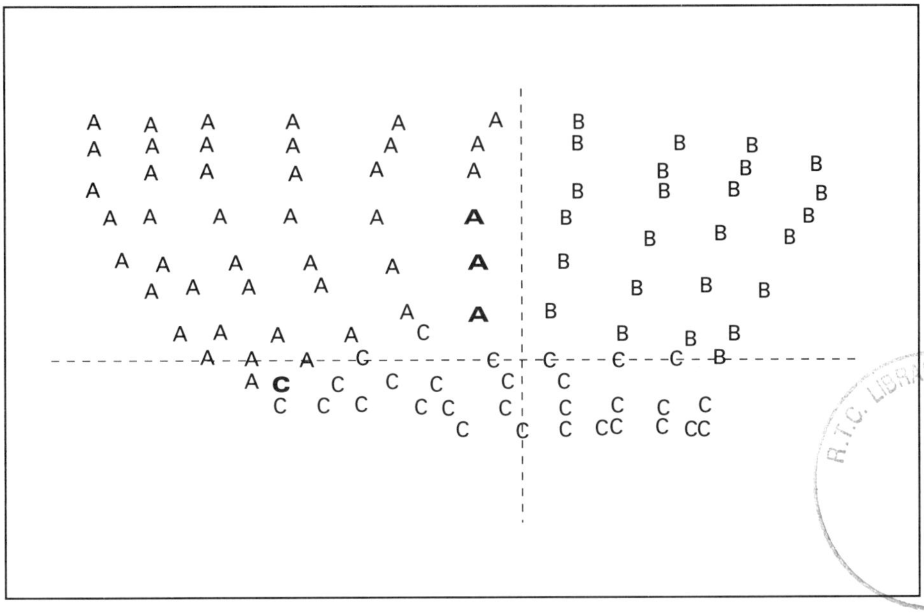

Figure 4.53 The weights after learning with the data of Figure 4.52 in a two-dimensional 10×10 neural net. A weight is labelled as 'A' if the first slave element is larger than the other two slave elements. Classes 'B' and 'C' are labelled similarly

classified as elements of class B. The bold face symbol **C** must be an A. A larger neural net would improve the results. ■

4.10 Application of the self-organizing neural net algorithm to the classification of handwritten digits

If we sample different pictures of a some class of pictures (in our case handwritten representations of some digit, see Figure 4.54) with the window introduced in Section 4.2, and present in a learning phase the observation vectors $\mathbf{v}$ obtained by that window to a self-organizing net, then the weight vectors will become similar to those observations that are common in all pictures in that class. In this way the topological features of a class of pictures will be stored in the weight vectors of the neural net.

We performed a classification experiment for handwritten digits. Figure 4.55 shows some examples of handwritten digits. We used the nearest-neighbour method discussed in Section 4.8.

We used ten two-dimensional self-organizing neural networks of 7×7 neurons, one network for each class of handwritten digits: '0', '1',..., '9'. Each handwritten

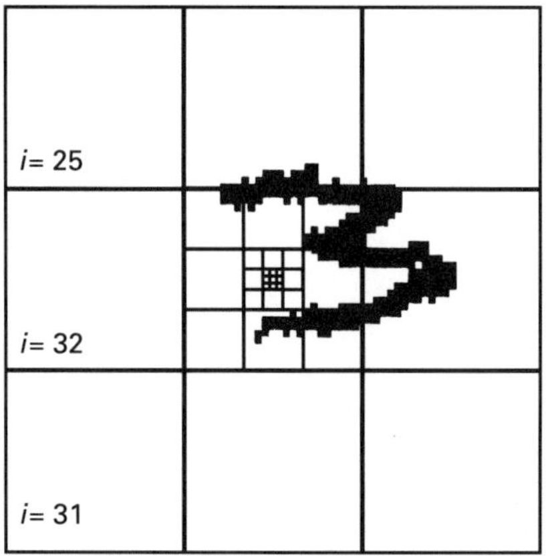

Figure 4.54 A handwritten '3' observed with the window

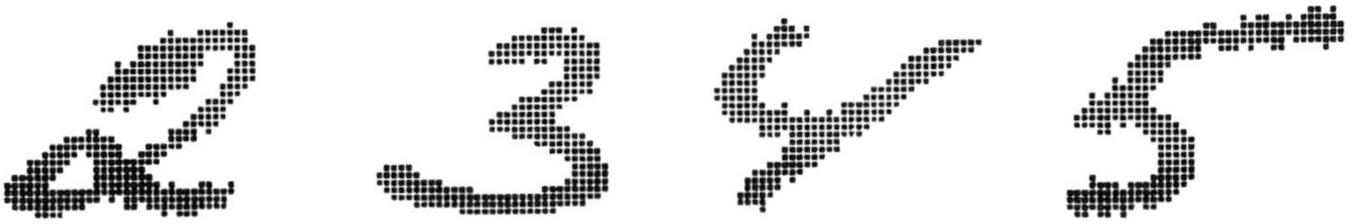

Figure 4.55 Examples of handwritten digits

digit was presented in a square of 30 × 40 pixels. The centre of the observation window can be placed at 30 × 40 different locations in a picture, giving 1200 different observation vectors $\mathbf{v}_i$ for each example of a digit. Each net was trained with 10 × 1200 observations from ten different handwritten examples of one type of digit. In the learning phase each observation vector was twice presented to the neural network. If a network was trained with examples of digit i we denote that network by N_i. After learning, fifteen new handwritten examples of each class were used as a test set. For each example 1200 observations were presented to the ten neural networks. An observation vector $\mathbf{v}$ obtained by sampling an example was assigned to a net i if $|\mathbf{v}-\mathbf{w}_i| = \min_j |v-\mathbf{w}_j|$, with $\mathbf{w}_i$ and $\mathbf{w}_j$ the weight vector of the winning neuron in neural net N_i, respectively N_j. If the majority of the 1200 observation vectors of one example was assigned to neural network N_k, then the example was classified as the digit k.

In Figure 4.56 we have given an outline of the classification procedure and the result of the classification for a handwritten representation of the digit '6'. (Note that we can use the typical distribution of the allocations of observation vectors of some digit to the different nets as a criterion for classification.)

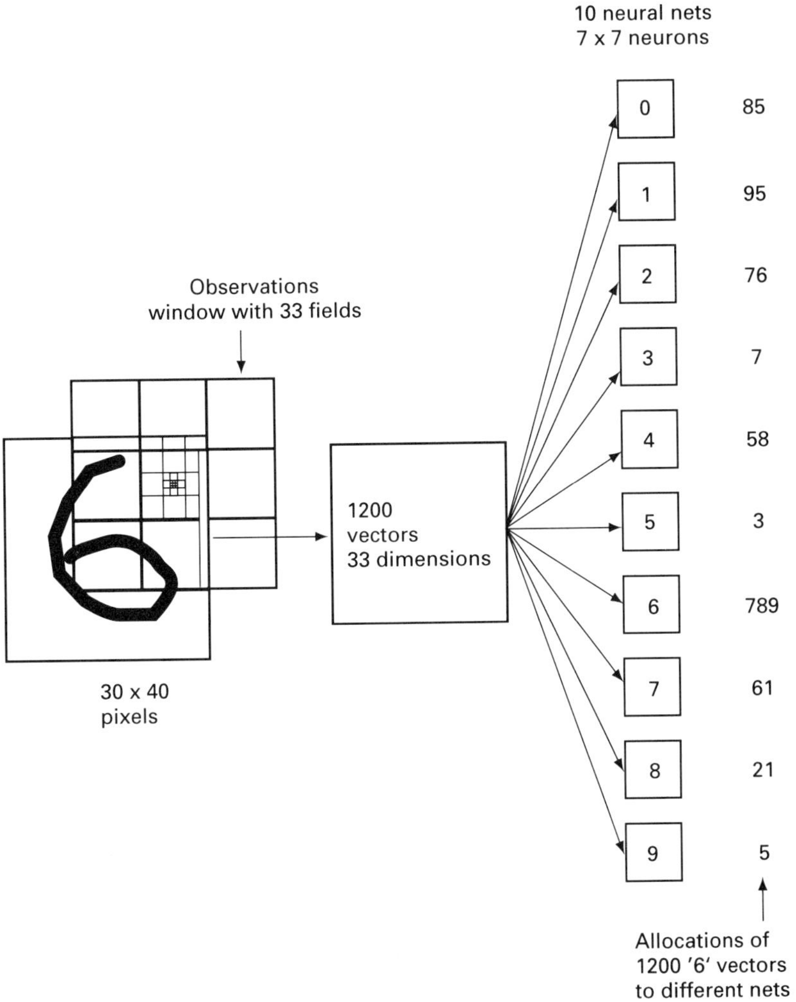

Figure 4.56 Outline of the classification procedure

From the $150 = 10 \times 15$ examples used in the test set, six examples were wrongly classified, i.e. a score of 96 per cent. To train the ten neural networks took 9 min on a HP 9000. The classification of one example in the test phase took 2 sec.

The same classification score was obtained when we only used twenty randomly selected observation vectors out of the 1200 possible observation vectors of some test digit. In this way we obtained a classification time of about 0.1 sec.

We see that if a class of pictures have some topological features in common, we can use the neural network for pattern recognition in a straightforward way.

4.11 Topology preservation with a self-organizing algorithm

In Section 4.5 we found that by using the adaptation rule:

$$\mathbf{w}_r(t+1)=\mathbf{w}_r(t)+\varepsilon(t)h(r, s, t)[\mathbf{v}_i(t)-\mathbf{w}_r(t)]$$

we are minimizing the error function:

$$E(W)=\frac{1}{2}\sum_{\mathbf{w}_r}\sum_{\mathbf{w}_s}\sum_{\mathbf{v}_i\in R(\mathbf{w}_s)} \varepsilon(t)p(\mathbf{v}_i)h(r, s, t)|\mathbf{v}_i-\mathbf{w}_r|^2$$

We could distinguish two different learning phases:

1. *The quantization phase* (final phase) In that phase we have for the neighbourhood function $h(r, s, t)=0$ for $r\neq s$ and $h(r, s, t)=1$ for $r=s$, and we are minimizing:

$$E(W)=\frac{1}{2}\sum_{\mathbf{w}_s}\sum_{\mathbf{v}_i\in R(\mathbf{w}_s)} p(\mathbf{v}_i)|\mathbf{v}_i-\mathbf{w}_s|^2$$

   The quantization phase is characterized by the property of *vector quantization*: an input data space of M d-dimensional vectors will be replaced by a smaller 'representative' set of N d-dimensional weight vectors of the neural net.
2. *The ordering phase* (initial phase) We are minimizing:

$$E_O(W, t)=\frac{1}{2}\sum_{\mathbf{w}_r}\sum_{\mathbf{w}_s} p(\mathbf{w}_s)g(r, s, t)|\mathbf{w}_s-\mathbf{w}_r|^2$$

   The ordering phase is characterized by the property that the *neural net will be ordered*: neighbouring neurons in the network will obtain similar weights. The net is *well-ordered* if neighbouring neurons have adjacent receptive fields.
   The approximation of the ordering error $E_O(W, t)$ by:

$$E_D(W)=\sum_{\delta_i\in\Delta}\sum_{d_L(\mathbf{i}_s, \mathbf{i}_r)=\delta_i} c_i|\mathbf{w}_s-\mathbf{w}_r|^2$$

   reveals more directly the property that we are *reducing the sum of weighted mutual distances between all weight vectors* in the ordering phase. The weight factor c_i is large for 'close neighbours' ($\delta_i=1$) of neurons in the neural net and will be small for 'distant neighbours' ($\delta_i\gg 1$).

The properties mentioned above deal mainly with the mapping of input vectors to weight vectors. Besides this *quantization mapping* ϕ from the space of input vectors V to the set of weight vectors W, there is a mapping ω (the *projection mapping*) from W to the lattice L of neurons, because with each weight vector there is associated a neuron in the neural net L. So we obtain a so-called *feature mapping* $\Psi=\phi\cdot\omega$ from V to the lattice L. In general the input vectors are obtained by observations (measurements) of some object space O (e.g. pictures or signals observed by some window). The *observation mapping* will be represented by the symbol α (see Figure 4.57). Frequently one is only interested in the representation of the input space V by

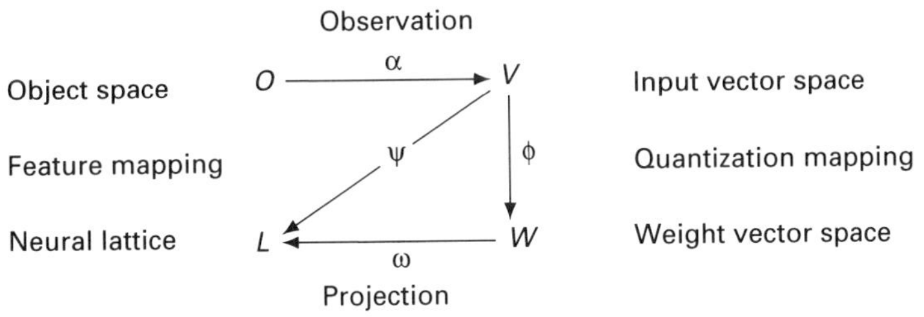

Figure 4.57 The set of interrelated mappings

the weight vectors of W, and one disregards the position of the neurons associated with the weight vectors.

In more sophisticated applications of the self-organizing neural net algorithm, the feature mapping ψ from the vector space V to the lattice L of neurons is used. We will say that an input vector $\mathbf{v}_i$ is represented by neuron u_j if u_j is the winning neuron when we present vector $\mathbf{v}_i$ to the neural net: $u_j = \psi(\mathbf{v}_i)$. It is frequently desired that similar input vectors are represented by the same neuron or by neighbouring neurons. This will not always be the case. If, for example, the training set D consists of many two-dimensional uniformly distributed vectors, and one is using a one-dimensional neural net (with two-dimensional weight vectors) with nine neurons, then the weight vectors will be uniformly distributed over the input space. The sequence of neurons u_i associated with the weight vectors $\mathbf{w}_i$ forms a chain through the input space. The input space will be divided by equally sized receptive fields $R(\mathbf{w}_i)$ (see, for example, Figure 4.58).

Similar input vectors on both sides of the border of, for example, $R(\mathbf{w}_2)$ and $R(\mathbf{w}_3)$, are represented by the neighbouring neurons u_2 and u_3, but similar input vectors on both sides of $R(\mathbf{w}_2)$ and $R(\mathbf{w}_9)$ are represented by neurons u_2 and u_9 that are not neighbours at all. If we had used instead a two-dimensional neural lattice, then similar input vectors would always have been represented by the same neuron or by neighbouring neurons with distance 1 (see Figure 4.59).

In Figure 4.59 the mutual relative position of two vectors $\mathbf{v}_i$ and $\mathbf{v}_j$ in the input space V is to a certain extent preserved by the mutual relative position of neurons $\Psi(\mathbf{v}_i)$ and $\Psi(\mathbf{v}_j)$. The *metric* of the input space is preserved.

A mapping from a metric space V with distance measure d_V to a metric space L with distance measure d_L is *metric preserving* if the triangular inequality property is preserved, i.e. if

$$d_V(\mathbf{v}_r, \mathbf{v}_t) \leqslant d_V(\mathbf{v}_r, \mathbf{v}_s) + d_V(\mathbf{v}_s, \mathbf{v}_t)$$

then

$$d_L[\psi(\mathbf{v}_r), \psi(\mathbf{v}_t)] \leqslant d_L[\psi(\mathbf{v}_r), \psi(\mathbf{v}_s)] + d_L[\psi(\mathbf{v}_s), \psi(\mathbf{v}_t)]$$

For metric preservation it is required that if the distance between two vectors from V is small, they will be represented by the same neuron or neighbouring neurons in

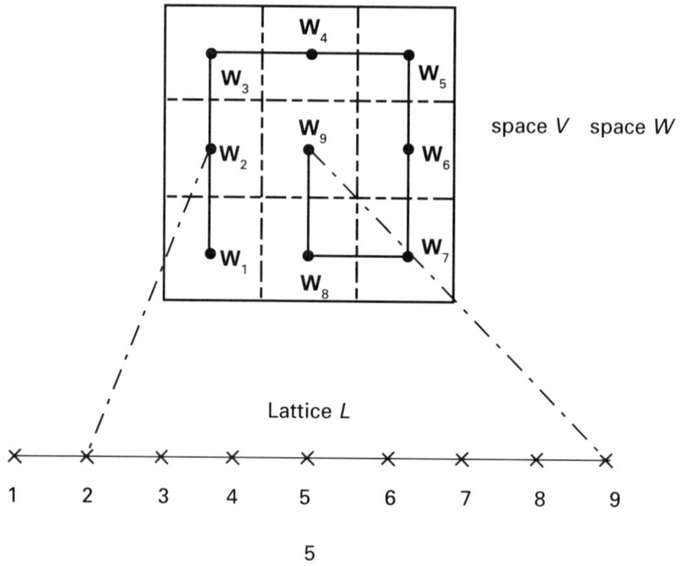

Figure 4.58 Two-dimensional data represented by two-dimensional weight vectors in a one-dimensional neural network

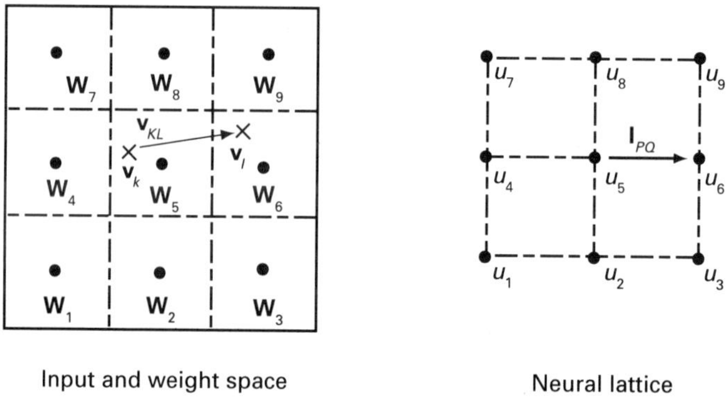

Figure 4.59 Metric preservation. Left figure: input space. Right figure: neural lattice with the projection u_5 of v_k and the projection of u_6 of v_1

the neural net. This requirement can be stated as follows. If two receptive fields $R(\mathbf{w}_i)$ and $R(\mathbf{w}_j)$ in V are adjacent (the common border $\partial R_{ij} = R(\mathbf{w}_i) \cap R(\mathbf{w}_j)$ is not empty), then neuron u_i with weight vector $\mathbf{w}_i$ is a neighbour of neuron u_j with weight vector $\mathbf{w}_j$.

If this property holds for all receptive fields, we say that the feature mapping is *topology preserving*. One may note that topology preservation is the complement of the well-ordering property. In Figure 4.58 there is no topology preservation while in Figure 4.59 there is topology preservation.

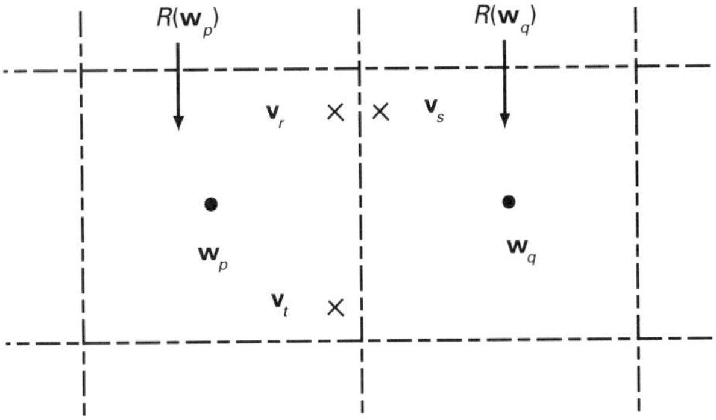

Figure 4.60 Two adjacent receptive fields

Because the self-organizing algorithm is minimizing the error function:

$$E(W, t) = \frac{1}{2} \sum_{\mathbf{w}_r} \sum_{\mathbf{w}_s} \sum_{\mathbf{v}_i \in R(\mathbf{w}_s)} \varepsilon(t) p(\mathbf{v}_i) h(r, s, t) |\mathbf{v}_i - \mathbf{w}_r|^2$$

the self-organizing algorithm will optimize the topology preservation. However, complete topology preservation is frequently not possible because the dimension of the input space V is not always the same as the dimension of the neural lattice L.

Although the feature mapping Ψ will try to preserve the topology of the input space in the neural lattice, it may happen that input vectors that are almost identical are represented by different neurons, whereas input vectors that are relatively more different are represented by the same neurons. In Figure 4.60 we have given two adjacent receptive fields $R(\mathbf{w}_p)$ and $R(\mathbf{w}_q)$ of two neighbouring neurons u_p and u_q with coordinate vectors $\mathbf{i}_p$ and $\mathbf{i}_q$ with distance $d_L(\mathbf{i}_p, \mathbf{i}_q) = 1$. The input vector $\mathbf{v}_r$ is almost identical to input vector $\mathbf{v}_s$, whereas $\mathbf{v}_r$ is more different from input vector $\mathbf{v}_t$. We have, however:

$$d_L[\psi(\mathbf{v}_r), \psi(\mathbf{v}_s)] = d_L(\mathbf{i}_p, \mathbf{i}_q) = 1$$

and

$$d_L[\psi(\mathbf{v}_r), \psi(\mathbf{v}_t)] = d_L(\mathbf{i}_p, \mathbf{i}_q) = 0$$

This phenomenon is due to the discontinuity of the feature mapping Ψ from V to L. The self-organizing algorithm will, however, try to minimize the discontinuity of the feature mapping.

If the dimension of the input space V is the same as the dimension of the neural lattice L and the learning set D_V contains many examples homogeneously distributed over some bounded area, then topology will be preserved for that area. This implies that if the distance between two input vectors approaches 0, then the distance between the corresponding winning neurons is at most equal to 1. If, however, the dimension m of the input space is larger than the dimension n of the neural lattice, then in general there will be no topology preservation, and if the distance of two input vectors approaches 0, then the distance between the corresponding winning neuron might become large ($\gg 1$) (see also Figure 4.58). A continuous path in V results in a corresponding track in the lattice L with in general large jumps between the successive winning neurons.

It might, however, occur that we can obtain topology preservation in a restricted sense if the training set D consists of m-dimensional vectors, whereas the neural lattice has a dimension n much smaller than m. This will be the case if the components of the m-dimensional vectors of the training set D are interrelated such that they can be represented by points in an n-dimensional space while preserving the topology for the elements of D. A trivial example is the set of three-dimensional vectors $\{[1, 1, 2], [2, 1, 4], [3, 1, 6], [4, 1, 8]\}$ that can be placed in a one-dimensional row by a mapping Ψ (if $\Psi(\mathbf{v}) = v_1$) while preserving the topology restricted to the set D, i.e. if

$$d_V(\mathbf{v}_r, \mathbf{v}_t) \leqslant d_V(\mathbf{v}_r, \mathbf{v}_s) + d_V(\mathbf{v}_s), (\mathbf{v}_t)$$

then

$$d_L[\psi(\mathbf{v}_r), \psi(\mathbf{v}_t)] \leqslant d_L[\psi(\mathbf{v}_r), \psi(\mathbf{v}_s)] + d_L[\psi(\mathbf{v}_s), \psi(\mathbf{v}_t)]$$

A less trivial example of this phenomenon was given in Section 4.2 where we showed that the topology of the two-dimensional pattern of the capital 'G' was preserved in a two-dimensional neural net. In that case we used a training set D of 1200 25-dimensional input vectors obtained by sampling the pattern with a special window. The observation window is constructed in such a way that the observation vectors can be mapped on points in a two-dimensional metric space with topology preservation between the data set D and that two-dimensional space. The dimension of the neural net used was two and the input space was restricted to D. We summarize our discussion in the next property:

Practical statement 4.1

Restricted topology preservation

If the *m*-dimensional vectors of the training set *D* can be mapped on a finite number of points of an *n*-dimensional metric space with preservation of topology, then we can use an *n*-dimensional neural net and preserve topology restricted to the elements of *D* (or restricted to vectors almost equal to elements of *D*).

4.12 Interpolation with the self-organizing algorithm

There is yet another property of the self-organizing neural net algorithm that might be profitable for some applications. If the neural lattice contains more neurons than there are input vectors in the training set, then, due to the vector quantization

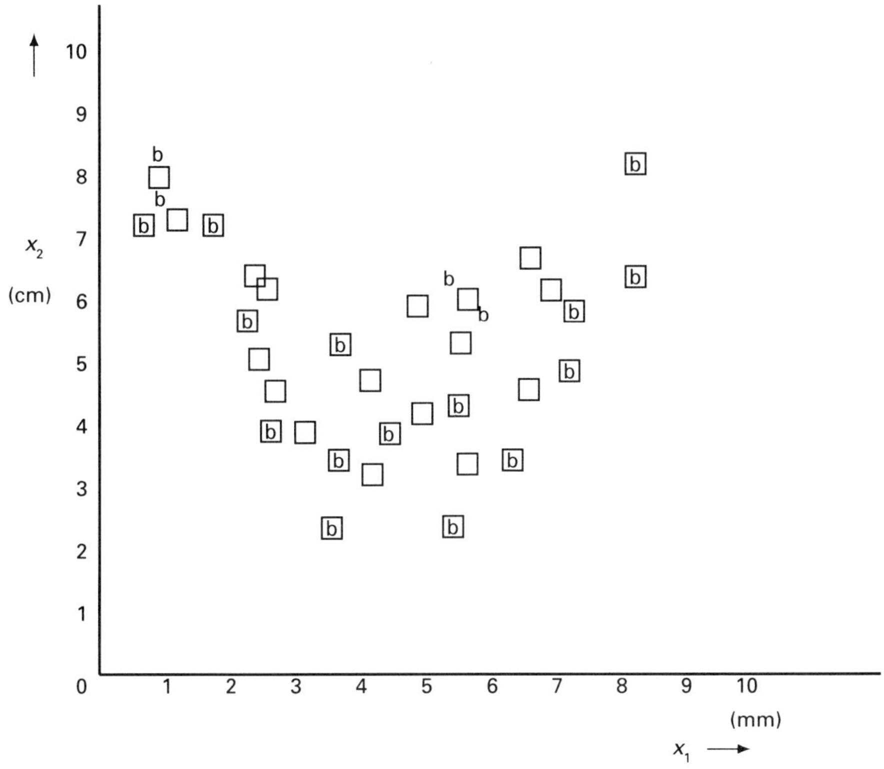

Figure 4.61 Input/weight space. The 'b's represent nineteen input vectors. The squares represent the weight vectors in a two-dimensional net with thirty-two neurons

property, there will be at the final phase of learning a set of weight vectors that will be copies of all the input vectors of the training set. During training of the neural net the redundant weights will also be adapted to the input vectors corresponding to the weight vectors of the surrounding winning neurons. In this case the redundant weight vectors will in this way obtain values that one would find by interpolation between the values of the training set.

Example 4.10

In Figure 4.61 we have given the result of training a two-dimensional neural net with thirty neurons and a training set with nineteen two-dimensional input vectors represented in Figure 4.61 by the letter 'b'. The final weights of the neurons are given by small squares. ■

We can summarize our discussion as follows:

Practical statement 4.2

Interpolation property

If the neural net contains more neurons than there are elements in the training set *D*, then the redundant weight vectors will be interpolated between the weight vectors that are copies (or almost copies) of the input vectors of the training set.

A frequently undesired interpolation of weight vectors between input vectors of the data set can, however, also occur if the number of elements in the data set is larger than the number of neurons but the elements of the data set D are separated by a relatively large empty area. We will explain this phenomenon shortly. In Section 4.5 we found that the algorithm tries to minimize the value of $|\mathbf{w}_i - \mathbf{w}_j|^2$ for all pairs of weight vectors $(\mathbf{w}_i, \mathbf{w}_j)$. By introducing a weight vector $\mathbf{w}_k$ between $\mathbf{w}_i$ and $\mathbf{w}_j$, if $|\mathbf{w}_i - \mathbf{w}_j|^2$ is large, a much smaller value of the replacement $|\mathbf{w}_i - \mathbf{w}_k|^2 + |\mathbf{w}_k - \mathbf{w}_j|^2$ can be obtained. The weight vector $\mathbf{w}_k$ is not, however, representing the input vectors.

Example 4.11

In Figure 4.62 we have given the result of training a two-dimensional network of 5×5 neurons with 1000 input vectors uniformly distributed on a circle. Small circles

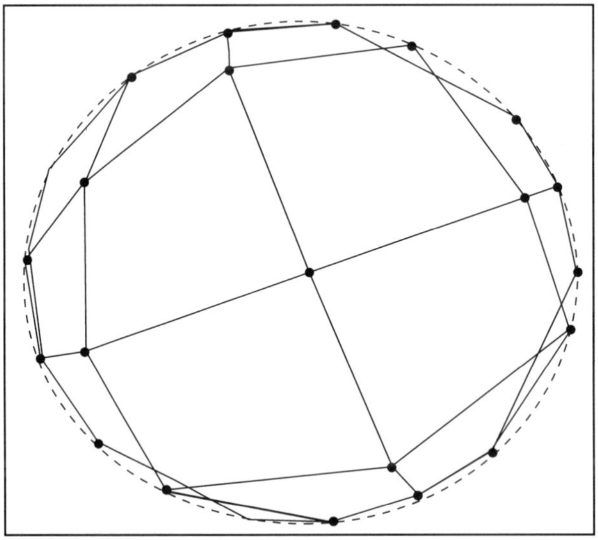

Figure 4.62 Representation of 1000 input vectors uniformly distributed on a circle by weight vectors in a 5×5 neural network

represent the final weight vectors. Weight vectors are connected with a straight line if the corresponding neurons are neighbours in the neural lattice. ■

4.13 Master–slave and multi-net decomposition of the self-organizing neural net algorithm

In this section we will discuss the decomposition of input vectors and the application of the self-organizing algorithm to the different parts of the decomposed input vector.

First we will discuss the *master–slave decomposition* as already applied in Section 4.9 on the Bayes classifier.

In several applications of the self-organizing algorithm the set of data vectors consists of pairs of vectors. For instance, in the case of function identification from samples, the training set contains pairs of argument vectors and function-value vectors. In the case of function identification we want generalization (or interpolation) from samples. To obtain a proper result the different parts of the vectors of the training set must be treated differently. We want a metric that preserves quantization of the argument values and a representation and interpolation of the function values.

One part of the data vectors will be treated by the self-organizing algorithm in the same way as was done in previous sections. We call that part of the input vector the *master input vector* $\mathbf{v}_i$. The second part of the input vector, called the *slave input vector*, denoted by $\tilde{\mathbf{v}}_i$, will *not* be used to find the winner in the neural net and will only be used to adapt the weight vectors in the neural net. The total input vector will be denoted by $\hat{\mathbf{v}}_i = [\mathbf{v}_i, \tilde{\mathbf{v}}_i]$.

In the same way we make a decomposition of the weight vectors of the neural lattice. One part is called the *master weight vector* $\mathbf{w}_i$ and the other part is called the *slave weight vector* $\tilde{\mathbf{w}}_i$. The total weight vector will be denoted by $\hat{\mathbf{w}}_i = [\mathbf{w}_i, \tilde{\mathbf{w}}_i]$. The algorithmic adaptation rule is then as follows.

Given some training vector $\hat{\mathbf{v}}(t) = [\mathbf{v}_i, \tilde{\mathbf{v}}_i]$:

1. Determine the winning neuron u_s for the *master vector* $\mathbf{v}(t)$, i.e.

$$d_V[\mathbf{w}_s(t), \mathbf{v}(t)] = \min_r d_V[\mathbf{w}_r(t), \mathbf{v}(t)]$$

2. Every weight vector $\hat{\mathbf{w}}_r(t) = [\mathbf{w}_r, \tilde{\mathbf{w}}]$ in the net will be changed to:

$$\hat{\mathbf{w}}_r(t+1) = \hat{\mathbf{w}}_r(t) + g(r, s, t)[\hat{\mathbf{v}}(t) - \hat{\mathbf{w}}_r(t)]$$

with $g(r, s, t)$ a scalar-valued adaptation function as discussed in Section 4.3.

From the algorithmic adaptation rule we conclude that the master input vectors and the master weight vectors are manipulated as in the previous section. The preceding theory will thus hold in the same way for the master vectors.

Due to the vector quantization property of the algorithm, the final complete weight vectors $\hat{\mathbf{w}}$ will become similar to the elements $\hat{\mathbf{v}}_i$ and thus the slave weight vectors will also become similar to the slave input vectors.

If the number of neurons in the neural lattice is larger than the number of input vectors, then the values of weight vectors of the redundant neurons will be interpolated between the weight vectors that are similar to the input vectors.

In the next two sections we apply the method of master–slave decomposition to function identification and control of a robot arm.

We found in Section 4.11 that it is preferable to have the dimension of the neural net equal to the dimension of the input space. Frequently the dimension of the input space is large and would thus require the neural lattice to be similarly large. But the required number of neurons will grow exponentially with the dimension of the neural lattice. If the number of neurons per dimension is equal to d and the dimension of the neural net is equal to m, then d^m neurons are required. This number may become larger than the number of elements in the training set and we cannot use the neural net for proper vector quantization.

Moreover, if the dimension of the input space is larger than the dimension of the neural net, the property of topology preservation is lost; the greater the difference in dimension, the greater the number of defects.

Therefore we frequently want the dimension m of the neural net to be low and equal (or close to) the dimension of the input vectors. A solution to this problem is to use the multi-net decomposition method.

The multi-net decomposition method is straightforward: we divide the input vector in some way into k parts and we use k different neural nets. The pth neural net is trained with the pth part of the vectors of the training set.

If we divide the original m-dimensional input vector into k equal parts of dimension m/k and use k neural networks of dimension m/k, and the number of neurons in each dimension of all subnetworks is d, then the number of neurons is reduced from d^m to $kd^{m/k}$.

In Section 4.16 we will apply the multi-net decomposition method to EEG analysis.

4.14 Application of the self-organizing algorithm to function identification

Assume we have several pairs of argument values and function values of some unknown function, and we want to know the functional relationship between arguments and function values. If we do not require a mathematical description of the functional relationship but are satisfied with a (hardware or software) realization of the function in a restricted domain, then we can use a neural network to approximate the unknown function. In Chapter 3 we have shown how we can use a continuous multi-layer Perceptron to identify an unknown function. We can, however, also use a self-organizing neural net algorithm for that purpose. The main difference will be that we obtain a quantized version of the unknown function.

At first glance one is inclined to make training vectors composed of pairs of argument and function values. If we use these training vectors we know that the final weight vectors will be copies of these training vectors, and if the number of neurons is larger than the number of training vectors we also obtain weight vectors by interpolation between the training vectors. The weight vectors are also composed of pairs of arguments and function values; the first part corresponds with an argument and the second part with a function value. After training, we present some argument of the function and determine the weight vector with an argument part with the minimal distance to the presented argument. Then we read in the pertinent weight vector the second part as the desired function value. However, this method will give incorrect results if there are weight vectors interpolated between the weight vectors that are copies of the training vectors. Interpolated weight vectors will be located in the area between the curves representing the functional relationship.

Example 4.12

In an experiment we applied 1000 training samples $[x_i, y_i]$ of the function $y=10x^2$ with x_i in the interval $[-0.3, +0.3]$ to the self-organizing algorithm. The neural net contained fifty neurons. The result is given in Figure 4.63 (the line represents the 1000 training samples and dots the $[x, y]$ value of the weight vectors). We observe

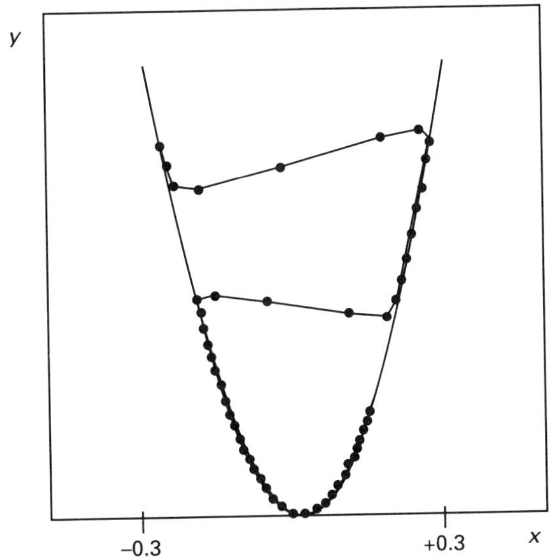

Figure 4.63 Representation of 1000 pairs of argument and function values of the function $y=10x^2$ by two-dimensional weight vectors (dots) in a one-dimensional neural net of fifty neurons. Input argument values in $[-0.3, +0.3]$

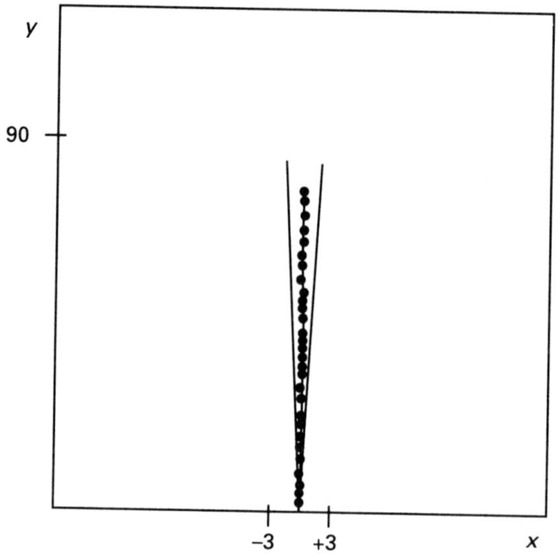

Figure 4.64 The representation of 1000 pairs of argument and function values of the function $y=10x^2$ by two-dimensional weight vectors (dots) in a one-dimensional neural net of fifty neurons. Input argument values in $[-3, +3]$

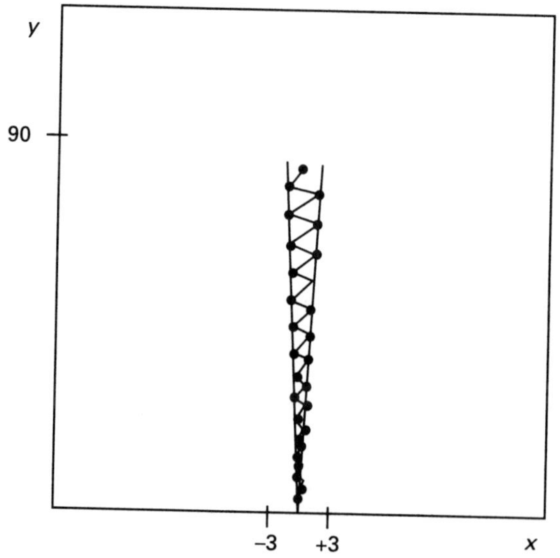

Figure 4.65 The representation of 10 000 pairs of argument and function values of the function $y=10x^2$ by two-dimensional weight vectors (dots) in a one-dimensional neural net of fifty neurons. Input argument values in $[-0.3, +0.3]$

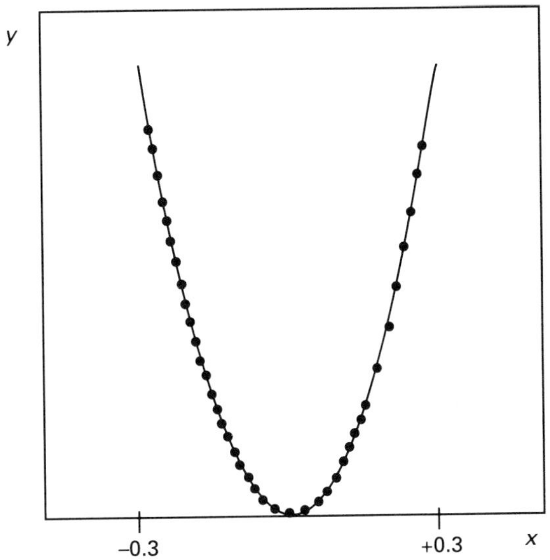

Figure 4.66 The master–slave method. The representation of 1000 pairs of argument and function values of the function $y=10x^2$ by two-dimensional master–slave weight vectors (dots) in a one-dimensional neural net of fifty neurons. Input argument values in $[-0.3, +0.3]$

that even in this case where the number of elements in the training set is much larger than the number of neurons, we obtain interpolated weight vectors that are wrongly located.

In a second experiment we used 1000 samples $[x_i, y_i]$ of the function $y=10x^2$ from the domain $[-3, +3]$. The result, given in Figure 4.64, is even worse. If we use more samples (i.e. 10 000) and extend the learning phase to 100 000 steps, then we obtain the result given in Figure 4.65. ■

Proper function identification with the self-organizing neural net algorithm can, however, be obtained by using the master–slave method presented in Section 4.13. If the argument **x** of the unknown function is m-dimensional, then we use an m-dimensional neural net and take **x** as the master vector. The corresponding n-dimensional **y** function value vector is used as the slave vector. If there are enough samples, then the **x** values of the weight vectors will be ordered in a regular m-dimensional lattice. The slave elements will be copies of the function values and if there are more neurons than samples, additional **y** values will be interpolated between the **y** values that are copies of the slave-training samples.

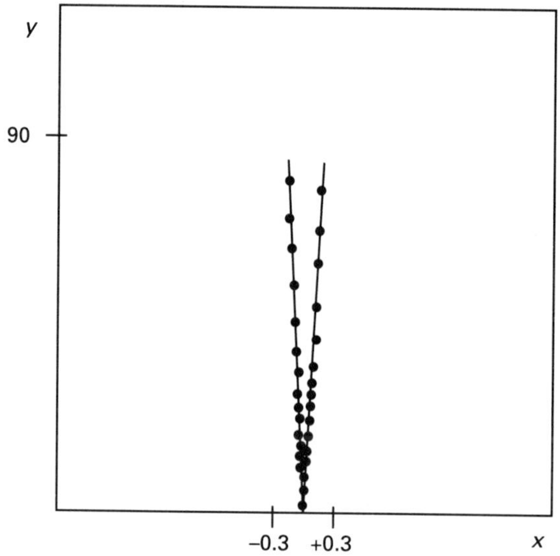

Figure 4.67 The master–slave method. The representation of 1000 pairs of argument and function values of the function $y = 10x^2$ by two-dimensional master–slave weight vectors (dots) in a one-dimensional neural net of fifty neurons. Input argument values in [3, +3]

Example 4.13

If we repeat the experiments mentioned in the previous example with the master–slave method, then we obtain under the same conditions the results respectively given in Figures 4.66–4.67. ■

4.15 Application of the self-organizing algorithm to robot arm control

Suppose we use a monitor connected to a camera to observe an object on a square table. The coordinates of the table top will be denoted by u and v. We want a robot to learn to grasp the object from the table given the x and y position of the object on the monitor screen (not u and v). For simplicity our robot consists of an arm with two parts moving in a horizontal plane (see Figure 4.68). With two servomotors we can control the two angles H_1 and H_2 in order to reach every point on the table.

In a training phase the object is placed somewhere on the table and we form a vector **v** with the observed values of x, y, H_1 and H_2 as the four components. Note

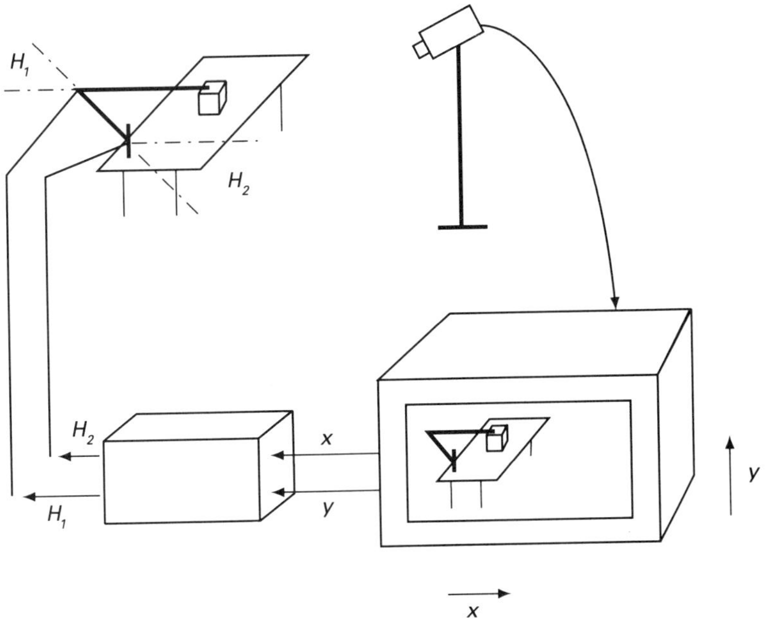

Figure 4.68 Simple outline of robot arm control

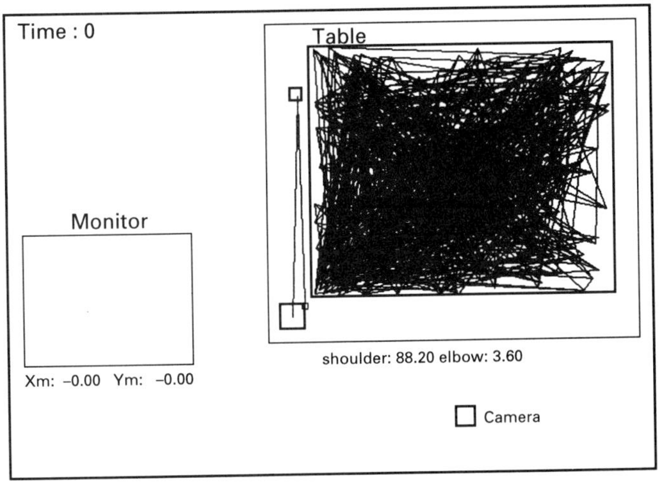

Figure 4.69 Initial representation of table coordinates by the two-dimensional master–slave weight vectors (dots in the right-hand figure) in a two-dimensional net of 20×20 neurons. Weight vectors are connected with a line if they belong to neighbouring neurons

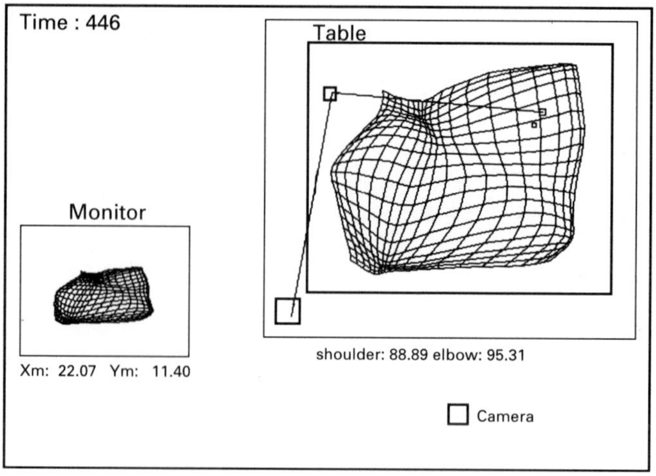

Figure 4.70 Representation of table coordinates by the two-dimensional master–slave weight vectors (dots in the right-hand figure) in a two-dimensional net of 20 × 20 neurons after 446 training examples. Weight vectors are connected with a line if they belong to neighbouring neurons

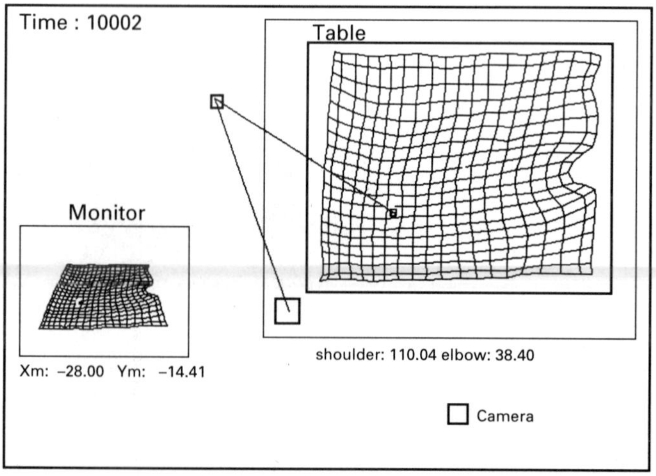

Figure 4.71 Representation of table coordinates by the two-dimensional master–slave weight vectors (dots in the right-hand figure) in a two-dimensional net of 20 × 20 neurons after 10 002 training examples. Weight vectors are connected with a line if they belong to neighbouring neurons

that the picture on the screen does not give the u and v position of the object on the table top but the x and y position of the object in a perspective view on the monitor.

In the training phase we place the object in 10 000 random positions on the table. In this way we obtain 10 000 training vectors. We use the master–slave decomposition method with $[x, y]$ the master vector and $[H_1, H_2]$ the slave component. We use a two-dimensional neural network with 20×20 neurons. After training the object is placed on the table and we observe the $[x, y]$ on the monitor. We apply this vector to the neural net and determine the winning neuron for this master vector. The corresponding slave vector will give us the information for the correct value of $[H_1, H_2]$ to grasp the object.

After learning, we have 400 weight vectors. For each slave weight vector $[H_1, H_2]$ there is a corresponding value of the table top coordinates $[u, v]$. These 400 values of $[u, v]$ are given by the corner points of the lattice depicted in Figures 4.69–4.71 after zero training examples, 446 training examples and 10 002 examples, respectively. We see that after 10 002 training steps 400 $[x, y]$ positions on the monitor are translated into 400 (almost) correct positions for the robot arm.

4.16 Application of the self-organizing algorithm to EEG signal analysis

In recording electroencephalograms (EEGs) of epileptic patients one frequently observes are irregular intervals certain short wave forms called spike–wave complexes (SWCs) in which a spike is followed by a slow wave (see Figure 4.72). The duration of such a SWC is about 0.5 sec, and SWCs can occur in sequences without interruption. EEGs are sometimes recorded over a 24-hour period, and it takes considerable time to screen such a recording for the occurrence of SWCs. Automatic detection and quantification of SWCs would be very useful.

We can use a self-organizing neural net algorithm after a training period for the

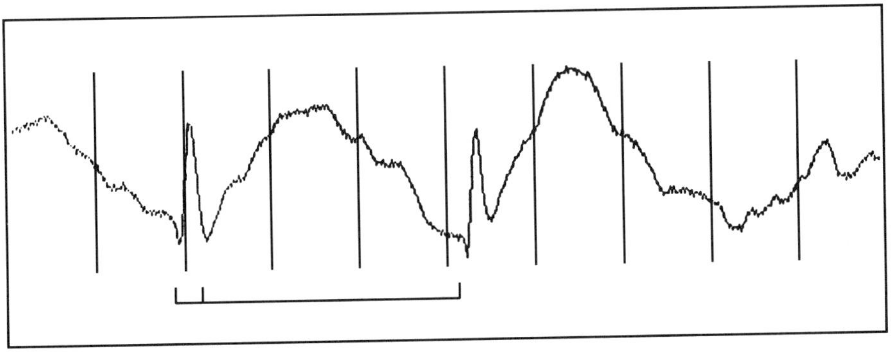

Figure 4.72 Two spike–wave complexes in an EEG signal

detection of the SWCs. The EEG signal is sampled first with a frequency of 366 Hz. We can observe a complete SWC with a *window* of about 180 samples. When we move the window along the EEG signal we obtain a large set of observation vectors composed of the values of the 180 consecutive samples of the window. If we move the window with a step size of one sample we obtain in this way 60×366 different observation vectors for a recording of 1 min.

Some observation vectors correspond to SWCs and others to observations of the ordinary EEG signals. We can use the self-organizing neural algorithm to make a vector quantization of the total set of observation vectors. In this way we can obtain weight vectors that are copies of clusters of observation vectors. Observation vectors obtained with the window located on SWCs will result in similar weight vectors (and interpolated weight vectors) in the neural lattice. After training, we label the weight vectors that are similar to observation vectors obtained from SWCs as elements of the SWC class. If after training we present an observation vector of a new recording to the neural net, we can determine whether or not it is similar to a weight vector of the class labelled as SWC vectors.

If we develop a classification system as described above we need, for a proper vector quantization and interpolation, a neural net with the same, or almost the same, dimension as the dimension of the input vectors. This would require a 180-dimensional neural lattice. The number of neurons will become very large (d^{180}, if d is the number of neurons per dimension) and the detection speed will be very low. We want to have the dimension of the neural lattice much lower so we have to reduce the dimension of the observation vector. We can achieve this goal by three methods: (i) reducing the length of the observation interval; (ii) preprocessing the observation vectors; (iii) using the multi-net decomposition method as described in Section 4.13.

1. We reduce the window length to 100 samples.
2. To detect the typical form of a spike we must observe the EEG very accurately. For the detection of the slow wave we can take the mean of the sample values in successive intervals. For our observation vector we take the first thirty samples for the first thirty components. The remaining interval of the window is divided into seven subintervals. The mean of the ten samples in the seven subintervals gives us the next seven components of the observation vector (see Figure 4.73). In this way the dimension of the observation vector becomes thirty-seven.
3. We divide the observation vector $\mathbf{v}_i$ into two vectors: $\mathbf{v}_{si}$ and $\mathbf{v}_{wi}$. The vector $\mathbf{v}_{si}$ contains the first thirty components of $\mathbf{v}_i$, and the vector $\mathbf{v}_{wi}$ contains the remaining seven components. We take two neural networks: one network, called the *spike network*, is trained with the vectors $\mathbf{v}_{si}$, the second network, the *wave network*, is trained with vectors $\mathbf{v}_{wi}$.

For the training we used an EEG recording of about 1 min duration. The recording contained forty-two SWCs. A neurologist indicated the beginning of nineteen out of these forty-two SWCs. The observation vectors were obtained by placing the window at 5000 random positions in the 1 min recording with the restriction that about 35

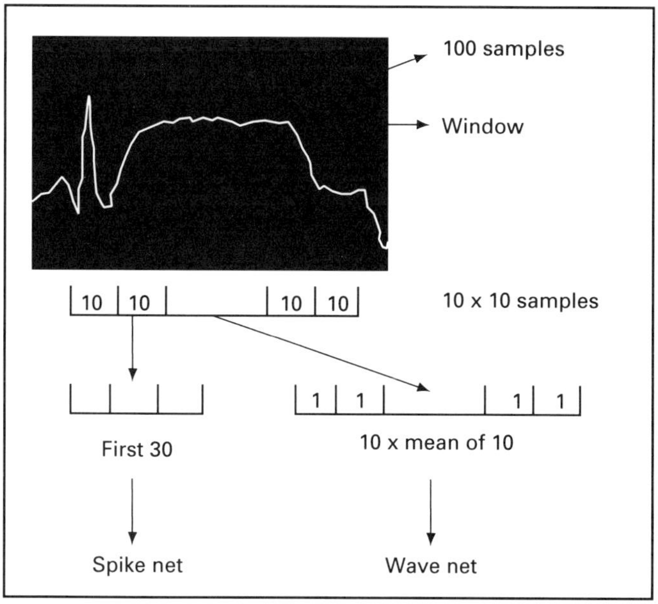

Figure 4.73 Outline for obtaining thirty-dimensional vectors for training the spike-net and ten-dimensional vectors for training the wave-net

per cent of the observations were obtained by locating the beginning of the window at the beginning of one of the nineteen marked SWCs.

We trained a two-dimensional 6×6 neural net (the spike network) with the vectors $\mathbf{v}_{si}$ and a two-dimensional 6×6 net with the vectors $\mathbf{v}_{wi}$. After learning, the weight vectors will represent (vector quantization) the set of training vectors. In Figure 4.74 we have given the reconstructed signal form of the weight vectors of the 6×6 spike neural net. Several of the weight vectors of the spike neural net represent the observation of a spike of an SWC and others represent the observation of an arbitrary party of the EEG signal that is not an SWC. The same holds for the waves of the SWC represented by the weight vectors of the wave network.

In order to know which weight vectors in the spike net represent spikes, we have to *label* the weight vectors. After learning, we take the nineteen observation vectors obtained by observing the marked SWCs in the registration. We determine for each of these observation vectors the weight vector with the smallest distance to the observation vector and label that weight vector as a *spike weight vector*. The same is done for the wave neural net.

After the training and labelling phase, we can use the neural net to detect SWCs in an EEG recording. We move with the window along the EEG signal and at each step we present the observation vector $\mathbf{v}_{si}$ to the spike net, and the vector $\mathbf{v}_{wi}$ to the wave net. If in the spike net and in the wave net the labelled weight vectors are

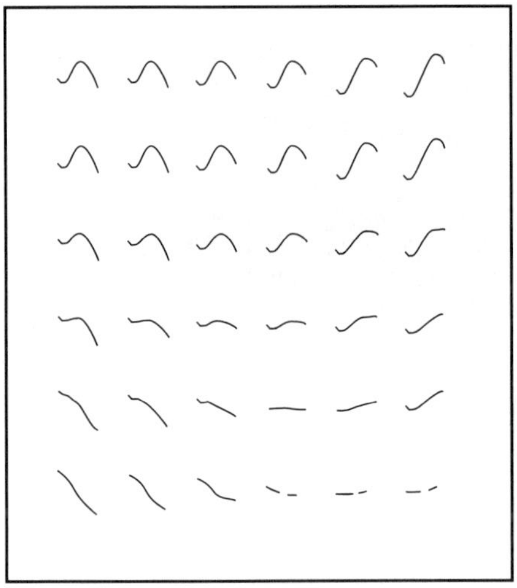

Figure 4.74 Signal segments corresponding to the weight vectors in the trained two-dimensional spike-net with 6×6 neurons

simultaneously the winning weight vectors, then the observed signal segment is classified as an SWC.

In the experimental setup we found that forty of the forty-two SWCs were detected and eight observations (signal forms resembling an SWC) were wrongly classified as an SWC.

4.17 Application of the self-organizing algorithm to speech recognition

The ultimate goal of a speech recognition system is the automatic conversion of recorded speech sound into corresponding written text. The most promising approach to large vocabulary automatic speech recognition is to build a recognizer for the smallest linguistic units that can occur in words: the phonemes. Subsequently one has to transform the resulting string of phonemes into words and at a still higher level into semantic knowledge. The number of phonemes is small (≈ 60) compared to the number of words in some vocabulary. In Table 4.1 we have given the phonemes used in the so-called TIMIT database, containing the sound recordings of ten sentences spoken by 630 speakers. The acoustic signal in the database is labelled with the corresponding sequences of phonemes. The speech signal was sampled at 16 kHz with a 16-bit analog/digital converter.

When we observe the acoustic signal corresponding with some phoneme (duration ≈ 0.1 sec), then the time-dependent frequency spectrum is to a certain extent

Table 4.1 List of phonemes

Phone	Example	Phone	Example	Phone	Example
/iy/	b**ea**t	/er/	b**ir**d	/z/	**z**oo
/ih/	b**i**t	/axr/	din**er**	/zh/	mea**s**ure
/eh/	b**e**t	/el/	bott**le**	/v/	**v**ery
/ae/	b**a**t	/em/	yes'**em**	/f/	**f**ie**f**
/ux/	b**eau**ty	/en/	butt**on**	/th/	**th**ief
/ix/	ros**e**s	/eng/	Washi**ng**ton	/s/	**s**i**s**
/ax/	th**e**	/m/	**m**o**m**	/sh/	**sh**oe
/ah/	b**u**tt	/n/	**n**o**n**	/hh/	**h**ay
/uw/	b**oo**t	/ng/	si**ng**	/hv/	Le**h**eigh
/uh/	b**oo**k	/ch/	**ch**ur**ch**	/pcl/	(p closure)
/ao/	b**ough**t	/jh/	**j**u**dge**	/tcl/	(t closure)
/aa/	c**o**t	/dh/	**th**ey	/kcl/	(k closure)
/ey/	b**ai**t	/b/	**b**o**b**	/qcl/	(q closure)
/ay/	b**i**te	/d/	**d**a**d**	/bcl/	(b closure)
/oy/	b**oy**	/dx/	(bu**tt**er)	/dcl/	(d closure)
/aw/	ab**ou**t	/nx/	(flapped n)	/gcl/	(g closure)
/ow/	b**oa**t	/g/	**g**a**g**	/epi/	(epi closure)
/l/	**l**ed	/p/	**p**o**p**	/h#/	(begin sil)
/r/	**r**ed	/t/	**t**o**t**	/#h/	(end sil)
/y/	**y**et	/k/	**k**i**ck**	/pau/	(between sil)
/w/	**w**et	/q/	(glottal stop)		

characteristic for that particular phoneme. This phenomenon can be captured by observing the speech sound through a window shorter (8 msec = 128 samples) than the smallest phoneme, shifting the window by discrete steps along the acoustic signal, and determining at each step the frequency spectrum of the signal in the window. The coefficients of the frequency spectrum (or some transformation thereof) corresponding with a window observation on some phoneme signal constitute a *spectral vector*.

We will now describe in a simplified experiment how we can use a self-organizing neural algorithm to learn to recognize a spoken sentence. The experiment will not reveal the quality of the recognition performance but will only reveal the idea of how a self-organizing neural can be used for speech recognition.

Learning with the multi-net method

We can use the self-organizing neural net algorithm to learn the time-dependent frequency characteristic of a phoneme. We use the multi-net method described in Section 4.13. For each phoneme we use a two-dimensional neural net of 5×5 neurons.

From the acoustic signal of a spoken sentence of some speaker (labelled with phonemes) we make a set of all spectral vectors obtained by placing the window somewhere on the acoustic signal. For each phoneme we collect all spectral vectors obtained by observing with our window the acoustic signal of that particular phoneme occurring in different positions in the sentence. Such a set will be called a phoneme spectral set. For each phoneme we train a separate neural net, called a phoneme net, by randomly selecting 10 000 times a vector from the pertinent phoneme spectral vector set. After training, due to the vector quantization property, each neural net will represent by its twenty-five weight vectors the most common spectral vectors of the particular phoneme.

Labelling the sequence of winning neurons

After training, we pass the observation window again step by step along the sentence, and for each phoneme interval we register the sequence of winning neurons (i.e. which neuron in which neural net) for the applied sequence of spectral vectors. Each time the window passes the same phoneme the sequence of winning neurons will be similar and almost all winning neurons will be located in the phoneme net that corresponds to the phoneme we are scanning.

For those acquainted with hidden Markov models, it should be noted that we could construct a phoneme characteristic hidden Markov model from those sequences of winning neurons. However, we can also use a cruder method to characterize a phoneme. When passing a phoneme with our window it turns out that the number of times the winning neuron will be located in the corresponding phoneme net exceeds a certain minimum. (There may be incidental interruptions that the winning neuron will be located in a phoneme net different from the phoneme we are scanning.) We can use this minimum score as a criterion for the detection of a certain phoneme.

Results of a simple experiment

We used the following single sentence spoken by one speaker to learn and to test the recognition by the self-organizing neural net: 'She had your dark suit in greasy wash water all year.' The acoustic signal was labelled by phonemes.

We used a simplified version of the set of phonemes, as should become clear from the sequence of phonemes as attached to the above spoken sentence, represented in Tables 4.2 and 4.3. In parentheses we give the sample numbers of the beginning and ending of some phoneme. (The time between two successive samples is about 0.06 msec.) Because we move the window in the test phase with steps of ten samples, we have also given the duration of a phoneme in intervals of ten samples. The final number in the row (after the arrow) gives the number of times the winning neuron was in the phoneme net that corresponds to the phoneme we are observing with steps of ten samples.

Table 4.2 Test result

sh as in **Sh**e	(00000–01666)	duration 166.6 × 10 samples	→146
iy as in Sh**e**	(01666–02626)	duration 96.0 × 10 samples	→103
hv as in **h**ad	(02626–03446)	duration 82.0 × 10 samples	→78
ae as in h**a**d	(03446–05285)	duration 183.9 × 10 samples	→179
*the phoneme (almost silence) at the beginning of the **d** in 'had'			
	(05285–05826)	duration 54.1 × 10 samples	→50
d as in ha**d**	(05826–06213)	duration 38.7 × 10 samples	→28
y as in **y**our	(06213–06642)	duration 42.9 × 10 samples	→29
er as in y**our**	(06642–07986)	duration 134.4 × 10 samples	→128
*the phoneme (almost silence) at the beginning of the **d** in 'dark'			
	(07986–09065)	duration 107.9 × 10 samples	→109
dd as in **d**ark	(09065–09266)	duration 20.1 × 10 samples	→0
aa as in d**ar**k	(09266–12159)	duration 289.3 × 10 samples	→274
*the phoneme (almost silence) at the beginning of the **k** in 'dark'			
	(12159–12866)	duration 70.7 × 10 samples	→64
kk as in dar**k**	(12866–13146)	duration 28.0 × 10 samples	→29
ss as in **s**uit	(13146–14997)	duration 185.1 × 10 samples	→182
uw as in s**ui**t	(14997–17051)	duration 204.5 × 10 samples	→188
*the phoneme (almost silence) at the beginning of the **t** in 'suit'			
	(17051–17306)	duration 25.5 × 10 samples	→23
tt as in sui**t**	(17306–17588)	duration 28.2 × 10 samples	→14
ix as in **i**n	(17588–18601)	duration 101.3 × 10 samples	→88
nn as in i**n**	(18601–19574)	duration 97.3 × 10 samples	→105
*the phoneme (almost silence) at the beginning of the **g** in 'greasy'			
	(19574–20546)	duration 97.2 × 10 samples	→72
g as in **g**reasy	(20546–21506)	duration 96.0 × 10 samples	→93
r as in g**r**easy	(21506–22013)	duration 50.7 × 10 samples	→49
iy as in gr**ea**sy	(22013–23026)	duration 101.3 × 10 samples	→100
s as in grea**s**y	(23026–25026)	duration 200.0 × 10 samples	→185
iy as in greas**y**	(25026–25943)	duration 91.7 × 10 samples	→77
w as in **w**ash	(25943–28199)	duration 225.6 × 10 samples	→222
ao as in w**a**sh	(28199–29828)	duration 162.9 × 10 samples	→150
sh as in wa**sh**	(29828–31373)	duration 154.5 × 10 samples	→141
*the stop at the end of wash			
	(31373–32130)	duration 75.7 × 10 samples	→74

We observe that for almost every phoneme, 90 per cent of the observations with our window result in a winning neuron in the correct phoneme net. An exception is the phoneme dd in '**d**ark' where the dd-phoneme net did not respond. A closer look reveals that the similar t-phoneme net responded eight times to the dd-phoneme.

Most spectral vectors of the observations of a phoneme are assigned to the same neural net and will traverse a path of winning neurons in that net. Occasionally the path may temporarily jump to another phoneme net as illustrated in Figure 4.75 for the twenty-eight observations on the phoneme /t/ in the word 'sui**t**'. In the figure the

Table 4.3 Continued test results

w as in **water**	(32130–32773)	duration 64.3 × 10 samples	→67
ao as in **water**	(32773–34186)	duration 141.3 × 10 samples	→130
d as in wa**t**er	(34186–34666)	duration 48.0 × 10 samples	→32
aa as in wat**er**	(34666–36226)	duration 156.0 × 10 samples	→140
q glottal stop at the end of water			
	(36226–37097)	duration 87.1 × 10 samples	→87
ao as in **all**	(37097–39729)	duration 263.2 × 10 samples	→235
l as in **all**	(39729–40689)	duration 96.0 × 10 samples	→94
y as in **year**	(40689–42008)	duration 131.9 × 10 samples	→130
ih as in **year**	(42008–43710)	duration 116.4 × 10 samples	→156
axr as in **year**	(43710–44874)	duration 116.4 × 10 samples	→117
*the phoneme representing silence at the end of the sentence from 44874 up to 57466			

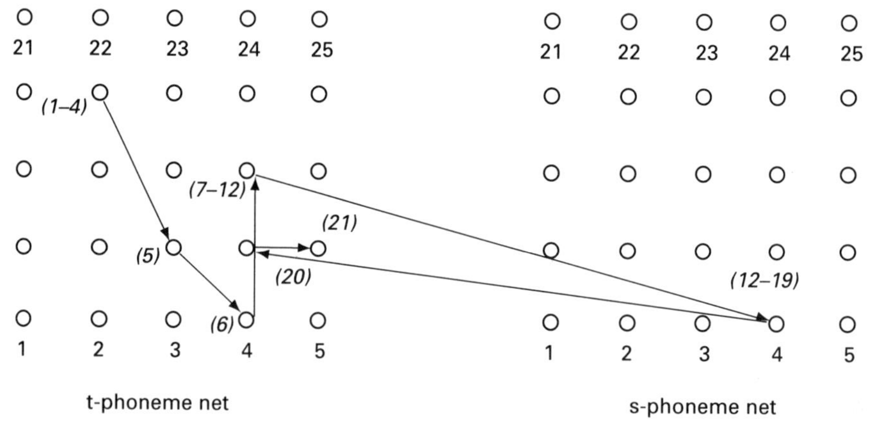

Figure 4.75 The path of winning neurons in the 't-phoneme net' and 's-phoneme net' by passing the window across the acoustic signal of the 't-phoneme'

t-phoneme net and the s-phoneme net are schematically drawn. In both nets the 5 × 5 neurons are represented by dots. For the first four observations, neuron 17 in the t-phoneme net is the winning neuron. For the fifth observation neuron 8 is the winning neuron. The next observation is assigned to neuron 4, then for six observations neuron 14 is the winning neuron. For the next seven observations we jump to the s-phoneme net where neuron 4 will be the winner, then we return again to the t-phoneme net where neurons 9 and 10 will be the winners, etc.

When we use a minimum score of fourteen successive winning neurons in the same phoneme net, with interrupting jumps to another phoneme net with at most a length of ten successive winning neurons, then we can detect the correct sequence of phonemes in the given sentence with one exception.

Note: We used some kind of preprocessing of the observation of the 128 samples in our window. The window sample vectors $\mathbf{x}_w$ were first multiplied with a Hanning window $\mathbf{w}_h$, $\mathbf{x}=\mathbf{x}_w \cdot \mathbf{x}_h$, with:

$$x_{hn}=\tfrac{1}{2}[1-\cos 2\pi n/(N-1)]$$

Next a discrete Fourier transform is applied to $\mathbf{x}$, resulting in a power spectrum vector $\mathbf{p}$. Finally we reduced the vector to sixteen components with the logarithmic frequency Melscale.

4.18 Selecting and scaling of training vectors

Suppose our data set $D_V \subseteq V$ of different d-dimensional data vectors $\mathbf{v}_i \in \mathbb{R}^d$ is generated in a sequence of steps. At each step a vector $\mathbf{v}_i$ is generated according to some probability density function $f(\mathbf{v}_i)$ in a d-dimensional input space V. The obtained sequence (length M) of vectors will be called the *training sequence* T_V and the consecutive elements will be indexed by a superscript: $T_V=\mathbf{v}^1, \mathbf{v}^2, \ldots, \mathbf{v}^M$. The same vector $\mathbf{v}_i$ may thus occur more than once in the training sequence T_V, say m_i times. Therefore we have $p(\mathbf{v}^i \in T_V)=1/M$ and $p(\mathbf{v}_i \in T_V)=m_i/M$ $(\approx f(\mathbf{v}_i)\,d\mathbf{v})$. We found in Section 4.4 that if we select at step t of the learning process a vector $\mathbf{v}(t)=\mathbf{v}_i$ with probability $p(\mathbf{v}_i)$ from the data set D_V and change the weight vectors according to the algorithmic adaptation rule, then we are minimizing the quantization error in the final phase of learning:

$$E(W)=\frac{1}{2}\sum_{\mathbf{w}_s} \sum_{\substack{\mathbf{v}_i \in R(\mathbf{w}_s)\\ \mathbf{v}_i \in D_V}} p(\mathbf{v}_i)|\mathbf{v}_i-\mathbf{w}_s|^2$$

During training, each vector $\mathbf{v}_i$ must be presented several times (preferably >10 times) to the neural algorithm. If, during training, the training sequence T_V contains enough input vectors to present a vector $\mathbf{v}_i$ (i.e. $m_i > 10$ for all i), the choice of an input vector from T_V at learning step t is not so difficult: choose $\mathbf{v}(t)=$'the tth element of T_V'. We must, however, be aware that at certain time T_E there will be almost no adaptation of weights in the neural net. If we use the function $\varepsilon(t)=(1+t)^{-1/2}$ for the learning rate $\varepsilon(t)$, then after 10 000 steps the adaptation of weight vectors will be 1 per cent; certainly there is no need to proceed with training to more than 100 000 steps. If we need a larger value of T_E we have to use another function for $\varepsilon(t)$.

A problem arises if $M \ll T_E$. In that case we draw randomly an element of T_V, $v(t)=\mathbf{v}^k$, with k an uniformly distributed integer $1 \leqslant k \leqslant M$. The ordering of weights in the neural net depends on the particular sequence of applied input vectors. To obtain a result independent of the applied sequence we must select the region adaptation function $h(r, s, t)$ such that the whole training set is presented several times before the quantization phase [$h(r, s, t)=1$ for $r=s$ and $h(r, s, t)=0$ otherwise] starts.

Another way of drawing input vectors is $\mathbf{v}(t)=\mathbf{v}^{k \bmod M}$ and as soon as $t=M$ the training set is permuted randomly. This method has the advantage that every input vector occurs p times if $T_E = p \times M$.

If the dimension of the input vectors is larger than the dimension of the neural net, the ordering and quantization is dominated by the components of the input vectors with the largest variance in the component values. In determining the 'winning neuron' during training, small differences in component values of **v** and **w** will be overruled by components with large differences. This implies that we have to scale the input vectors in such a way that the range of all components is almost the same. This can be done by the following method:

1. Determine for each component i of the input vectors $\mathbf{x}_j$ of the data set: $x_{i_{\max}} = \max_{\mathbf{x}_j}\{x_{ji}\}$ and $x_{i_{\min}} = \min_{\mathbf{x}_j}\{x_{ji}\}$.
2. Determine for each i: $d_i = \max\{x_{i_{\max}} - x_{i_{\min}}\}$.
3. Determine $d_{\max} = \max_i\{d_i\}$.
4. Multiply the ith component x_{ji} of all vectors $\mathbf{x}_j$ by $d_{\max}/d_i$.

This procedure will bring all components in the same range $d_{\max}$. If after training we need the values of the weight vectors, we have to multiply the ith component of a weight vector by $d_i/d_{\max}$.

4.19 Some practical measures of performance of the self-organizing neural net algorithm

In order to evaluate the performance of vector quantization by the self-organizing algorithm, we can calculate, after learning, the *energy of quantization noise* defined as follows (see Section 4.4):

$$E_{DW} = \frac{1}{M} \sum_{\mathbf{w}_s} \sum_{\substack{\mathbf{v}_i \in R(\mathbf{w}_s) \\ \mathbf{v}_i \in D_V}} |\mathbf{v}_i - \mathbf{w}_s|^2$$

with M the number of elements in D_V.

The quantization noise is defined by:

$$Q_{DW} = \frac{1}{M} \sum_{\mathbf{w}_s} \sum_{\substack{\mathbf{v}_i \in R(\mathbf{w}_s) \\ \mathbf{v}_i \in D_V}} |\mathbf{v}_i - \mathbf{w}_s|$$

If we use the training sequence T_V, as defined in Section 4.18, to evaluate E_{DW} and Q_{DW}, we obtain respectively:

$$E_{DW} = \frac{1}{M} \sum_{\mathbf{w}_s} \sum_{\substack{\mathbf{v}_i \in R(\mathbf{w}_s) \\ \mathbf{v}_i \in D_V}} p(\mathbf{v}_i)|\mathbf{v}_i - \mathbf{w}_s|^2$$

with M the number of elements in T_V, and

$$Q_{DW}=\frac{1}{M}\sum_{\mathbf{w}_s}\sum_{\substack{\mathbf{v}_i\in R(\mathbf{w}_s)\\ \mathbf{v}_i\in D_V}} p(\mathbf{v}_i)|\mathbf{v}_i-\mathbf{w}_s|$$

Furthermore the vector quantization of the self-organizing algorithm will 'try' to preserve topology (see Section 4.11). Thus we also want to have a measure to evaluate the topology preservation.

For topology preservation we want the mutual relative position of two vectors $\mathbf{v}_i$ and $\mathbf{v}_j$ in the input space V to be to a certain extent preserved by the mutual relative position of the corresponding winning neurons $\Psi(\mathbf{v}_i)$ and $\Psi(\mathbf{v}_j)$ in the neural lattice.

A mapping from a metric space V with distance measure d_V to a metric space L with distance measure d_L is *metric preserving* if the triangular inequality property is preserved; that is, if

$$d_V(\mathbf{v}_r, \mathbf{v}_t)\leqslant d_V(\mathbf{v}_r, \mathbf{v}_s)+d_V(\mathbf{v}_s, \mathbf{v}_t)$$

then

$$d_L[\psi(\mathbf{v}_r), \psi(\mathbf{v}_t)]\leqslant d_L[\psi(\mathbf{v}_r), \psi(\mathbf{v}_s)]+d_L[\psi(\mathbf{v}_s), \psi(\mathbf{v}_t)]$$

For metric preservation it is required that if the distance between two vectors from V is small, they will be represented by the same neuron or by neighbouring neurons in the neural net. This requirement can be stated as follows: if two receptive fields $R(\mathbf{w}_i)$ and $R(\mathbf{w}_j)$ in V are adjacent [the common border $\partial R_{ij}=R(\mathbf{w}_i)\cap R(\mathbf{w}_j)$ is not empty], then neuron u_i with weight vector $\mathbf{w}_i$ is a neighbour of neuron u_j with weight vector $\mathbf{w}_j$. If this property holds for all receptive fields, we say that the feature mapping is *topology preserving.*

If ω is the projection from the weight space W to the neural lattice L, then we can consider for a practical measure of topology preservation the distance in the neural lattice between $\omega(\mathbf{w}_i)$ and $\omega(\mathbf{w}_j)$ if the corresponding receptive fields $R(\mathbf{w}_i)$ and $R(\mathbf{w}_j)$ are adjacent (i.e. they have a common border). For optimal topology preservation this distance would always be 1. If there is a distortion of topology preservation, then the distance $\omega(\mathbf{w}_i)$ and $\omega(\mathbf{w}_j)$ will be larger than 1. So we can use the following *estimation of topological energy* as a practical measure for topology preservation:

$$T_{EW}=\frac{1}{b}\sum_i\sum_{\substack{j=i+1\\ R(\mathbf{w}_i)\text{ adjacent }R(\mathbf{w}_j)}} d_L^2\{\omega(\mathbf{v}_i), \omega(\mathbf{v}_j)\}$$

with b the number of borders of adjacent receptive fields in the input space V.

The *topological noise* is defined as:

$$J_{EW}=\frac{1}{b}\sum_i\sum_{\substack{j=i+1\\ R(\mathbf{w}_j)\text{ adjacent }R(\mathbf{w}_j)}} d_L\{\omega(\mathbf{v}_i), \omega(\mathbf{v}_j)\}$$

A simple (but not perfect way) to determine whether two receptive fields $R(\mathbf{w}_i)$ and

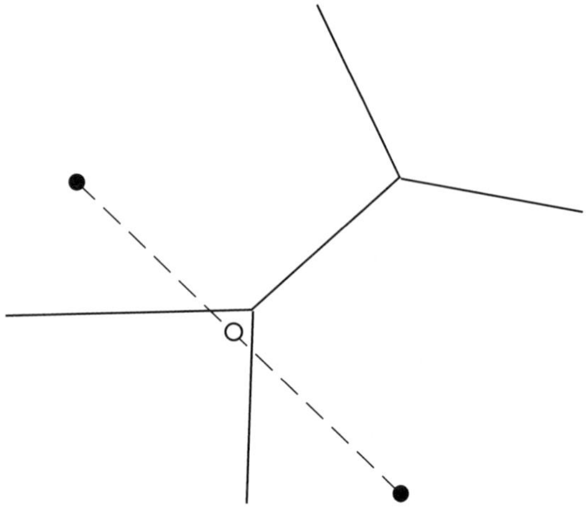

Figure 4.76 Two adjacent receptive fields where $(\mathbf{w}_i+\mathbf{w}_j)/2$ is not in $R(\mathbf{w}_i)\cup R(\mathbf{w}_j)$

$R(\mathbf{w}_j)$ are adjacent or not is to check whether:

$$\frac{\mathbf{w}_i+\mathbf{w}_j}{2}\in R(\mathbf{w}_i)\cap R(\mathbf{w}_j)$$

It may occur, however, that two receptive fields are adjacent for a 'small' common border and the expression above is not true, as illustrated in Figure 4.76.

Example 4.14

As a training sequence 1000 two-dimensional input vectors were randomly chosen from an input space V with a uniform distribution of vectors in the domain $[0, 1]^2$. The neural lattice was one-dimensional with 100 neurons, and so perfect topology preservation is not possible. The number of training steps was: $T_E=2^{17}$. In Figure 4.77 we have given the quantization noise Q_{DW} as a function of the learning steps. The initial quantization noise is low because an initial random distribution of weight vectors is already a good representation of randomly distributed input vectors. After two steps the quantization noise is a monotone decreasing function of t.

In Figure 4.78 we have given the value of the topological noise J_{EW} as a function of the training steps. We observe an initial decrease of J_{EW} from initial disorder to perfect topology preservation at $t=2^4$. The topology noise increases again starting at $t=2^7$. The reason for this increase can be explained as follows. In the first relatively short period, the ordering phase, ordering and topology preservation takes place. In the quantization phase [$h(r, s, t)=1$ for $r=s$ and $h(r, s, t)=0$ for $r\neq s$] the ordering is overruled by the quantization (see also Sections 4.4 and 4.11). The quantization noise is decreasing whereas the topology noise is increasing. ■

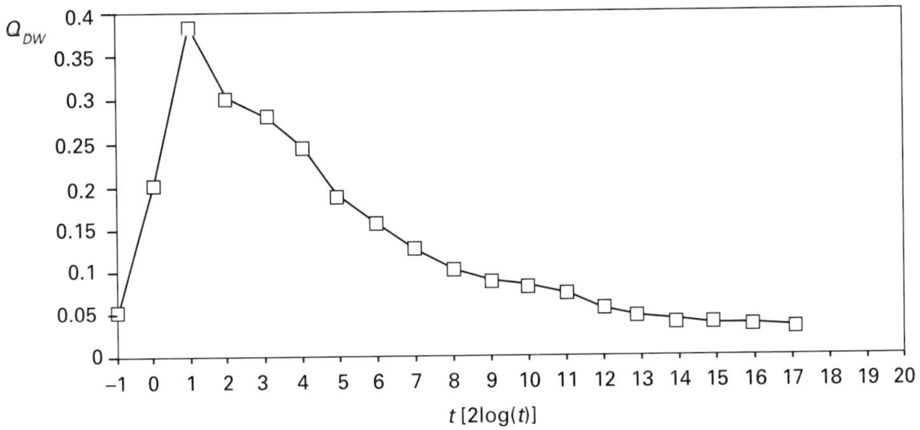

Figure 4.77 Quantization noise during training of a one-dimensional net with 100 neurons with 1000 two-dimensional uniformly distributed input vectors

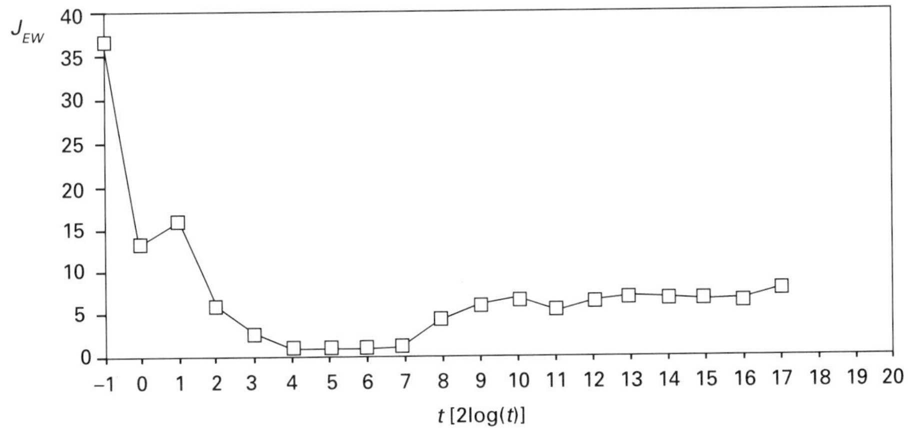

Figure 4.78 Topological noise during training of a one-dimensional net with 100 neurons with 1000 two-dimensional uniformly distributed input vectors

Other simple measures of performance are as follows:

- *Number of geometric neighbours l* (lines): the number of pairs of neurons in the lattice with distance $d_L = 1$.
- *Number of effective close neighbours c* (connections): the number of pairs of adjacent receptive fields which belong to neurons with distance $d_L = 1$.

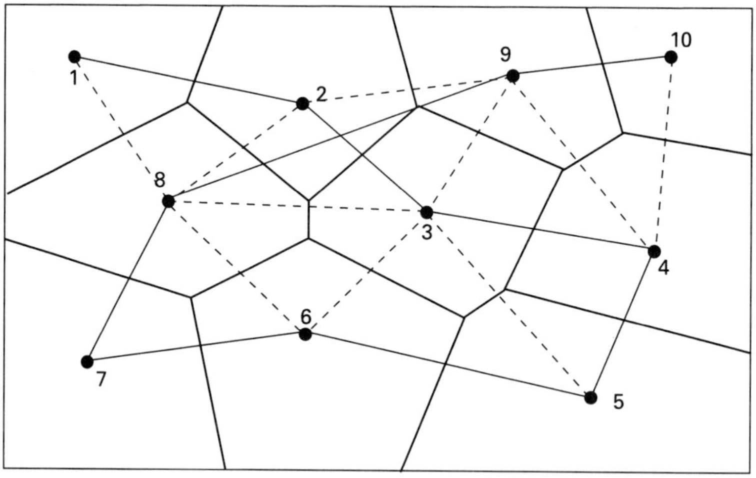

Figure 4.79 Two-dimensional input/weight space for a one-dimensional neural net of ten neurons. Areas are the receptive fields

- *Number of conflicting close neighbours d* (defects): the number of pairs of receptive fields that are not adjacent which belong to neurons with distance $d_L=1$.
- *Number of effective neighbours b* (borders): the number of pairs of adjacent receptive fields.
- *Number of effective distant neighbours j* (jumps): the number of pairs of adjacent receptive fields which belong to neurons with a distance $d_L>1$.

Note: $l=c+d$ and $b=j+c$.

Example 4.15

In Figure 4.79 we have given a two-dimensional input space V. The dots represent the weights of a one-dimensional neural lattice with 10 neurons. We find for Figure 4.79: $l=9$, $c=8$, $d=1$, $b=18$ and $j=10$. ■

In Section 4.11 we found that if the dimension of the input space V is the same as the dimension of the neural lattice L, then topology will be preserved. This implies that if the distance between two input vectors approaches zero, then the distance between the corresponding winning neurons is at most equal to 1. If, however, the dimension m of the input space is larger than the dimension n of the neural lattice, then in general there will be no topology preservation, and if the distance between two input vectors approaches zero, then the distance between the corresponding winning neurons might become large ($\gg 1$). A continuous path in V results in a corresponding track in the lattice L with in general large jumps between the successive winning neurons.

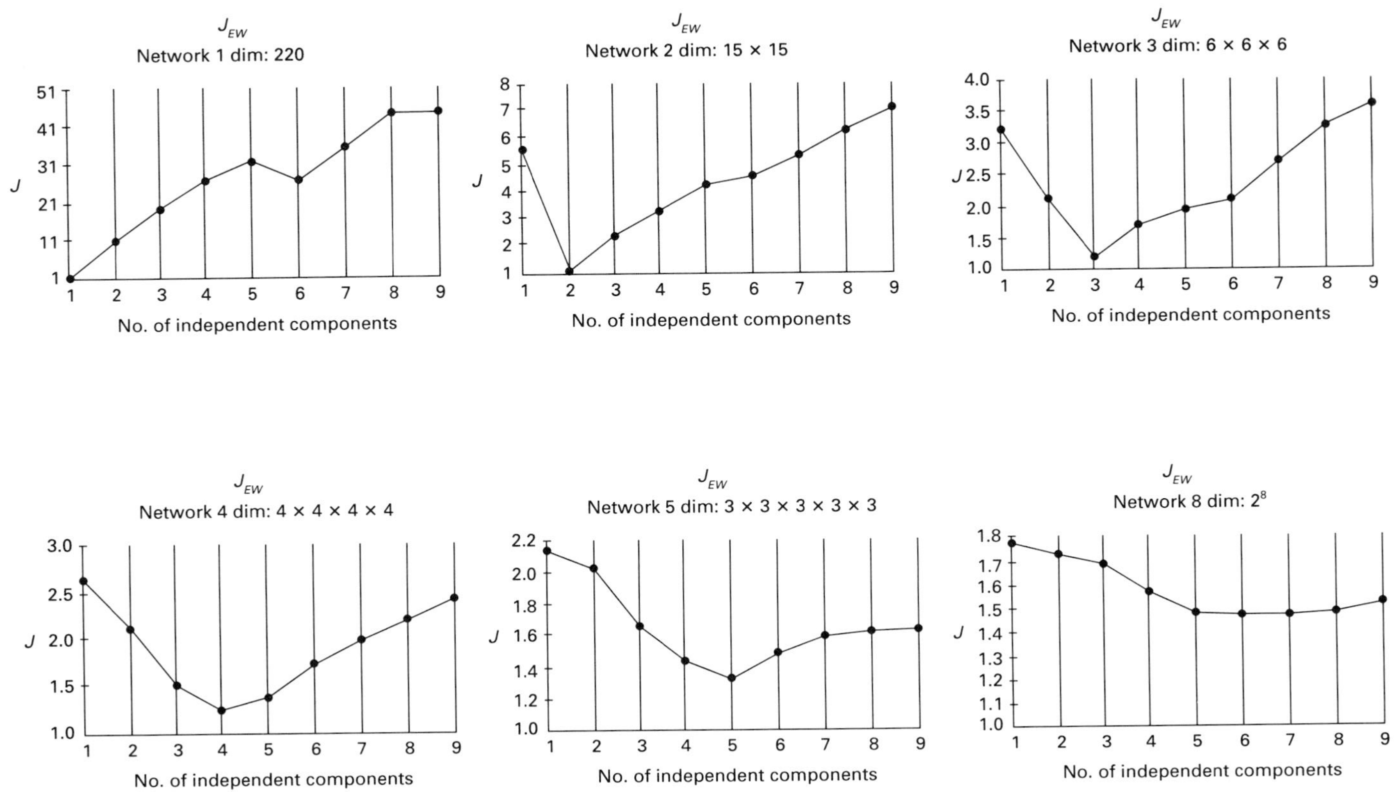

Figure 4.80 Topological noise as function of the number of independent input vector components for different dimensions of the neural network

We found, however, that we can obtain topology preservation in a restricted sense if the training set D consists of m-dimensional vectors, whereas the neural lattice has a dimension n much smaller than m. This will be the case if the components of the m-dimensional vectors of the training set D are interrelated such that they can be represented by points in an n-dimensional space while preserving the topology for the elements of D.

With the measure of J_{EW} of topological noise we demonstrate this mechanism by a simple experiment.

Example 4.16

In an experiment we used nine-dimensional training vectors. A number of the components of the training vector were independent by selecting for each of them a random number in the interval [0, 1]. The dependent components were just copies of the independent elements. We used six different neural network dimensions: one-dimensional (220 neurons), two-dimensional (15×15 neurons), three-dimensional ($6 \times 6 \times 6$ neurons), four-dimensional ($4 \times 4 \times 4 \times 4$ neurons), five-dimensional ($3 \times 3 \times 3 \times 3 \times 3$ neurons), eight-dimensional ($2 \times 2 \times 2 \times 2 \times 2 \times 2 \times 2 \times 2$ neurons).

In Figure 4.80 we have given the results for the topological noise J_{EW} for the different neural nets. We observe that a minimum of the topological noise is obtained if the number of independent components is equal to the dimension of the neural network. Note that for optimal topology preservation, the value of topological noise is $J_{EW} = 1$. ■

4.20 Application of the self-organizing algorithm to signature identification

We will end this book with a nice and powerful application of the self-organizing neural net algorithm.

In many situations the signature of a person will authorize some official certificate. However, after some training, the pattern of a signature can be duplicated. To avoid the falsification of signatures many corporations would like to have means to check the authenticity of a signature.

From the static pattern of a signature alone one cannot deduce its authenticity. However, the dynamic pattern of writing down a signature is far more difficult to copy. The speed of making the curves in a signature is quite characteristic for a person and hard to imitate.

We can use the self-organizing algorithm to solve this problem. We will only give a global outline of the implementation because the concept will be patented.

First we have to store the dynamic and static characteristics of a signature. For that purpose a person is asked to write down his or her signature on an electronic notepad several times. For each signature the position of the pencil is sampled at

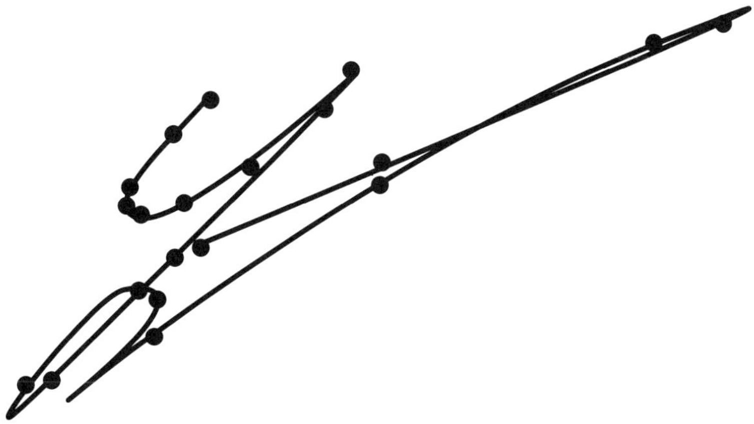

Figure 4.81 The author's signature with sample observation points

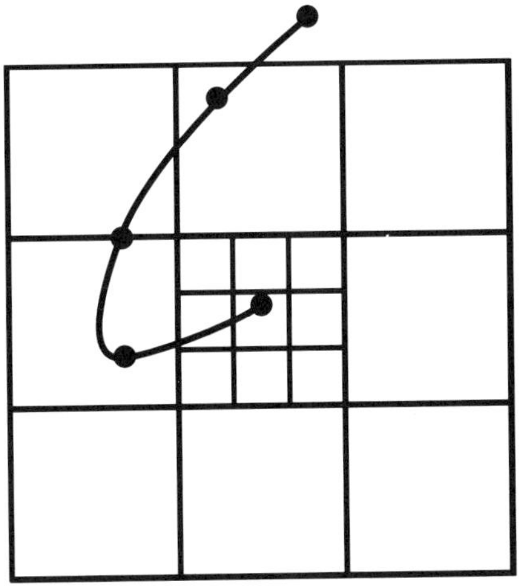

Figure 4.82 The observation of the signature with a window at the fifth sample point

small equidistant moments of time ($\approx$0.1 sec). Figure 4.81 presents the signature of the author; the twenty dots give the positions of the sample points.

Now we use the observation window introduced in Section 4.2, as was used for character recognition in Section 4.10. At each sample point we observe through the window the pattern of the signature but only for that part of the signature as is actually written down at the moment of sampling. In Figure 4.82 we observe the first part of the signature at the fifth sample point. In Figure 4.83 we observe a larger part of the signature at the tenth sample point.

Each observation will give us a observation vector of seventeen components. Each component reflects the mean grey value in a field of the window (see Sections 4.2

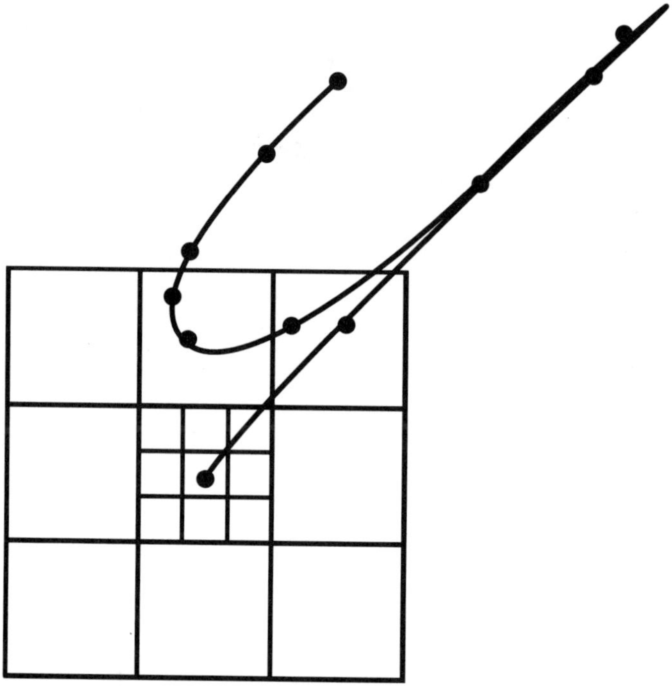

Figure 4.83 The observation of the signature with a window at the tenth sample point

and 4.10). For the signature of Figure 4.81 we obtain twenty different observation vectors. If we have ten examples of the signature, we obtain in this way 200 observations. We train a self-organizing two-dimensional neural net with 5×5 neurons with the 200 observation vectors. Because of the vector quantization performed by the self-organizing algorithm, we obtain twenty-five weight vectors representing clusters of similar observations. The examples of the signature are now used again to determine the dynamics of the signature. For each signature we have a sequence of twenty observation vectors. We present the sequence of observation vectors of a signature again to the neural net and determine the corresponding sequence of 'winning neurons'. The sequence of winning neurons is stored in a file. The same will be done for the remaining set of nine examples of signatures. The obtained (similar) sequences of winning neurons will reflect the dynamics of a signature. If, for example, a part of the signature is written very slowly, the same neuron will be the winning neuron for several consecutive observations, because almost the same observation is presented several times to the neural net. The transition from a winning neuron to the next neuron represents the movement of the pencil from one part of the signature to the next part.

We can do some statistical analysis on the obtained sequences of winning neurons in order to determine the probability of a sequence of winning neurons for that particular signature.

After this learning phase we can use the neural net to identify with some probability the authenticity of a signature written down on an electronic notepad.

4.21 Exercises

1. Assume we have a two-dimensional neural network with nine neurons u_j. The coordinate vectors $\mathbf{i}_j$ of the neurons in the lattice are: $\mathbf{i}_1=[0,0]$, $\mathbf{i}_2=[1,0]$, $\mathbf{i}_3=[2,0]$, $\mathbf{i}_4=[0,1]$, $\mathbf{i}_5=[1,1]$, $\mathbf{i}_6=[2,1]$, $\mathbf{i}_7=[0,2]$, $\mathbf{i}_8=[1,2]$, $\mathbf{i}_9=[2,2]$.
   At a certain time step t the weight vectors $\mathbf{w}_j$ are: $\mathbf{w}_1=[4,5]$, $\mathbf{w}_2=[5,3]$, $\mathbf{w}_3=[4,6]$, $\mathbf{w}_4=[1,1]$, $\mathbf{w}_5=[2,2]$, $\mathbf{w}_6=[0,0]$, $\mathbf{w}_7=[6,4]$, $\mathbf{w}_8=3,5]$, $\mathbf{w}_9=[5,4]$.
   At time t we have an input vector $\mathbf{v}=[3,3]$. What will be the value of the weights at time $t+1$ if we apply the algorithmic adaptation rule, if $\varepsilon(t)=1$ and $h(r,s,t)=1$ if $|\mathbf{i}_r-\mathbf{i}_s|=0$, $h(r,s,t)=0.5$ if $|\mathbf{i}_r-\mathbf{i}_s|=1$ and $h(r,s,t)=0$ otherwise?
2. What will be the value of the root mean squared error $E(W)$:
$$E(W)=\frac{1}{2}\sum_{\mathbf{w}_r}\sum_{\mathbf{w}_s}\sum_{\mathbf{v}_i\in R(\mathbf{w}_s)} p(\mathbf{v}_i)\varepsilon(t)h(r,s,t)|\mathbf{v}_i-\mathbf{w}_r|^2$$
   for the data set $D=\{[3,3]\}$ with $p([3,3])=1$ before and after adaptation for the self-organizing neural net of exercise 1 if $\varepsilon(t)=\varepsilon(t+1)$ and $h(r,s,t)=h(r,s,t+1)$?
3. How could we arrange the training process such that the solution route of the travelling salesman problem (see Section 4.6) will start in a given town?
4. We have a two-dimensional input data set D_V with data uniformly distributed in the domain $[0.5, 3.5]\times[0.5, 3.5]$. We use a two-dimensional self-organizing neural with nine neurons. The coordinate vectors $\mathbf{i}_j$ of the neurons in the lattice are: $\mathbf{i}_1=[0,0]$, $\mathbf{i}_2=[1,0]$, $\mathbf{i}_3=[2,0]$, $\mathbf{i}_4=[0,1]$, $\mathbf{i}_5=[1,1]$, $\mathbf{i}_6=[2,1]$, $\mathbf{i}_7=[0,2]$, $\mathbf{i}_8=[1,2]$, $\mathbf{i}_9=[2,2]$.
   What could be the final value of the weights $\mathbf{w}_j$ of all neurons if the algorithm performs a perfect vector quantization?
5. The same problem as in exercise 4 but now with a one-dimensional neural net. The coordinate vectors $\mathbf{i}_j$ of the neurons in the lattice are: $\mathbf{i}_1=[1]$, $\mathbf{i}_2=[2]$, $\mathbf{i}_3=[3]$, $\mathbf{i}_4=[4]$, $\mathbf{i}_5=[5]$, $\mathbf{i}_6=[6]$, $\mathbf{i}_7=[7]$, $\mathbf{i}_8=[8]$, $\mathbf{i}_9=[9]$.

Bibliography

Baker, M. A. and Loeb, M. (1973) Implications of measurement of eye fixations for a psychophysics of form perception, *Perception and Psychophysics*, **13**(2), 185–192.

Biswas, N. N. and Kumar, R. (1990) A new algorithm for learning representations in Boolean neural networks, *Current Science*, **59**, 595–600.

Cover, T. M. (1965) Geometrical and statistical properties of systems of linear inequalities with applications to pattern recognition, *IEEE Transactions on Electronic Computers*, **14**, 326–334. [The paper deals with the problems discussed in Section 2.2.]

Duda, R. O. and Hart, P. E. (1973) *Pattern Classification and Scene Analysis*, Wiley. [A good and pleasant to read book on pattern recognition with an excellent treatment of the problems discussed in Sections 2.2 and 2.3 of this book.]

Gerrits, H. J. M. and Vendrik, A. J. H. (1972) Eye movement necessary for continuous perception during stabilization of retinal images, *Biblitheca Ophthalmologica*, **82**, 339–347.

Gersho, A. and Cuperman, V. (1983) Vector quantization: a pattern matching technique for speech coding, *IEEE Communication Magazine*, December.

Gray, D. L. and Michel, A. N. (1992) A training algorithm for binary feedforward neural networks, *IEEE Transactions on Neural Networks*, **3**(2), 176–194.

Hebb, D. O. (1949) *Organization of Behavior*. Wiley.

Hecht Nielsen, R. (1990) *Neurocomputing*, Addison-Wesley.

Hertz, J., Krogh, A. and Palmer, R. G. (1991) *Introduction to the Theory of Neural Networks*. Addison-Wesley. [A very good, more advanced introduction to neural networks.]

Hopfield, J. J. (1984) Neurons with graded response have collective computational properties like those of two state neurons, *Procedures of the Natural Academy of Science USA*, **81**, 3088–3092.

Jannarone, R. J., Yu, K. F. and Takefuji, Y. (1988) Conjuntoids: statistical learning modules for binary events, *Neural Networks*, **1**, 325–337.

Kleene, S. C. (1956) Representation of events in nerve nets and finite automata, *Automata Studies*, 3–44.

Kohonen, T. (1982) Analysis of a simple self organizing process, *Biological Cybernetics*, **44**, 135–140.

Kohonen, T. (1984) *Self-organization and Associative Memory*, Springer Verlag. [A good introductory book from the founder of the self-organizing neural network.]

Kohonen, T. (1988) The neural phonetic typewriter, *IEEE Computer*, March, 11–22. [A short introductory paper with application to the recognition of phonetic units.]

Kosko, B. (1992) *Neural Networks and Fuzzy Systems*, Prentice Hall.

Martinetz, T. (1992) Selbstorganisierende neurale Netzwerkmodelle zur Bewegungskontrolle, Dissertation, Technische Universität Munchen, Theoretische Physik. [A more advanced introduction to self-organizing neural networks with application to robot control.]

McCullogh, W. S. and Pitts, W. (1943) A logical calculus of the ideas immanent in neurons activity, *Bulletin of Mathematical Biophysics*, **5**, 115–133.

Minsky, M. and Papert, S. (1969) *Perceptrons*, MIT Press. [The book gives a critical analysis of the performance of Perceptrons. There is a close relation with Section 2.6 of this book.]

Piaget, J. and Inhelder, B. (1969) *The Psychology of the Child*, Basic Books.

Ritter, H. and Kohonen, T. (1989) Self-organizing semantic maps, *Biological Cybernetics*, **61**, 241–254. [A short introductory paper with application to the recognition of semantic relationships in data.]

Ritter, H. and Schulten, K. (1986) On the stationary state of Kohonen's self-organizing sensory mapping, *Biological Cybernetics*, **54**, 99–106.

Rosenblatt, F. (1962) *Principles of Neurodynamics*, Spartan Books. [One of the first books on neural networks. It deals mainly with the reinforcement learning rule.]

Rummelhart, D. E. and McClelland, J. L. (1988) *Parallel Distributed Processing*, vol. 1, MIT Press. [A pleasant readable introduction into the field of neural networks.]

Senden von M. (1932) *Raum und Gestalltauffassung bei Operierten Blindgeboren vor und nach der Operation*, Barth, Leipzig.

Stassen, H. P. W. (1980) Measurement and analysis of eye movement and their role in the process of visual brightness perception, Ph.D. Thesis, University of Nijmegen.

Tromp, E. (1993) Practical implications of a theoretical analysis of the behaviour of a multilayer neural network, *Report no. 93N063*, University of Twente, Department of Engineering.

Veelenturf, L. P. J. (1970) Pattern recognition by syntactic analysis, Technical Report, University of Twente.

Veelenturf, L. P. J. (1978) Interference of sequential machines from sample computations. *IEEE Transactions on Computers*, **C-27** (2), 167–170.

Veelenturf, L. P. J. (1981) An automata theoretical approach to developing learning neutral networks, *Cybernetics and Systems*, **12**, 179–202.

Widrow, B. and Hoff, M. E. (1960) Adaptive switching circuits, *IRE WESCON Convention Record*, part 4, 96–104.

INDEX